Studies in Classification, Data Analysis, and Knowledge Organization

Titles in the Series:

E. Diday, Y. Lechevallier, M. Schader, P. Bertrand, and B. Burtschy (Eds.)
New Approaches in Classification and Data Analysis. 1994 (out of print)

W. Gaul and D. Pfeifer (Eds.)
From Data to Knowledge. 1995

H.-H. Bock and W. Polasek (Eds.)
Data Analysis and Information Systems. 1996

E. Diday, Y. Lechevallier, and O. Opitz (Eds.)
Ordinal and Symbolic Data Analysis. 1996

R. Klar and O. Opitz (Eds.)
Classification and Knowledge Organization. 1997

C. Hayashi, N. Ohsumi, K. Yajima, Y. Tanaka, H.-H. Bock, and Y. Baba (Eds.)
Data Science, Classification, and Related Methods. 1998

I. Balderjahn, R. Mathar, and M. Schader (Eds.)
Classification, Data Analysis, and Data Highways. 1998

A. Rizzi, M. Vichi, and H.-H. Bock (Eds.)
Advances in Data Science and Classification. 1998

M. Vichi and O. Opitz (Eds.)
Classification and Data Analysis. 1999

W. Gaul and H. Locarek-Junge (Eds.)
Classification in the Information Age. 1999

H.-H. Bock and E. Diday (Eds.)
Analysis of Symbolic Data. 2000

H. A. L. Kiers, J.-P. Rasson, P. J.F. Groenen, and M. Schader (Eds.)
Data Analysis, Classification, and Related Methods. 2000

W. Gaul, O. Opitz, and M. Schader (Eds.)
Data Analysis. 2000

R. Decker and W. Gaul (Eds.)
Classification and Information Processing at the Turn of the Millenium. 2000

S. Borra, R. Rocci, M. Vichi, and M. Schader (Eds.)
Advances in Classification and Data Analysis. 2001

W. Gaul and G. Ritter (Eds.)
Classification, Automation, and New Media. 2002

K. Jajuga, A. Sokołowski, and H.-H. Bock (Eds.)
Classification, Clustering and Data Analysis. 2002

M. Schwaiger and O. Opitz (Eds.)
Exploratory Data Analysis in Empirical Research. 2003

M. Schader, W. Gaul, and M. Vichi (Eds.)
Between Data Science and Applied Data Analysis. 2003

H.-H. Bock, M. Chiodi, and A. Mineo (Eds.)
Advances in Multivariate Data Analysis. 2004

D. Banks, L. House, F.R. McMorris, P. Arabie, and W. Gaul (Eds.)
Classification, Clustering, and Data Mining Applications. 2004

D. Baier and K.-D. Wernecke (Eds.)
Innovations in Classification, Data Science, and Information Systems. 2005

M. Vichi, P. Monari, S. Mignani and A. Montanari (Eds.)
New Developments in Classification and Data Analysis. 2005

D. Baier, R. Decker, and L. Schmidt-Thieme (Eds.)
Data Analysis and Decision Support. 2005

C. Weihs and W. Gaul (Eds.)
Classification the Ubiquitous Challenge. 2005

M. Spiliopoulou, R. Kruse, C. Borgelt, A. Nürnberger and W. Gaul (Eds.)
From Data and Information Analysis to Knowledge Engineering. 2006

Vladimir Batagelj · Hans-Hermann Bock
Anuška Ferligoj · Aleš Žiberna
Editors

Data Science and Classification

With 67 Figures and 42 Tables

Prof. Dr. Vladimir Batagelj
Department of Mathematics, FMF
University of Ljubljana
Jadranska 19
1000 Ljubljana, Slovenia
vladimir.batagelj@fmf.uni-lj.si

Prof. Dr. Anuška Ferligoj
Faculty of Social Sciences
University of Ljubljana
Kardeljeva pl. 5
1000 Ljubljana, Slovenia
anuska.ferligoj@fdv.uni-lj.si

Prof. Dr. Hans-Hermann Bock
Institute of Statistics
RWTH Aachen University
52056 Aachen, Germany
bock@stochastik.rwth-aachen.de

Aleš Žiberna
Faculty of Social Sciences
University of Ljubljana
Kardeljeva pl. 5
1000 Ljubljana, Slovenia
ales.ziberna@fdv.uni-lj.si

ISSN 1431-8814
ISBN-10 3-540-34415-2 Springer Berlin Heidelberg New York
ISBN-13 978-3-540-34415-5 Springer Berlin Heidelberg New York

Springer · Part of Springer Science+Business Media
springer.com

Printed in Germany

Production: LE-TEX Jelonek, Schmidt & Vöckler GbR, Leipzig
Softcover-Design: Erich Kirchner, Heidelberg

SPIN 11759263 43/3100/YL – 5 4 3 2 1 0 – Printed on acid-free paper

Preface

This volume contains a refereed selection of papers presented during the 10th Jubilee Conference of the International Federation of Classification Societies (IFCS) on Data Science and Classification held at the Faculty of Social Sciences of the University of Ljubljana in Slovenia, July 25-29, 2006. Papers submitted for the conference were subjected to a careful reviewing process involving at least two reviewers per paper. As a result of this reviewing process, 37 papers were selected for publication in this volume.

The book presents recent advances in data analysis, classification and clustering from methodological, theoretical, or algorithmic points of view. It shows how data analysis methods can be applied in various subject-specific domains. Areas that receive particular attention in this book are similarity and dissimilarity analysis, discrimination and clustering, network and graph analysis, and the processing of symbolic data. Special sections are devoted to data and web mining and to the application of data analysis methods in quantitative musicology and microbiology. Readers will find a fine selection of recent technical and application-oriented papers that characterize the current developments in data science and classification. The combination of new methodological advances with the wide range of real applications collected in this volume will be of special value for researchers when choosing appropriate newly developed analytical tools for their research problems in classification and data analysis.

The editors are grateful to the authors of the papers in this volume for their contributions and for their willingness to respond so positively to the time constraints in preparing the final versions of their papers. Without their work there would be no book. We are especially grateful to the referees – listed at the end of this book – who reviewed the submitted papers. Their careful reviews helped us greatly in selecting the papers included in this volume.

We also thank Dr. Martina Bihn and the staff of Springer-Verlag, Heidelberg for their support and dedication for publishing this volume in the series *Studies in Classification, Data Analysis, and Knowledge Organization.*

Vladimir Batagelj, Ljubljana
Hans–Hermann Bock, Aachen
Anuška Ferligoj, Ljubljana
Aleš Žiberna, Ljubljana

May 2006

Preface

The 10th IFCS Conference – a Jubilee

The International Federation of Classification Societies (IFCS) was founded July 4, 1985 in Cambridge (UK) at a time when classification problems were frequently being encountered in such diverse fields as biology, marketing, social sciences, pattern recognition, picture processing, information retrieval, and library science. These often massive problems had to be solved by quantitative or computerized methods based on data and measurements. In fact, the IFCS founding agreement paved the way for an intensive bilateral and multilateral cooperation and scientific as well as personal contacts among the members of the six 'national' classification societies from United Kingdom (BCS), North America (CSNA), Germany (GfKl), Japan (JCS), France (SFC), and Italy (SIS) that formed the IFCS in those times.

A main activity of IFCS is the organization of a biennial conference series. The first one with the title 'Classification, Data Analysis, and Related Methods' was held in Aachen (Germany) from June 29 to July 1, 1987 with about 300 participants from 25 countries and more than 180 scientific papers. Since that time, eight more IFCS conferences have taken place at different locations around the world (see the table below), always with a broad international participation from inside and outside the IFCS. Typically, a selection of scientific papers was published in a Proceedings volume in the Springer series 'Studies in Classification, Data Analysis, and Knowledge Organization' so that the results became available worldwide.

The biennial IFCS conferences

Year	*Place*	*Hosting Society*	*Organizer*
1987	Aachen (Germany)	GfKl	H.-H. Bock
1989	Charlottesville (USA)	CSNA	H. Bozdogan
1991	Edinburgh (UK)	BCS	D. Wishart, A. Gordon
1993	Paris (France)	SFC	E. Diday
1996	Kobe (Japan)	JCS	Ch. Hayashi
1998	Rome (Italy)	SIS	A. Rizzi, M. Vichi
2000	Namur (Belgium)	SFC, VOC	J.-P. Rasson, H. Kiers
2002	Cracow (Poland)	SKAD	A. Sokolowski
2004	Chicago (USA)	CSNA	F.R. McMorris
2006	Ljubljana (Slovenia)	SDS	A. Ferligoj, V. Batagelj

As a consequence, more and more national groups or societies working in the classification and data analysis field joined the IFCS: the VOC from Belgium/Netherlands, the CLAD from Portugal, the Polish society (SKAD), the KCS from Korea, the Irish Pattern Recognition and Classification Society (IPRCS), and finally the Central American and Caribbean society (SoCCAD).

The 10th IFCS conference is being hosted by the Statistical Society of Slovenia (SDS) at the University of Ljublajana (Slovenia), in July 2006, with Anuska Ferligoj and Vladimir Batagelj chairing. It will, without any doubt, be a new highlight in the history of IFCS, provide a challenging marketplace for scientific and applied research results, and foster further cooperation and contacts within the worldwide classification community.

This Jubilee Conference is certainly an occasion for recalling the history and achievements of the 20 years of IFCS's life. But it also marks the beginning of another decade of tasks and activities for IFCS: with new challenges for research and application, with interesting scientific conferences, with an intensive cooperation among IFCS members, and hopefully also a large impact on the worldwide development of our favorite domains: data analysis and classification.

May 2006 *Hans-Hermann Bock*

Contents

Part VII. Analysis of Music Data

Part VIII. Gene and Microarray Analysis

Part I

Similarity and Dissimilarity

A Tree-Based Similarity for Evaluating Concept Proximities in an Ontology

Emmanuel Blanchard, Pascale Kuntz, Mounira Harzallah, and Henri Briand

Laboratoire d'informatique de Nantes Atlantique
Site École polytechnique de l'université de Nantes
rue Christian Pauc
BP 50609 - 44306 Nantes Cedex 3
emmanuel.blanchard@univ-nantes.fr

Abstract. The problem of evaluating semantic similarity in a network structure knows a noticeable renewal of interest linked to the importance of the ontologies in the semantic Web. Different semantic measures have been proposed in the literature to evaluate the strength of the semantic link between two concepts or two groups of concepts within either two different ontologies or the same ontology. This paper presents a theoretical study synthesis of some semantic measures based on an ontology restricted to subsumption links. We outline some limitations of these measures and introduce a new one: the Proportion of Shared Specificity. This measure which does not depend on an external corpus, takes into account the density of links in the graph between two concepts. A numerical comparison of the different measures has been made on different large size samples from WordNet.

1 Introduction

With a long history in psychology (Tversky (1977), Rosch (1975)), the problem of evaluating semantic similarity in a network structure knows a noticeable renewed interest linked to the development of the semantic Web. In the 70's many researches on categorization were influenced by a theory which states that, from an external point of view, the categories on a set of objects were organized in a taxonomy according to an abstraction process. Describing proximity relationships between domain concepts by a hierarchy, or more generally a graph, remains a common principle of the current knowledge representation systems, namely the ontologies associated with the new languages of the semantic Web –in particular OWL (Bechhofer et al. (2004)). As defined by Gruber (1993), "an ontology is a formal, explicit specification of a shared conceptualization" and "a conceptualization is an abstract, simplified view of the world that we wish to represent for some purpose".

From an operational point of view, the development and the exploitation of an ontology remains a complex task in a global process of knowledge engineering. Upstream, extracting and structuring large sets of concepts with increasing sizes represents one of the major difficulties. Downstream, retrieving subsets of concepts requires approaches that are not time-consuming and are efficient in terms of semantic relevance of the results. To make these tasks easier, some proposals resort to an estimation of a semantic similarity between the pairs of concepts.

Generally speaking, a "semantic similarity" $\sigma(o_i, o_j)$ between two objects o_i and o_j is related to their commonalities and sometimes their differences. Most of the definitions considered in the literature suppose that the objects are associated with their extension –a collection of elements or a set of descriptors. In this case, the semantic similarities can be roughly classified in three main classes depending on the object description:

1. The extensions are simply subsets of elements. Then, the similarity $\sigma(o_i, o_j)$ between two objects o_i and o_j is a function of the subsets of the elements common (set intersection) to and the elements different (set symmetric difference) for o_i and o_j. This class includes similarities well-known in the taxonomic literature such as the Jaccard's coefficient and the Dice's coefficient.
2. Each element in the domain is taken to be a dimension in a vector space (Salton and McGill (1983)), and each object o_i can be described by a vector whose components describe the elements in the collection. This representation popular in the information retrieval domain makes use of the usual Cosine-Similarity, and many adaptations have been proposed (Salton and McGill (1983)).
3. A hierarchical domain structure is given. The leaves are the elements of the collection and nodes represents an organization of concepts. Different similarities –often generalizations of case 2– have been proposed to exploit the hierarchical structure (see Ganesan et al. (2003) for a recent review). Very recently, the problem has been extended to graphs (Maguitman et al. (2005)).

When considering ontologies, the concepts are not necessarily described by their extension ; the internal organization of a domain ontology is often the product of a consensus of experts (Guarino (1995)). Hence, defining a semantic similarity $\sigma(c_i, c_j)$ between a pair c_i, c_j of concepts sets specific problems depending on the information at our disposal. Some measures only depend on the concept structuring –often hierarchical– ; others, in addition, require textual corpus of the domain.

This paper presents a review of the main measures of the literature available for comparing pairs of concepts within a domain ontology. Our description in a unified framework allows to highlight their commonalities and their differences. In particular, we show that they can be defined as functions of a restricted set of parameters, and that only one of the measures based on a corpus exploits all the parameters. To overcome this limitation, we propose a new measure, the Proportion of Shared Specificity (PSS) which is an adaptation of the Dice's coefficient to a hierarchical structure. We show that the Wu and Palmer's coefficient is a particular case of our measure. And, we present experimental comparisons on different large size samples extracted from the semantic network WordNet 2.0.

2 Graph-based similarities for domain ontologies

Formally, an ontology can be modeled by a graph where nodes represent concepts and arcs represent labeled relationships. Here, like often in the literature, we restrict ourselves to the hyperonymy and hyponymy relationships associated to the relationship of subsumption (is-a). For instance: "a dog is an animal" implies that dog is an hyponym of animal and animal is an hyperonym of dog. This relationship is common to every ontology, and different researches have been confirmed that it is the more structuring one (e.g. Rada et al. (1989)).

Let O be a domain ontology with a set $C = \{c_1, c_2, \ldots, c_n\}$ of concepts. The relationships "hyperonymy" (generalization) and "hyponymy" (specialization) are dual : for any pair $(c_i, c_j) \in C \times C$ if $hyperonymy(c_i, c_j)$ then $hyponymy(c_j, c_i)$, and vice-versa. And, each concept c_i has no more than one hyperonym. Moreover, to maintain the connectivity, it is common to add a virtual concept c_0 ("thing" or "entity"). Consequently, O can be modeled by a rooted tree $T_O(C, A)$ with the root c_0 and so that each arc $(c_i, c_j) \in A$ represents an hyponymy relationship between c_i and c_j (figure 1). By construction of an ontology, the deeper is a concept in $T_O(C, A)$, the more specific it is. We adopt this restricted framework of a rooted tree in the following of this paper.

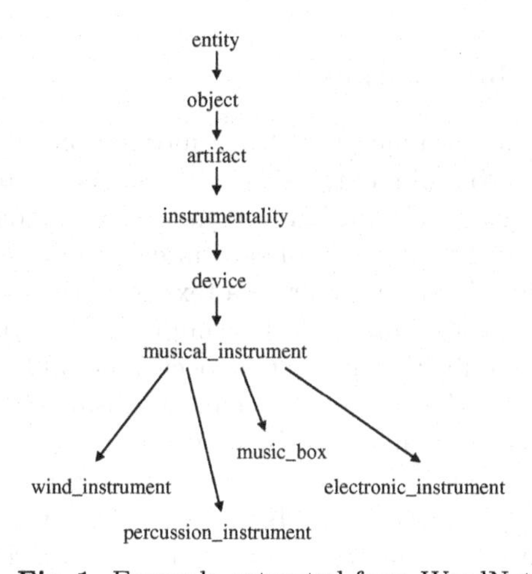

Fig. 1. Example extracted from WordNet.

The graph-based measures –also called structural measures– can be regrouped in two main classes : the functions of combinatorial properties, and the measures which incorporate an additional knowledge source from a corpus analysis. Note that, depending on the authors, the definitions have been originally given on the form of a dissimilarity δ or a similarity σ ; as the transformation of σ in δ is not unique, we here retain this diversity.

2.1 Approaches using combinatorial properties

A comparative study of the different measures based on combinatorial properties shows that their definitions depends on four main components: *(i)* the length (edge number) $len(c_i, c_j)$ of the path between two concepts c_i, c_j in $T_O(C, A)$, *(ii)* the most specific subsumer $mscs(c_i, c_j)$ (the lowest common ancestor in the tree) of c_i and c_j, *(iii)* the depth $h(c_i)$ (length from the root to c_i) of a concept c_i in $T_O(C, A)$.

The simplest dissimilarity, proposed by Rada et al. (1989), is $\delta_r(c_i, c_j) = len(c_i, c_j)$. Despite its simplicity, experiments in information retrieval have shown that, when the paths are restricted to is-a links like here, this measure could contribute to good results. A normalization has been introduced by Leacock and Chodorow (1998): $\sigma_{lc}(c_i, c_j) = -\log \frac{len(c_i,c_j)}{2 \cdot \max_{c \in C}[h(c_O)]}$.

The Wu and Palmer similarity takes into account the specificity of the subsumer of c_i and c_j: $\sigma_{wp}(c_i, c_j) = \frac{2 \cdot h(mscs(c_i,c_j))}{len(c_i,c_j) + 2 \cdot h(mscs(c_i,c_j))}$.

Based on linguistic properties, Sussna (1993) have introduced a weight function for the relationships : $w_r(c_i, c_j)$ for the hyperonymy (resp. $w_o(c_j, c_i)$ for the hyponymy). When the concepts c_i, c_j are adjacent in $T_O(C, A)$, the dissimilarity is a scaled sum of the weights : $\delta_s(c_i, c_j) = \frac{w_r(c_i,c_j) + w_o(c_i,c_j)}{2 \cdot \max(h(c_i), h(c_j))}$. For two arbitrary concepts, it is computed as the sum of the dissimilarities between the pairs of adjacent concepts along the path connecting them.

2.2 Approaches using a corpus

These approaches introduce a measure of the information shared by a pair of concepts. One criterion to evaluate the similarity between two concepts is the extent to which they share information in common. The required additional notion is the probability $P(c_i) \in [0, 1]$ of encountering an occurrence of c_i. In practice, this probability is estimated from a text corpus S by the occurrence frequency of c_i in S $num(c_i)/num(c_O)$. To compute $num(c_i)$, Resnik (1995) proposes to count not only the number of occurrences of c_i, but also the number of occurrences of the concepts which are subsumed by c_i.

For Resnik, the information shared by two concepts c_i and c_j is the "information content" of their most specific common subsumer : $\sigma_r(c_i, c_j) = -\log P(mscs(c_i, c_j))$. Lin (1998) and Jiang and Conrath (1997) moderate this value by the quantity of information which distinguishes the two concepts : $\sigma_l(c_i, c_j) = \frac{2 \cdot \log P(mscs(c_i,c_j))}{\log P(c_i) + \log P(c_j)}$ and $\delta_{jc}(c_i, c_j) = 2 \cdot \log P(mscs(c_i, c_j)) - (\log P(c_i) + \log P(c_j))$.

3 A new measure: the Proportion of Shared Specificity

Each of the previous measures attempts to exploit the information contained in the ontology at best to evaluate the similarity between the pairs of concepts

(c_i, c_j). The Rada's coefficient is the simplest ; it takes only into account the length of the paths joining c_i and c_j. The Lin's coefficient is the more complex ; it takes into account the common information shared by the concepts and, via the estimation of $P(c_i)$, the density of concepts between the root c_0 and c_i and c_j. When the density is high, this means that the hyponyms are more specific, and consequently the Lin's similarity is higher. However, the computation of the Lin's coefficient requires a corpus in addition to the ontology. And this latter should be significant enough to provide a "good" estimation of $P(c_i)$. As this condition may be very restrictive for numerous real-life applications, we have developed a new measure which shares some properties of the Lin's coefficient but which only depends on $T_O(C, A)$.

When each concept c_i is described by a set of unitary characteristics $cha(c_i)$, one of the most commonly used measure is the Dice coefficient (Ganesan et al. (2003)):

$$\sigma_d(c_i, c_j) = \frac{2 \cdot |cha(c_i) \bigcap cha(c_j)|}{|cha(c_i)| + |cha(c_j)|} \tag{1}$$

In this model, two concepts are all the more similar since they share numerous common characteristics and few different ones. We here propose to adapt this formula to measure the (dis)similarity between a pair of concepts in an ontology.

Intuitively, $|cha(c_i)|$ is an indicator of a known part of the information carried by the concept c_i. In the approaches using a corpus (2.2) the quantity of information carried by a concept is evaluated by $-\log(P(c_i))$. Let us remark, that when replacing in (1) $|cha(c_i)|$ (resp. $|cha(c_j)|$) by $-\log(P(c_i))$ (resp. $-\log(P(c_j))$) and $|cha(c_i) \cap cha(c_j)|$ by $-\log(P(mscs(c_i, c_j)))$ we rediscover the Lin's formula.

When no additional information is available, we can exploit the hierarchical structure of the ontology $T_O(C, A)$. Our reasoning rests on an analogy with the case where we dispose of a set of instances. In an ontological model, each concept is not necessarily described by a collection of characteristics, and the associated instances are not explicitly given. Nevertheless, by construction, we can consider that the probability of an instance to be associated with the more general concept c_0 (the root) is equal to 1. If c_0 specializes in k hyponyms $c_1^1, c_1^2, \ldots, c_1^k$, then $\sum_{i=1}^{k} P(c_1^i) = 1$. And, if the distribution of the instance number is supposed to be uniform for each concept, then $P(c_1^i) = \frac{1}{k}$. Then, when this assumption is available for all the levels of $T_O(C, A)$, for any concept c_i deeper in T_O, we have a series of specialisations in $k_1, k_2, \ldots, k_{h(c_i)}$ concepts. Then $P(c_i) = \frac{1}{k_1 \cdot k_2 \cdot \cdots \cdot k_{h(c_i)}}$. Consequently, the quantity of information associated with the specificity of c_i is measured by $\log(k_1) + \log(k_2) + \cdots + \log(k_{h(c_i)})$.

Moreover, the quantity of information shared by two concepts c_i and c_j can be measured, like in the Lin's coefficient, by the quantity of information of their most specific subsumer. Hence, the Proportion of Shared Specificity is defined by:

$$\sigma_{pss}(c_i, c_j) = \frac{2 \cdot \log\Big(P(mscs(c_i, c_j))\Big)}{\log\Big(P(c_i)\Big) + \log\Big(P(c_j)\Big)} \tag{2}$$

Let us remark, that when we consider the simplest case where all the nodes have the same degree k, then $P(c_i) = k^{-h(c_i)}$ for any concept c_i at depth $h(c_i)$. Consequently, the quantity of information associated with the specificity of c_i is measured by $h(c_i) \cdot \log k$ and we obtain the Wu and Palmer's measure.

4 Experimental results

This section presents an analysis of the behavior of the previous coefficients on samples of concepts drawn from the semantic network WordNet 2.0 (Fellbaum (1998)). Let us briefly recall that, inspired from the psycholinguistic theories on the lexical human memory, WordNet was created as an attempt to model the lexical knowledge of a native English speaker. The lexical entities (e.g. nouns, verbs, adjectives) are organized into synonym sets that are interlinked with different relations. We here restrict ourselves to nouns and to the subsumption hierarchy (hyperonymy/hyponymy). This hierarchy, which contains 146690 nodes, constitutes the backbone of the noun subnetwork accounting for close to 80% of the links (Budanitsky (1999)). It can be properly represented by the tree model $T_O(C, A)$ described in section 2, and consequently, for our experiments we do not enter into the discussion between experts concerning the ontological nature of WordNet. The computations have been performed with the Perl modules of Pedersen et al. (2004).

We have randomly drawn a sample of 1000 concepts from T_O, and computed the coefficient values for each pair. Figure 2 shows the distributions associated with the different measures. Due to the random draw and the size of T_O, numerous pairs contain concepts without semantic links, and it is not surprising to find numerous null values (around 60%). Contrary to other combinatorial measures, the PSS coefficient allows a more subtle differentiation of the concepts.

Due to the theoretical relationships between σ_{pss}, σ_l (Lin) and σ_{wp} (Wu & Palmer) we have analyzed the values of the rank correlation of Spearman on the pair set of the sample : $\rho(wp, l) = 0.813$, $\rho(l, pss) = 0.835$ and $\rho(pss, wp) = 0.970$. Obviously, it is interesting to find a strong correlation between the PSS coefficient and the Lin's coefficient which requires an additional corpus. However, a part of these high values can be explained by the great number of distant pairs. Nevertheless, when we restrict ourselves to the subset which contains 1% of the most similar pairs both for σ_{pss} and σ_l, the correlation is still significant ($\rho(pss, l) = 0.53$).

To go deeper in the analysis, we have computed the correlations on different subtrees of T_O associated with identified sets of themes (e.g. insect, tree,

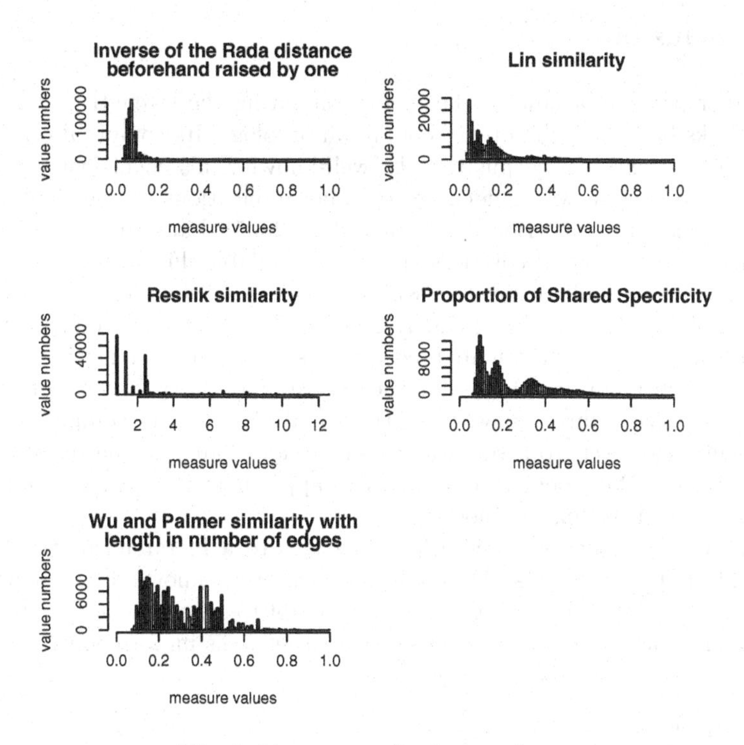

Fig. 2. histograms of value numbers.

musical instrument). Table 1 gives the different correlations. As WordNet is homogeneous (the number of hyponyms is barely constant), the correlation between Wu Palmer and PSS measures is naturally strong. It is important to note that $\rho(pss, l) > \rho(w, l)$ for all our experiments.

Table 1. Correlation on three WordNet subsets

root	insect	tree	musical_instrument
number of concepts	157	454	1013
number of pairs	12246	102831	512578
median number of hyponyms	3	2	3
$\rho(pss, l)$	0.65	0.53	0.30
$\rho(pss, wp)$	0.91	0.85	0.90
$\rho(w, l)$	0.63	0.52	0.27

5 Conclusion

This paper presents a new similarity for evaluating the strength of the semantic links between pairs of concepts in an ontology. Its computation does not require an external corpus like the well-known Lin's coefficient. At the origin, our objective was guided by real-life applications, in particular in knowledge management (Berio and Harzallah (2005), Laukkanen and Helin (2005)), where additional corpuses are rarely available. From the Dice's coefficient, we have built a measure which exploits the structural properties of the ontology. Our numerical experiments on WordNet have confirmed its discriminant behavior and highlighted its links with other coefficients of the literature. The main arguments for the choice of this experimental framework are its size –which allows statistical analysis– and its computational accessibility. However, to assess the semantic significance of the results obtained with the PSS coefficient, we plan to apply it in the near future to a professional environment ontology.

Moreover, we have here restricted ourselves to a hierarchical structure deduced from the "is-a" link. Although this structure is known to be the most structuring of a real-life ontology, we now attempt to generalize our approach to a graph structure to simultaneously take other links into account.

References

BECHHOFER, S., VAN HARMELEN, F., HENDLER, J., HORROCKS, I., MCGUINNESS, D. L., PATEL-SCHNEIDER, P. F., AND STEIN, L. A. (2004): Owl web ontology language reference. http://www.w3.org/TR/2004/REC-owl-ref-20040210/.

BERIO, G. AND HARZALLAH, M. (2005): Knowledge management for competence management. *Universal Knowledge Management*, 0.

BUDANITSKY, A. (1999): Lexical semantic relatedness and its application in natural language processing. Technical report, Computer Systems Research Group – University of Toronto.

FELLBAUM, C., editor (1998): *WordNet: An electronic lexical database.* MIT Press.

GANESAN, P., GARCIA-MOLINA, H., AND WIDOM, J. (2003): Exploiting hierarchical domain structure to compute similarity. *ACM Trans. Inf. Syst.*, 21(1):64–93.

GRUBER, T. R. (1993): A translation approach to portable ontology specifications. *Knowledge Acquisition*, 5(2):199–220.

GUARINO, N. (1995): Formal ontology, conceptual analysis and knowledge representation. *Human-Computer Studies*, 43(5/6):625–640.

JIANG, J. J. AND CONRATH, D. W. (1997): Semantic similarity based on corpus statistics and lexical taxonomy. In *Proc. of Int. Conf. on Research in Comp. Linguistics*.

LAUKKANEN, M. AND HELIN, H. (2005): Competence management within and between organizations. In *Proc. of 2nd Interop-EMOI Workshop on Enterprise*

Models and Ontologies for Interoperability at the 17th Conf. on Advanced Information Systems Engineering, pages 359–362. Springer.
LEACOCK, C. AND CHODOROW, M. (1998): Combining local context and wordnet similarity for word sense identification. In Fellbaum, C., editor, *WordNet: An electronic lexical database*, pages 265–283. MIT Press.
LIN, D. (1998): An information-theoretic definition of similarity. In *Proc. of the 15th Int. Conf. on Machine Learning*, pages 296–304. Morgan Kaufmann.
MAGUITMAN, A. G., MENCZER, F., ROINESTAD, H., AND VESPIGNANI, A. (2005): Algorithmic detection of semantic similarity. In *Proc. of the 14th int. conf. on World Wide Web*, pages 107–116. ACM Press.
PEDERSEN, T., PATWARDHAN, S., AND MICHELIZZI, J. (2004): Wordnet::similarity – measuring the relatedness of concepts. In *Proc. of the Fifth Annual Meeting of the North American Chapter of the Association for Comp. Linguistics*, pages 38–41.
RADA, R., MILI, H., BICKNELL, E., AND BLETTNER, M. (1989): Development and application of a metric on semantic nets. *IEEE Transactions on Systems, Man, and Cybernetics*, 19(1):17–30.
RESNIK, P. (1995): Using information content to evaluate semantic similarity in a taxonomy. In *Proc. of the 14th Int. Joint Conf. on Artificial Intelligence*, volume 1, pages 448–453.
ROSCH, E. (1975): Cognitive representations of semantic categories. *Experimental Psychology: Human Perception and Performance*, 1:303–322.
SALTON, G. AND MCGILL, M. J. (1983): *Introduction to modern information retrieval*. McGraw-Hill.
SUSSNA, M. (1993): Word sense disambiguation for free-text indexing using a massive semantic network. In *Proc. of the Sec. Int. Conf. on Information and Knowledge Management*, pages 67–74.
TVERSKY, A. (1977): Features of similarity. *Psychological Review*, 84(4):327–352.

Improved Fréchet Distance for Time Series

Ahlame Chouakria-Douzal[1] and Panduranga Naidu Nagabhushan[2]

[1] TIMC-IMAG, Université Joseph Fourier Grenoble 1,
F-38706 LA TRONCHE Cedex, France
Ahlame.Douzal@imag.fr

[2] Dept. of Studies in Computer Science, University of Mysore
Manasagangothri, Mysore, Karnataka- 570 006, India
pn@amrita.edu

Abstract. This paper focuses on the Fréchet distance introduced by Maurice Fréchet in 1906 to account for the proximity between curves (Fréchet (1906)). The major limitation of this proximity measure is that it is based on the closeness of the values independently of the local trends. To alleviate this set back, we propose a dissimilarity index extending the above estimates to include the information of dependency between local trends. A synthetic dataset is generated to reproduce and show the limited conditions for the Fréchet distance. The proposed dissimilarity index is then compared with the Fréchet estimate and results illustrating its efficiency are reported.

1 Introduction

Time series differ from "non-temporal" data due to the interdependence between measurements. This work focuses on the distances between time series, an important concept for time series clustering and pattern recognition tasks. The Fréchet distance is one of the most widely used proximity measure between time series. Fréchet distance uses time distortion by acceleration or deceleration transformations to look for a mapping that minimizes the distance between two time series. We show in section 4, that the Fréchet distance ignores the interdependence among the occurring values; proximity is only based on the closeness of the values; which can lead to irrelevant results. For this reason, we propose a dissimilarity index extending this classical distance to include the information of dependency between local trends. The rest of this paper is organized as follows: the next section presents the definitions and properties of the conventional Fréchet distance. Section 3, discusses the major limitations of such proximity estimate, then gives the definition and properties of the new dissimilarity index. Section 4, presents a synthetic dataset reproducing limited conditions for this widely used time series proximity measure, then perform a comparison between the proposed dissimilarity index and the Fréchet distance before concluding.

2 The Fréchet distance between time series

The success of a distance, intended to distinguish the events of a time series that are similar from those that are different, depends on its adequacy with respect to the proximity concept underlying the application domain or the experimental context.

The Fréchet distance was introduced by Maurice Fréchet in 1906 (Fréchet (1906)) to estimate the proximity between continuous curves. We present a discrete variant of this distance. An in-depth study of the Fréchet distance is provided by Alt (Alt and Godau (1992)) and an interesting comparison of the different distance theories can be found in Eiter and Mannila (1994). The popular and highly intuitive Fréchet distance definition is: "A man is walking a dog on a leash. The man can move on one curve, the dog on another. Both may vary their speed independently, but are not allowed to go backwards. The Fréchet distance corresponds to the shortest leash that is necessary". Let's provide a more formal definition.

We define a mapping $r \in M$ between two time series $S_1 = (u_1, ..., u_p)$ and $S_2 = (v_1, ..., v_q)$ as the sequence of m pairs preserving the observation order:

$$r = ((u_{a_1}, v_{b_1}), (u_{a_2}, v_{b_2}), ..., (u_{a_m}, v_{b_m}))$$

with $a_i \in \{1, .., p\}$, $b_j \in \{1, .., q\}$ and satisfying for $i \in \{1, .., m-1\}$ the following constraints:

$$a_1 = 1,\ a_m = p \qquad b_1 = 1, b_m = q \tag{1}$$

$$a_{i+1} = a_i \text{ or } a_i + 1 \qquad b_{i+1} = b_i \text{ or } b_i + 1 \tag{2}$$

We note $|r| = \max_{i=1,..,m} |u_{a_i} - v_{b_i}|$ the mapping length representing the maximum span between two coupled observations. The Fréchet distance $\delta_F(S_1, S_2)$ is then defined as:

$$\delta_F(S_1, S_2) = \min_{r \in M} |r| = \min_{r \in M}(\max_{i=1,..,m} |u_{a_i} - v_{b_i}|) \tag{3}$$

Graphically, a mapping between two time series $S_1 = (u_1, ..., u_p)$ and $S_2 = (v_1, ..., v_q)$ can be represented by a path starting from the corner $(1, 1)$ and reaching the corner (p, q) of a grid of dimension (p, q). The value of the square (i, j) is the span between the coupled observations (u_i, v_j). The path length corresponds to the maximum span reached through the path. Then, the Fréchet distance between S_1 and S_2 is the minimum length through all the possible grid paths. We can easily check that δ_F is a metric verifying the identity, symmetry and triangular inequality properties (a proof can be found in Eiter and Mannila (1994)).

According to δ_F two time series are similar if there exists a mapping between their observations, expressing an acceleration or a deceleration of the occurring observation times so that the maximum span between all coupled observations is close.

Note that the Fréchet distance is very useful when only the occurring events, not their occurring times, are determinant for the proximity evaluation. This explains the great success of Fréchet distance in the particular domain of voice processing where only the occurring syllables are used to identify words; the flow rate being specific to each person.

3 Fréchet distance extension for time series proximity estimation

Generally, the interdependence among the occurring values, characterizing the local trends in the time series, is determinant for the time series proximity estimation. Thus, Fréchet distance fails as it ignores such main information. Section 4 illustrates two major constraints in the Fréchet measure: ignorance of the temporal structure and the sensitivity to global trends. To alleviate these drawbacks in the classical Fréchet estimate we propose a dissimilarity index extending Fréchet distance to include the information of dependency between the time series local trends. The dissimilarity index consists of two components. The first one estimates the closeness of values and is based on a normalized form of the conventional proximity measure. The second component, based on the temporal correlation Von Neumann (1941-1942)), Geary (1954) and (Chouakria-Douzal (2003), estimates the dependency between the local trends.

3.1 Temporal correlation

Let's first recall the definition of the temporal correlation between two time series $S_1 = (u_1, ..., u_p)$ and $S_2 = (v_1, ..., v_p)$:

$$\text{CORT}(S_1, S_2) = \frac{\sum_{i=1}^{p-1}(u_{(i+1)} - u_i)(v_{(i+1)} - v_i)}{\sqrt{\sum_{i=1}^{p-1}(u_{(i+1)} - u_i)^2 \sum_{i=1}^{p-1}(v_{(i+1)} - v_i)^2}}$$

The temporal correlation coefficient CORT $\in [-1, 1]$ estimates how much the local trends observed simultaneously on both times series, are positively/negatively dependent. By dependence between time series we mean a stochastic linear dependence: if we know at a given time t the growth of the first time series then we can predict, through a linear relationship, the growth of the second time series at that time t. Similar to the classical correlation coefficient, a value of CORT $= 1$ means that, at a given time t, the trends observed on both time series are similar in direction and rate of growth, a value of -1 means that, at a given time t, the trends observed on both time series are similar in rate but opposite in direction and finally, a value of 0 expresses that the trends observed on both time series are stochastically linearly independent.

Contrary to the classical correlation coefficient, the temporal correlation estimates locally not globally the dependency between trends; indeed, two time series may be highly dependent through the classical correlation and linearly independent through the temporal correlation (illustrated in section 4). Finally, contrary to classical correlation, the temporal correlation is global trend effect free. Let's now present the new dissimilarity index as an extension of the Fréchet distance.

3.2 The dissimilarity index

The proposed dissimilarity index consists in the combination of two components. The first one, estimates the closeness of values and is based on a normalized form of the Fréchet distance. The second one is based on the temporal correlation introduced above. Many functions could be explored for such combination function. To illustrate well the additive value of the temporal correlation to account for local trends dependency, we limit this work to a linear combination function. Let's note $DisF$ the dissimilarity index extending δ_F:

$$DisF(S_1, S_2) = \alpha \frac{\delta_F(S_1, S_2)}{\max_{S_i, S_j \in \Omega_S} \delta_F(S_i, S_j)} + (1 - \alpha) \frac{1 - \textsc{cort}(S_1, S_2)}{2}$$

where $DisF(S_1, S_2) \in [0, 1]$, Ω_S is the set of the observed time series, $\alpha \in [0, 1]$ determines the weight of each component in the dissimilarity evaluation and CORT the temporal correlation defined above.

Note that for $\alpha = 1$, $DisF$ corresponds to the normalized δ_F and the proximity between two time series is only based on taken values, considered as independent observations. For $\alpha = 0$, $DisF$ corresponds to CORT and the proximity is based solely on the dependency between local trends. Finally for $0 < \alpha < 1$, $DisF$ implies a weighted mean of the normalized δ_F and CORT, the proximity between time series includes then, according to their weights, both the proximity between occurring values and the dependency between local trends.

4 Applications and results

In this section, we first present the time series synthetic dataset which reproduces the limited conditions for Fréchet distance. Then we explore and compare the distribution of the temporal and classical correlations between the synthetic dataset time series. Finally, the proposed dissimilarity index is compared to the conventional estimate.

4.1 Synthetic dataset

To reproduce the limited conditions for the widely used conventional distances, we consider a synthetic dataset of 15 time series divided into three classes of functions. The first five time series are of class F_1, the next five are of class F_2 and the last five are of the class F_3; where, F_1, F_2 and F_3 are defined as follows:

$$F_1 = \{f_1(t) \mid f_1(t) = f(t) + 2t + 3 + \epsilon\}$$
$$F_2 = \{f_2(t) \mid f_2(t) = \mu - f(t) + 2t + 3 + \epsilon\}$$
$$F_3 = \{f_3(t) \mid f_3(t) = 4f(t) - 3 + \epsilon\}$$

$f(t)$ is a given discrete function, $\mu = E(f(t))$ is the mean of $f(t)$ through the observation period, $\epsilon \rightsquigarrow N(0,1)$ is a zero mean gaussian distribution and 2t + 3 describes a linear upward trend tainting F_1 and F_2 classes. Figure 1 represents simultaneously these three classes through 15 synthetic time series. Note that F_1 and F_3 show similar local tendencies, they increase (respectively

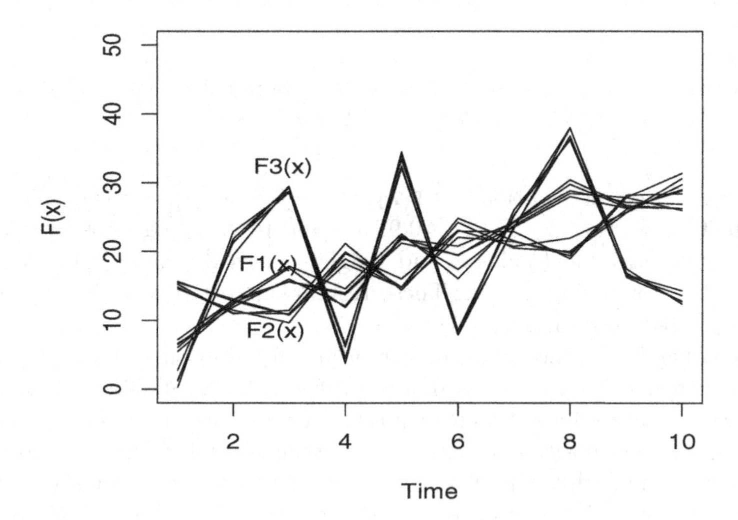

Fig. 1. Three classes of synthetic time series

decrease) simultaneously. On the contrary, F_2 shows local tendencies opposite to those of F_1 and F_3, when F_2 increases (respectively decreases) F_1 and F_3 decreases (respectively increases). Finally, F_1 and F_2 are the closest in values.

4.2 Time series temporal correlation vs classical correlation

Let's explore in figure 2 the distribution of the temporal and classical correlations among the times series into F_1, F_2 and F_3 classes. On the one hand, the

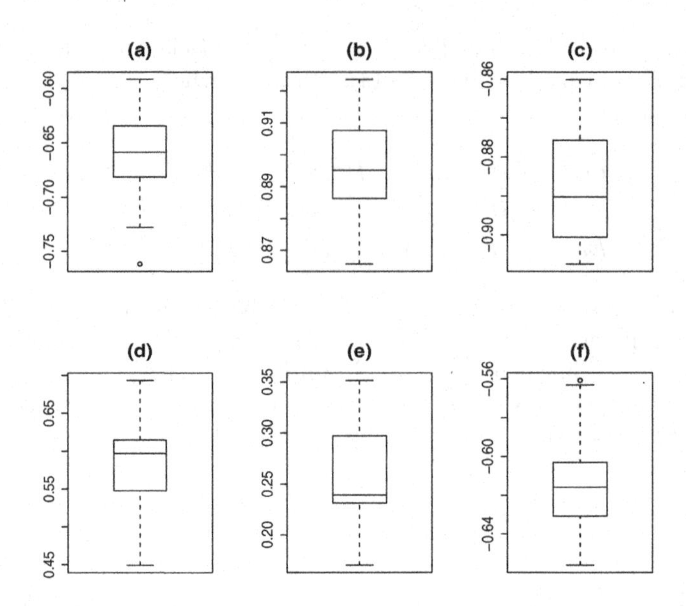

Fig. 2. **(a)** $\text{CORT}(F_1(x), F_2(x))$ **(b)** $\text{CORT}(F_1(x), F_3(x))$ **(c)** $\text{CORT}(F_2(x), F_3(x))$ **(d)**$\text{COR}(F_1(x), F_2(x))$ **(e)** $\text{COR}(F_1(x), F_3(x))$ **(f)** $\text{COR}(F_2(x), F_3(x))$

temporal correlation distribution $\text{CORT}(F_1, F_3) \in [0.87, 0.92]$, $\text{CORT}(F_1, F_2) \in [-0.73, -0.60]$ and $\text{CORT}(F_2, F_3) \in [-0.91, -0.86]$) reveal a high positive dependency between F_1 and F_3 classes and a high negative dependency between F_2 and the two remaining classes. These results supported well the dependencies illustrated above in figure 1.

On the other hand, the classical correlation distribution $\text{COR}(F_1, F_3) \in [0.15, 0.35]$, $\text{COR}(F_1, F_2) \in [0.45, 0.70]$ and $\text{COR}(F_2, F_3) \in [-0.66, -0.56]$) indicates a weak (nearly independence) positive dependency between F_1 and F_3 classes and a high positive dependency between F_1 and F_2 classes. These results illustrate well that the classical correlation estimates globally (not locally) the dependency between tendencies of time series. Indeed, F_1 and F_2 which are not locally but globally dependent, due to the linear upward trend tainting them, are considered as highly dependent; whereas F_1 and F_3 which are dependent locally not globally are considered as very weakly dependent. Note that contrary to classical correlation, the temporal correlation is global-trend effect free.

4.3 Comparative analysis

To compare the above proximity measures, we estimate first the proximity matrices between the 15 synthetic time series, according to $DisF$ and δ_F. $DisF$ is evaluated with $\alpha = 0.5$ and $\alpha = 0$. For $\alpha = 1$, results are similar to those obtained from δ_F. A hierarchical cluster analysis is then performed on the obtained proximity matrices. Figure 3 illustrates the obtained dendograms. Note first that the three above proximity measures (δ_F,

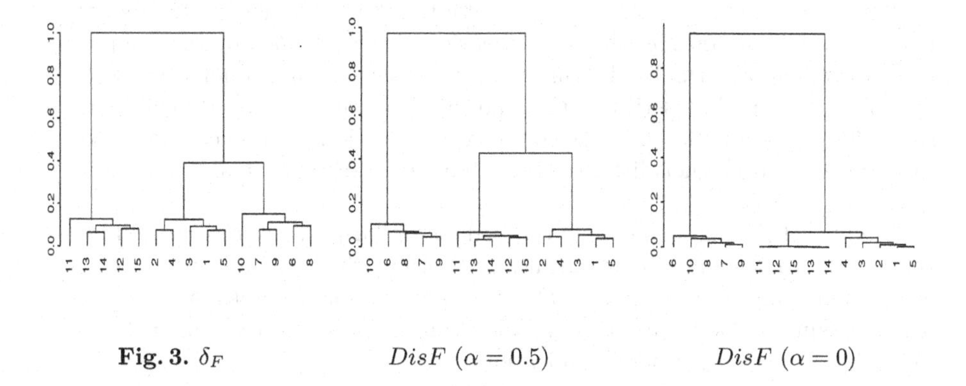

Fig. 3. δ_F $DisF$ $(\alpha = 0.5)$ $DisF$ $(\alpha = 0)$

$DisF(\alpha = 0.5)$ and $DisF(\alpha = 0)$) divide the 15 time series on the well expected three classes F_1 (from 1 to 5), F_2 (from 6 to 10) and F_3 (from 11 to 15). In addition, on the one hand, δ_F dendogram works out the time series of the classes F_1 and F_2 as the closest. Indeed, for δ_F, after stretching each class to match well an other class, the proximity evaluation is based solely on the taken values, which are close on F_1 and F_2.

On the other hand, $DisF$ for $\alpha = 0.5$ and $\alpha = 0$ determines successfully the classes F_1 and F_3 as the closest. Note particularly that for $\alpha = 0.5$ $DisF$ still provides three classes with a high proximity between F_1 and F_3; whereas for $\alpha = 0$ F_1 and F_3 are nearly merged and the respective dendogram comes out with only two main classes. Indeed, for $\alpha = 0$ the proximity evaluation are based solely on the dependency between time series which is very high between F_1 and F_3.

5 Discussion and conclusion

This paper focuses on the Fréchet distance between time series. We have provided the definitions and properties of this conventional measure. Then we illustrated the limits of this distance. To alleviate these limits, we propose a new dissimilarity index based on the temporal correlation to include the information of dependency between the local trends.

Note that, as this paper introduces the benefits of the temporal correlation for time series proximity estimation, and mainly for clarity reasons, we limit our work on two points. First we restrict the combination function to a linear function to show clearly, by varying the parameter α, the additive value of the temporal correlation. Secondly, we restrict the illustration of the proposed index to a synthetic dataset which reproduces the limited conditions for the conventional Fréchet distance.

Future works, on the one hand, will study other combination functions. For instance, if we consider the two dimensional space defined by the components CORT and a normalized form of δ_F, then we can define a new euclidean distance between time series as their norm vector in such two dimensional space. On the second hand, these combination functions will be compared to the conventional Fréchet distance through a wide range of a real datasets.

Finally, let's remark that the proposed dissimilarity index $DisF$ could be very useful for time series classification problem, where the aim consists in determining the most adaptable $DisF$ by looking for the optimal value of α maximizing a classification rate. This is an interesting direction to study through a priori time series classification.

References

GEARY, R.C. (1954): The contiguity ratio and statistical mapping. *The Incorporated Statistician, 5/3, 115-145.*

VON NEUMANN, J. (1941): Distribution of the ratio of the mean square successive difference to the variance. *The Annals of Mathematical Statistics, 12/4.*

VON NEUMANN, J., KENT, R.H., BELLINSON, H.R. and HART, B.I. (1942): The mean square successive difference to the variance. *The Annals of Mathematical Statistics. 153-162.*

FRÉCHET, M. (1906): Sur quelques points du calcul fonctionnel. Rendiconti del Circolo Mathematico di Palermo, 22, 1-74.

GODAU, M. (1991): A natural metric for curves - computing the distance for polygonal chains and approximation algorithms. In Proc. 8th Sympos. Theor. Aspects of Comp. STACS, LNCS 480, 127-136.

ALT, H. and GODAU, M. (1992): Measuring the resemblance of polygonal curves. In Proc. 8th Annu. ACM Sympos. Comput. Geom. 102-109.

EITER T. and MANNILA, H. (1994): Computing Discrete Fréchet distance, Technical Report CD-TR 94/64, Christian Doppler Laboratory for Expert Systems. TU Vienna, Austria.

CHOUAKRIA-DOUZAL, A. (2003): Compression Technique Preserving Correlations of a Multivariate Temporal Sequence. In: M.R. Berthold, H-J Lenz, E. Bradley, R. Kruse, C. Borgelt (eds.). *Advances in Intelligent Data Analysis, 5, Springer, 566-577.*

Comparison of Distance Indices Between Partitions

Lucile Denœud[1,2] and Alain Guénoche[3]

[1] École nationale supérieure des télécommunications, 46, rue Barrault, 75634 Paris cedex 13 (e-mail: denoeud@infres.enst.fr)
[2] CERMSEM CNRS-UMR 8095, MSE, Université Paris 1 Panthéon-Sorbonne, 106-112, boulevard de l'Hôpital, 75647 Paris cedex 13
[3] Institut de Mathématiques de Luminy, 163, avenue de Luminy, 13009 Marseille (e-mail: guenoche@iml.univ-mrs.fr)

Abstract. In this paper, we compare five classical distance indices on $\mathcal{P}_n$, the set of partitions on n elements. First, we recall the definition of the *transfer distance* between partitions and an algorithm to evaluate it. Then, we build sets $\mathcal{P}_k(P)$ of partitions at k transfers from an initial partition P. Finally, we compare the distributions of the five index values between P and the elements of $\mathcal{P}_k(P)$.

1 Introduction

The comparison of partitions is a central topic in clustering, as well for comparing partitioning algorithms as for classifying nominal variables. The literature abounds in distances (small values when partitions are close) or indices (large values when partitions are close) defined by many authors to compare two partitions P and Q defined on the same set X. The most commonly used are: the Rand index (1971), the Jaccard index and the Rand index corrected for chance (Hubert and Arabie, 1985). We also test two indices derived from lattice distances on partitions. The definitions of these indices are given in Section 4. The comparison of these indices is only interesting (from a practical point of view) if we consider close partitions, which differ randomly one from each others as mentioned by Youness and Saporta (2004). They generate such partitions from an Euclidian representation of the elements of X. Here, we develop a more general approach, independent of the representation space for X.

In 1964, Régnier used a distance between partitions which fits this type of study. It is the minimum number of transfers of one element from its class to another (eventually empty) to turn P into Q (see Section 2). The computational aspect has been largely studied by Day (1981) and we have recently bounded the distance between two partitions with respect to their number of classes (Charon et al., 2005). Here we compare the distributions of the similarity indices mentioned above on partitions at k transfers from P. If k is small enough, these partitions represent a small percentage α of all the partitions of X. This permits to define a value k_α of the maximum number of transfers allowed (see Section 3), and to build sets of random partitions at k transfers from P for $k < k_\alpha$. The results of the comparison of the indices for some given initial partition P are given in Section 5.

2 The transfer distance

Let P and Q be two partitions on a set X of n elements with respectively $|P| = p$ and $|Q| = q$ classes (also called cells in the literature); we will assume that $p \leq q$.

$$P = \{P_1, .., P_p\} \text{ and } Q = \{Q_1, .., Q_q\}.$$

The minimum number of transfers to turn P into Q, denoted $\theta(P, Q)$, is obtained by establishing a matching between the classes of P and those of Q keeping a maximum number of elements in the matched classes, those that do not need to be moved. Consequently, we begin to add $q - p$ empty cells to P, so that P is considered also as a partition with q classes.

Let Υ be the mapping from $P \times Q \longrightarrow \mathbb{N}$ which associates to any pair of classes the cardinal of their intersection. Classically, $n_{i,j} = |P_i \cap Q_j|$ and $p_i = |P_i|$ and $q_j = |Q_j|$ denote the cardinals of the cells. Let Δ be the mapping which associates to each pair of classes (P_i, Q_j) the cardinal of their symmetrical difference, noted $\delta_{i,j}$. We have $\delta(i,j) = p_i + q_j - 2 \times n_{i,j}$. We consider the complete bipartite graph whose vertices are the classes of P and Q, with edges weighted either by Υ (denoted $(K_{q,q}, \Upsilon)$) or by Δ ($(K_{q,q}, \Delta)$).

Proposition 1 (Day 1981) *The bijection minimizing the number of transfers between two partitions P and Q with q classes corresponds to a matching of maximum weight w_1 in $(K_{q,q}, \Upsilon)$ or, equivalently, to a perfect matching of minimum weight w_2 in $(K_{q,q}, \Delta)$; moreover, $\theta(P, Q) = n - w_1 = \frac{w_2}{2}$.*

Establishing the bipartite graph can be done in $O(n^2)$. The weighted matching problem in a complete bipartite graph can be solved by an assignment method well-known in operational research. The algorithm has a polynomial complexity in $O(q^3)$. We will not go into further details, given for instance in J. Van Leeuwen (1990). We just develop an example of the computation of the transfer distance.

Example 1 *We consider the two partitions* $P = (1, 2, 3|4, 5, 6|7, 8)$ *and* $Q = (1, 3, 5, 6|2, 7|4|8)$. *The following two tables correspond to the intersections and to the symmetrical differences between classes of P and Q. Two extreme matchings are edited in bold. Each one gives* $\theta(P, Q) = 4$.

To the maximum weighted matching in the table Υ may correspond the series of 4 transfers: $(1, 2, 3|4, 5, 6|7, 8) \rightarrow (1, 3|4, 5, 6|2, 7, 8) \rightarrow (1, 3, 5|4, 6|2, 7, 8) \rightarrow (1, 3, 5, 6|4|2, 7, 8) \rightarrow (1, 3, 5, 6|4|2, 7|8)$.

To the minimum weighted perfect matching in the table Δ may correspond another optimal series: $(1, 2, 3|4, 5, 6|7, 8) \rightarrow (1, 2, 3, 7|4, 5, 6|8) \rightarrow (2, 3, 7|1, 4, 5, 6|8) \rightarrow (2, 7|1, 3, 4, 5, 6|8) \rightarrow (2, 7|1, 3, 5, 6|8|4)$.

Υ	1,3,5,6	2,7	4	8	Δ	1,3,5,6	2,7	4	8
1,2,3	**2**	1	0	0		3	**3**	4	4
4,5,6	2	0	**1**	0		**3**	5	2	4
7,8	0	**1**	0	1		6	2	3	**1**
$\emptyset$	0	0	0	**0**		4	2	**1**	1

3 Close partitions with respect to transfers

We note $\mathcal{P}_n$ the set of partitions on a set of n elements and $\mathcal{P}_k(P)$ the set of partitions at k transfers from P and $\mathcal{P}_{\leq k}(P)$ the set of partitions at at most k transfers from P.

$$\mathcal{P}_{\leq k}(P) = \{Q \in \mathcal{P}_n \text{ such that } \theta(P,Q) \leq k\} = \bigcup_{0 \leq i \leq k} \mathcal{P}_i(P)$$

Statistically, we consider that a partition Q is close to P at threshold α if there are less than α percent of the partitions that are closer to P than Q with respect to transfers. The matter is then to know how many partitions are within a k radius from P. For $k = 0$, there is just one partition, P itself, otherwise θ would not be a distance. We can easily give a formula for $\mathcal{P}_1(P)$, but for larger k it becomes more difficult. For a given partition P, we call *critical value* at threshold α, the greatest number of transfers k_α such as

$$\frac{|\mathcal{P}_{\leq k_\alpha}(P)|}{|\mathcal{P}_n|} \leq \alpha.$$

While $n \leq 12$, we can enumerate all the partitions in $\mathcal{P}_n$ to compute $|\mathcal{P}_k(P)|$. For that, we use the procedure *NexEqu* in Nijenhuis and Wilf (1978). The algorithm builds the next partition with respect to the lexicographic order, starting from the partition with a single class.

For $n > 12$, there are too many partitions to achieve an exhaustive enumeration. Then we select at random a large number of partitions to be compared to P, to estimate $|\mathcal{P}_{\leq k}(P)|/|\mathcal{P}_n|$. To obtain a correct result, these partitions must be equiprobable; the book by Nijenhuis and Wilf provides also such a procedure (*RandEqu*).

Thus we measure a frequency f in order to estimate a proportion p. We want to find a confidence interval around $p = 0.1$ for a risk ρ fixed ($\rho = 5\%$) and a gap δ between f and p judged as acceptable ($\delta = 0.01$). For these values, we can establish the size of the sample S using the classical formula

$$t(\rho)\sqrt{\frac{f(1-f)}{|S|}} \leq \delta$$

in which $t(\rho)$ is given by the normal Gauss distribution (Brown et al 2002). We obtain that 3600 trials should be carried out, which is quite feasible. Note that this number decreases with p (when $p < 0.5$) and it is independent of n.

Example 2 *For $n = 12$, there are $|\mathcal{P}_{12}| = 4213597$ partitions that can be compared to P in order to establish the distribution of $|\mathcal{P}_k(P)|$ according to k. For $P = \{1,2,3,4|5,6,7|8,9|10,11|12\}$, as for all the partitions with classes having the same cardinality, the number of partitions at $0,\ldots,8$ transfers from P are respectively* 1, 57, 1429, 20275, 171736, 825558, 1871661, 1262358, 60522. *The cumulated proportions in % are respectively* 0.0, 0.0, 0.0, 0.5, 4.6, 24.2, 68.6, 99.6, *and* 100. *For $\alpha = 5\%$ the critical value is 4; indeed there are just 4.6% of the partitions that are at 4 transfers or less from P, while for 5 transfers, there are 24.2%. The cumulated frequencies computed from P and 5000 random partitions are:* $0.0, 0.0, 0.1, 0.5, 4.4, 23.9, 68.7, 98.3$ *and* 100. *Thus the critical value computed by sampling is also equal to* 4.

4 Distance indices between partitions

Classically, the comparison of partitions is based on the pairs of elements of X. Two elements x and y can be joined together or separated in P and Q. Let r be the number of pairs simultaneously joined together, s the number of pairs simultaneously separated, and u (resp. v) the number of pairs joined (resp. separated) in P and separated (resp. joined) in Q. Also classical is the following partial order relation on $\mathcal{P}_n$ providing a lattice structure. A partition P is finer than Q ($P \prec Q$) if and only if all the classes of P are included in those of Q (dually, Q is coarser than P):

$$P \prec Q \text{ iff } \forall (i,j)\ P_i \cap Q_j \in \{P_i, \emptyset\}.$$

In this lattice, the finest partition P^n has n singletons and the coarsest P^0 has only one class. Let us recall the meet ($\wedge$) and the join ($\vee$) definitions on partitions: The classes of partition $P \wedge Q$ are the intersections of those of P and Q ; the classes of $P \vee Q$ are the unions of classes having a non-empty intersection or the transitive closure of the relation "to be joined" in P or Q. We will note $\pi(P)$ the set of joined pairs in P, that is to say $|\pi(P)| = \sum_{i\in\{1,\ldots,p\}} \frac{n_i(n_i-1)}{2}$. According to the previous notations, we have

$$r = \sum_{i,j} \frac{n_{i,j}(n_{i,j}-1)}{2} = \pi(P \wedge Q).$$

We compare five of the most usual indices used in the literature. One can find a larger sample in Arabie & Boorman (1973) or Day (1981).

4.1 The Rand index

The Rand index, denoted R, is simply the fraction of pairs for which P and Q agree. It belongs to $[0,1]$ and $1 - R(P,Q)$ is the symmetrical difference distance between $\pi(P)$ and $\pi(Q)$.

$$R(P,Q) = \frac{r+s}{n(n-1)/2}.$$

4.2 The Jaccard index

In the Rand index, the pairs simultaneously joined or separated are counted in the same way. However, partitions are often interpreted as classes of joined elements, separations being the consequences of this clustering. We use then the Jaccard index (1908), denoted J, which does not take the s simultaneous separations into account:

$$J(P,Q) = \frac{r}{r+u+v}.$$

4.3 The corrected Rand index

In their famous paper of 1985, Hubert and Arabie noticed that the Rand index is not *corrected for chance* so that its expectation is equal to zero for random partitions having the same *type*, that is the same number of elements in their respective classes. They introduced the corrected Rand index, whose expectation is equal to zero, denoted here HA, to pay homage to the authors.

The corrected Rand index is based on three values: the number r of common joined pairs in P and Q, the expected value $Exp(r)$ and the maximum value $Max(r)$ of this index, among the partitions of the same type as P and Q. It leads to the formula

$$HA(P,Q) = \frac{r - Exp(r)}{Max(r) - Exp(r)}$$

with $Exp(r) = \frac{|\pi(P)| \times |\pi(Q)|}{n(n-1)/2}$ and $Max(r) = \frac{1}{2}(|\pi(P)| + |\pi(Q)|)$. This maximum value is questionable since the number of common joined pairs is necessarily bounded by $\inf\{|\pi(P)|, |\pi(Q)|\}$, but $Max(r)$ insures that the maximum value of HA is equal to 1 when the two partitions are identical. Note that this index can take negative values.

4.4 The Johnson index

This index is another normalization of the number of pairs simultaneously joined in P and in Q. It has been essentially suggested by Johnson in 1968, and therefore denoted Jo:

$$Jo(P,Q) = \frac{2r}{\pi(P) + \pi(Q)}.$$

A similar index has been proposed by Wallace (1983) using the geometrical average instead of the arithmetical one ; it gives almost identical results.

4.5 The Boorman index

This index has been proposed in 1973 and is denoted herein B.

$$B(P,Q) = 1 - \frac{|P| + |Q| - 2|P \vee Q|}{n-1}.$$

It is not based on the pairs of elements but on the number of classes of P, Q and $P \vee Q$; for any partition P there exists at least one partition Q such that $P \vee Q = P^n$, so one can divide the distance by $n-1$:

5 Comparison of indices

Let P be a partition on X with p classes. When $n = |X| \leq 12$, we enumerate the sets $\mathcal{P}_k(P)$, then we evaluate the minimum and maximum values of each index above between P and any Q belonging to $\mathcal{P}_k(P)$. We consider the results obtained for $n = 10$ and $P = (1,2,3,4,5|6,7,8,9,10)$ in Table 1. The partitions being at at most 3 transfers represent 1.7% of the 115975 partitions on 10 elements.

Table 1. Distribution of the number of partitions at k transfers from $P = (1,2,3,4,5|6,7,8,9,10)$ and extreme values of the distance indices.

Nb. of transfers	1	2	3	4	5	6	7	8
Nb. of partitions	20	225	1720	9112	31361	54490	17500	1546
R min	.80	.64	.53	.47	.44	.44	.44	.44
R max	.91	.87	.82	.78	.69	.64	.60	.56
J min	.64	.43	.32	.22	.15	.08	.04	0.0
J max	.80	.70	.60	.50	.44	.21	.10	0.0
HA min	.60	.28	.06	-.08	-.12	-.17	-.19	-.22
HA max	.82	.72	.63	.53	.32	.22	.11	0.0
Jo min	.78	.60	.49	.36	.27	.14	.08	0.0
Jo max	.89	.82	.75	.67	.62	.34	.18	0.0
B min	.78	.67	.56	.44	.33	.22	.11	0.0
B max	.89	.89	.78	.78	.89	.67	.56	.44

The lowest index value obtained for partitions at 3 transfers are quite small and do not reveal their closeness to P; it is particularly true for the corrected Rand index that can only be slightly positive. *A contrario*, the Rand and the Boorman indices can keep large values for partitions at maximum transfer distance. More serious, for each index the maximum value obtained for partitions at 5 transfers are greater than the minimum value obtained for 2 transfers, suggesting that partitions at 5 transfers from P can be closer to P than partitions at 2 transfers.

When $n > 12$, we cannot enumerate $\mathcal{P}_n$ anymore, nor select randomly equiprobable partitions at a small transfer distance, that are too rare. So, we build a set of partitions from P, by selecting recursively and at random one element and its new class number between 1 and $p+1$ (the number of classes is eventually updated). The number of iterations varies in order to obtain a set $\mathcal{Q}_k(P)$ of partitions for each value of $k \in \{1, ..., k_{5\%}\}$. We then compute the values of each index and the average, standard deviation, minimum and maximum values for any k.

Here we restrict ourselves to two partitions of 100 elements; P_1 with 5 balanced classes and P_2 having type (40, 30, 20, 5, 5). Both critical values at 5% are here equal to 83. The averages of the different indices are represented in Figure 1 which fits for both (we have considered here 170 000 partitions nearly uniformly distributed in $\mathcal{Q}_k(P)$ for $k \leq 80$).

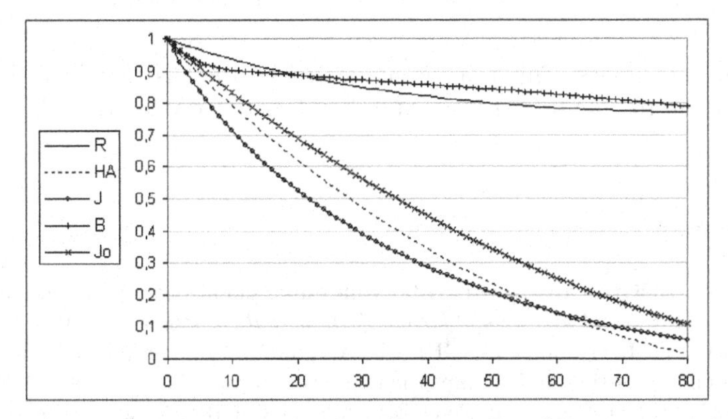

Fig. 1. Averages of the distance indices between P and partitions of $\mathcal{Q}_k(P)$

The indices of Jaccard, corrected Rand and Johnson have approximately the same behavior: they are high when k is small and decrease regularly near to 0 when $k = k_{5\%}$. Remarkably, the corrected Rand index becomes negative for larger transfer values. The Rand index and the Boorman index stay near 0.8 whatever is k ; consequently they seem unadapted for the comparison of partitions. Standard deviations of all the indices are similar in the range $[0; 0.03]$. Therefore we focus again on the extreme values of each index: We have computed in Table 5 the largest interval of transfer distance $[\theta^-; \theta^+]$ for partitions having the same index value; that is to say that two partitions being respectively at θ^- and θ^+ transfers from P can have the same index value.

Contrary to the case $n = 10$, the values of $\theta^+ - \theta^-$ are reasonably small for the indices of HA, J and Jo. That is to say that high values of these indices cannot be obtained by distant partitions in terms of transfer.

Table 2. Largest interval of transfer for partitions having the same index value

	R	HA	J	Jo	B
θ^-	36	45	32	32	9
θ^+	80	62	42	42	75
$\theta^+ - \theta^-$	44	17	10	10	66

We have done the same observations with several other initial partitions. We can conclude, according to the studied partitions, that the Jaccard and the Johnson indices are the most accurate to compare close partitions. The corrected Rand index comes after according to table 5. We also illustrate that these classical indices are correlated with the small values of the transfer distance, but only when n is large enough. For small n, the transfer distance is a much appropriate measure of the closeness of partitions.

Acknowledgements This work is supported by the CNRS ACI IMP-Bio. We also want to thank B. Fichet and J.P. Barthélemy for their help and advices.

References

ARABIE, P., BOORMAN, S.A. (1973): Multidimensional Scaling of Measures of Distance between Partitions, *Journal of Mathematical Psychology*, 10, 148–203.

BROWN, L., CAL, T. and DASGUPTA, A. (2002): Confidence intervals for a binomial proportion and asymptotic expansions, *Ann. Statist.*, 160–201.

CHARON, I., DENŒUD, L., GUÉNOCHE, A. and HUDRY, A. (2005): Comparing partitions by element transferts, submitted.

DAY, W. (1981): The complexity of computing metric distances between partitions, *Mathematical Social Sciences*, 1, 269-287.

HUBERT, L. and ARABIE, P. (1985): Comparing partitions, *J. of Classification*, 2, 193-218.

NIJENHUIS, A. and WILF, H. (1978): *Combinatorial algorithms*, Academic Press, New-York.

RAND, W.M. (1971): Objective criteria for the evaluation of clustering methods, *J. Amer. Statist. Assoc.*, 66, 846-850.

VAN LEEUWEN, J. (1990): *Handbook of thoretical computer science*, Vol A, Elsevier, Amsterdam.

WALLACE, D.L. (1983): Comment, *J. of the Am. Stat. Association*, 78, 569-579.

YOUNESS, G. and SAPORTA, G. (2004): Une méthodologie pour la comparaison des partitions, *Revue de Statistique Appliquée*, 52 (1), 97-120.

Design of Dissimilarity Measures: A New Dissimilarity Between Species Distribution Areas

Christian Hennig[1] and Bernhard Hausdorf[2]

[1] Department of Statistical Science, University College London
Gower St, London WC1E 6BT, United Kingdom
[2] Zoologisches Museum der Universität Hamburg
Martin-Luther-King-Platz 3, 20146 Hamburg, Germany

Abstract. We give some guidelines for the choice and design of dissimilarity measures and illustrate some of them by the construction of a new dissimilarity measure between species distribution areas in biogeography. Species distribution data can be digitized as presences and absences in certain geographic units. As opposed to all measures already present in the literature, the geco coefficient introduced in the present paper takes the geographic distance between the units into account. The advantages of the new measure are illustrated by a study of the sensitivity against incomplete sampling and changes in the definition of the geographic units in two real data sets.

1 Introduction

We give some guidelines for the choice and design of dissimilarity measures (in Section 2) and illustrate some of them by the construction of a new dissimilarity measure between species distribution areas in biogeography.

Species distribution data can be digitized as presences and absences in certain geographic units, e.g., squares defined by a grid over a map. In the so-called R-mode analysis in biogeography, the species are the objects between which the dissimilarity is to be analyzed, and they are characterized by the sets of units in which they are present.

More than 40 similarity and dissimilarity measures between distribution areas have already been proposed in the literature (see Shi, 1993, for 39 of them). The choice among these is discussed in Section 3.

Somewhat surprisingly, none of these measures take the geographic distance between the units into account, which can provide useful information, especially in the case of incomplete sampling of species presences. Section 4 is devoted to the construction of a new dissimilarity coefficient which incorporates a distance matrix between units. In most applications this will be the geographic distance.

Some experiments have been performed on two data sets of species distribution areas, which explore the high stability of the new measure under incomplete sampling and change of the grid defining the units. They are explained in detail in Hennig and Hausdorf (2005). An overview of their results is given in Section 5.

2 Some thoughts on the design of dissimilarity measures

In many situations, dissimilarities between objects cannot be measured directly, but have to be constructed from some known characteristics of the objects of interest, e.g. some values on certain variables.

From a philosophical point of view, the assumption of the objective existence of a "true" but not directly observable dissimilarity value between two objects is highly questionable. Therefore we treat the dissimilarity construction problem as a problem of the choice or design of such a measure and not as an estimation problem of some existing but unknown quantities.

Therefore, subjective judgment is necessarily involved, and the main aim of the design of a dissimilarity measure is the proper representation of a subjective or intersubjective concept (usually of subject-matter experts) of similarity or dissimilarity between the objects. Such a subjective concept may change during the process of the construction - the decisions involved in such a design could help the experts to re-think their conceptions. Often the initial expert's conceptions cannot even be assumed to be adequately representable by formulae and numbers, but the then somewhat creative act of defining such a representation may still have merits. It enables the application of automatic data analysis methods and can support the scientific discussion by making the scientist's ideas more explicit ("objectivating" them in a way).

Note that Gordon (1990) discussed the problem of finding variable weights in a situation where the researchers are able to provide a dissimilarity matrix between the objects but not a function to compute these values from the variables characterizing the objects, in which case the design problem can be formalized as a mathematical optimization problem. Here we assume that the researchers cannot (or do not want) to specify all dissimilarity values directly, but rather are interested in formalizing their general assessment principle, which we think supports the scientific discourse better than to start from subjectively assigned numbers.

The most obvious subjective component is the dependence of a dissimilarity measure on the research aim. For example, different similarity values may be assigned to a pair of poems depending on whether the aim is to find poems from the same author in a set of poems with unknown author or to assess poems so that somebody who likes a particular poem will presumably also like poems classified as similar. This example also illustrates a less subjective aspect of dissimilarity design: the quality of the measure with respect to the research aim can often be assessed by observations (such as the analysis of dissimilarities between poems which are known to be written by the same author). Such analyses, as well as the connection of the measure to scientific knowledge and common sense considerations, improve the scientific acceptability of a measure.

A starting point for dissimilarity design is the question: "how can the researcher's (or the research group's) idea of the similarity between objects given a certain research aim be translated into a formally defined function of the observed object characteristics?"

This requires at first a basic identification of how the observed characteristics are related to the researcher's concept. For species distribution areas, we start with the idea that similarity of two distribution areas is the result of the origin of the species in the same "area of endemism", and therefore distribution areas should be treated as similar if this seems to be plausible. Eventually, the dissimilarity analysis (using techniques like ordination and cluster analysis) could provide us with information concerning the historic process of the spciation (Hausdorf and Hennig, 2003).

It is clear that the dissimilarity measure should become smaller if (given constant sizes of the areas) the number of units in which both species are present becomes larger. Further, two very small but disjunct distribution areas should not be judged as similar just because the number of units in which both species are not present is large, while we would judge species present at almost all units as similar even if their few non-occurrences don't overlap. This suggests that the number of common absences is much less important (if it has any importance at all) for dissimilarity judgments than the number of common presences. The species distribution area problem is discussed further in the next section.

Here are some further guidelines for the design and "fine-tuning" of dissimilarity measures.

- After having specified the basic behaviour of the dissimilarity with respect to certain data characteristics, think about the importance weights of these characteristics. (Note that variable weights can only be interpreted as importance weights if the variables are suitably standardized.)
- Construct exemplary (especially extreme) pairs of objects in which it is clear what value the dissimilarity should have, or at least how it should compare with some other exemplary pairs.
- Construct sequences of pairs of objects in which one characteristic changes while others are held constant, so that it is clear how the dissimilarity should change.
- Think about whether and how the dissimilarity measure could be disturbed by small changes in the characteristics, what behaviour in these situations would be adequate and how a measure could be designed to show this behaviour.
- Think about suitable invariance properties. Which transformations of the characteristics should leave the dissimilarities unchanged (or only changed in a way that doesn't effect subsequent analyses, e.g. multiplied

by a constant)? There may be transformations under which the dissimilarities can only be expected to be approximately unchanged, e.g. the change of the grid defining the geographic units for species areas.
- Are there reasons that the dissimilarity measure should be a metric (or have some other particular mathematical properties)?
- The influence of monotone characteristics on the dissimilarity should not necessarily be linear, but can be convex or concave (see the discussion of the function u below).
- If the measure should be applied to a range of different situations, it may be good to introduce tuning constants, which should have a clear interpretation in terms of the subject matter.

3 Jaccard or Kulczynski coefficient?

We denote species areas as sets A of geographic units, which are subsets of the total region under study $R = \{r_1, \ldots, r_k\}$ with k geographic units. $|A|$ denotes the number of elements in A (size of A).

The presumably most widely used dissimilarity measure in biogeography is the Jaccard coefficient (Jaccard, 1901)

$$d_J(A_1, A_2) = 1 - \frac{|A_1 \cap A_2|}{|A_1 \cup A_2|}.$$

This distance has a clear direct interpretation as the proportion of units present in A_1 or A_2, but not in both of them. It does not depend on the number of common absences, which is in accord with the above discussion.

However, there is an important problem with the Jaccard distance. If a smaller area is a subset of a much larger area, the Jaccard distance tends to be quite large, but this is often inadequate. For example, if there are $k = 306$ units (as in an example given below), $A \subset B$, $|A| = 4, |B| = 20$, we have $d_J(A, B) = 0.8$, though both species may have originated in the same area of endemism. A may only have a worse ability for dispersal than B. We would judge A as more similar (in terms of our research aims) to B than for example a species C with $|C| = 20$, $|B \cap C| = 10$, but $d_J(B, C) = 0.67$. The reason is that the Jaccard denominator $|A_1 \cup A_2|$ is dominated by the more dispersed species which therefore has a higher influence on the computation of d_J.

Giving both species the same influence improves the situation, because $|A \cap B|$ is small related to $|B|$, but large related to $|A|$. This takes into account differences in the sizes of the species areas to some extent (which is desirable because very small species areas should not be judged as very similar to species occupying almost all units), but it is not dominated by them as strongly as the Jaccard distance. This leads to the Kulczynski coefficient (Kulczynski, 1927)

$$d_K(A_1, A_2) = 1 - \frac{1}{2}\left(\frac{|A_1 \cap A_2|}{|A_1|} + \frac{|A_1 \cap A_2|}{|A_2|}\right),$$

for which $d_K(A, B) = 0.4$ and $d_K(B, C) = 0.5$ while the good properties of the Jaccard coefficient mentioned above are preserved. However, the Jaccard coefficient is a metric (Gower and Legendre, 1986) while the triangle inequality is not fulfilled for the Kulczynski coefficient. This can be seen as follows. Consider $D \subset B$, $|D| = 4$, $|D \cap A| = 0$. Then $d_K(D, B) + d_K(B, A) = 0.8 < d_K(A, D) = 1$. But this makes some sense. Using only set relations and ignoring further geographical information, the dissimilarity between A and D should be the maximal value of 1 because they are disjunct. On the other hand, for the reasons given above, it is adequate to assign a small dissimilarity to both pairs A, B and B, D, which illustrates that our subject matter concept of dissimilarity is essentially non-metric. Therefore, as long as we do not require the triangle inequality for any of the subsequent analyses, it is more adequate to formalize our idea of dissimilarity by a non-metric measure. Actually, if we apply a multidimensional scaling algorithm to embed the resulting dissimilarity matrix in the Euclidean space, such an algorithm will essentially reduce the distance between A and D in the situation above, which is satisfactory as well, because now the fact that the common superset B exists can be taken into account to find out that A and D may have more in common than it seems from just looking at $A \cap D$. For example, they may be competitors and therefore not share the same units, but occur in the same larger area of endemism.

Note that the argument given above is based on the fact that $|B| = 20$ is much smaller than the whole number of units. This suggests that a more sophisticated approach may further downweight the relation of $|A_1 \cap A_2|$ to the size of the larger area, dependent on the number of common absences (an extreme and for our aims certainly exaggerated suggestion is the consideration of $1 - \frac{|A_1 \cap A_2|}{|A_1|}$ where A_1 is the smaller area, see Simpson, 1960).

4 Incorporating geographic distances

Assume now that there is a distance d_R defined on R, which usually will be the geographic distance. Obviously this distance adds some useful information. For example, though A and D above are disjunct, the units of their occurrence could be neighboring, which should be judged as a certain amount of similarity in the sense of our conception.

Furthermore, small intersections (and therefore large values of both d_J and d_K) between seemingly similar species areas may result from incomplete sampling or very fine grids.

The motivation for the definition of our new geco coefficient (the name comes from "geographic distance and congruence") was that we wanted to maintain the equal weighting of the species of the Kulczynski coefficient while incorporating the information given by d_R.

The general definition is

$$d_G(A_1, A_2) = \frac{1}{2}\left(\frac{\sum_{a\in A_1} \min_{b\in A_2} u(d_R(a,b))}{|A_1|} + \frac{\sum_{b\in A_2} \min_{a\in A_1} u(d_R(a,b))}{|A_2|}\right),$$

where u is a monotone increasing transformation with $u(0) = 0$. To motivate the geco coefficient, consider for a moment u as the identity function. Then, d_G is the mean of the average geographic distance of all units of A_1 to the respective closest unit in A_2 and the average geographic distance of all units of A_2 to the respective closest unit in A_1. Thus, obviously, $d_G(A_1, A_1) = 0$, $d_G(A_1, A_2) \geq 0$, $d_G(A_1, A_2) = d_G(A_2, A_1)$ and $d_G(A_1, A_2) \leq \max u(d_R)$. If $u(d_R(a, b)) > 0$ for $a \neq b$, then $d_G(A_1, A_2) > 0$ for $A_1 \neq A_2$. d_G reduces to the Kulczynski coefficient by taking $d_R = \delta$ with $\delta(a, a) = 0$, $\delta(a, b) = 1$ if $a \neq b$ and u as the identity function, because

$$|A_1 \cap A_2] = \sum_{a\in A_1} \min_{b\in A_2} (1 - \delta(a, b)) = \sum_{b\in A_2} \min_{a\in A_1} (1 - \delta(a, b)).$$

It follows that d_G is not generally a metric, though it may become a metric under certain choices of u and d_R (δ is a metric, which shows that demanding d_R to be a metric does not suffice). Given that A and D from the example of the previous Section are far away from each other and B is present at both places, the violation of the triangle inequality may still be justified.

Note that for general d_R, $\sum_{a\in A} \min_{b\in B}(1-d_R(a, b)) = \sum_{b\in B} \min_{a\in A}(1-d_R(a, b))$ does not hold, and therefore it is favorable for the aim of generalization that $|A \cap B|$ appears in the definition of the Kulczynski coefficient related to $|A|$ *and* $|B|$. A corresponding generalization of the Jaccard distance would be less intuitive.

The identity function may be reasonable as a choice for u in particular situations, but often it is not adequate. Consider as an example d_R as geographic distance, and consider distribution areas A, B, C and D all occupying only a single geographic unit, where the unit of A is 10 km distant from B, 5000 km distant from C and 10000 km distant from D. Then, if u is the identity function, the geco distances from A to B, C and D are 10, 5000 and 10000, thus distribution area D is judged as twice as different from A than C. But while in many circumstances a small geographic distance is meaningful in terms of the similarity of distribution areas (because species may easily get from one unit to another close unit and there may be similar ecological conditions in close units, so that species B is in fact similar to A), the differences between large distances are not important for the similarity between species areas and units which are 5000 and 10000 km away from A may both simply not be in any way related to the unit of A. Thus, we suggest for geographical distances a transformation u that weights down the differences between large distances. A simple choice of such a transformation is the following:

$$u(d) = u_f(d) = \begin{cases} \frac{d}{f*\max d_R} & : d \leq f * \max d_R \\ 1 & : d > f * \max d_R \end{cases}, 0 \leq f \leq 1.$$

That is, u_f is linear for distances smaller than f times the diameter (maximum geographical distance) of the considered region R, while larger geographical distances are treated as "very far away", encoded by $u_f = 1$. This yields $\max d_G = \max u(d_R) = 1$, makes the geco coefficient independent of the scaling of the geographical distances (kilometers, miles etc.) and directly comparable to the Kulczynski distance. In fact, $f = 0$ (or f chosen so that $f*\max d_R$ is smaller than the minimum nonzero distance in R) yields the Kulczynski distance, and $f = 1$ is equivalent to u chosen as the identity function scaled to a maximum of 1. f should generally be chosen so that $f*\max d_R$ can be interpreted as the minimal distance above which differences are no longer meaningful with respect to the judgment of similarity of species. We suggest $f = 0.1$ as a default choice, assuming that the total region under study is chosen so that clustering of species may occur in much smaller subregions, and that relevant information about a particular unit (e.g., about possible incomplete sampling) can be drawn from a unit which is in a somewhat close neighborhood compared to the whole area of the region. $f = 0.1$ has been used in both experiments below. A larger f may be adequate if the region under study is small, a smaller f may be used for a very fine grid.

There are alternatives to the choice of u that have a similar effect, e.g., $u(d) = \log(f * d + 1)$. However, with this transformation, f would be more difficult to choose and to interpret.

The geco coefficient may be used together with more sophisticated measures d_R quantifying for example dissimilarities with respect to ecological conditions between units or "effective distances" taking into account geographical barriers such as mountains.

5 Experiments with the geco coefficient

We carried out two experiments to explore the properties of the geco coefficient and to compare it with the Kulczynski coefficient. Full descriptions and results can be found in Hennig and Hausdorf (2005).

The first experiment considers the sensitivity against incomplete sampling. The data set for this experiment includes the distribution of 366 land snail species on 306 grid squares in north-west Europe. The data set has been compiled from the distribution maps of Kerney et al. (1983). These maps are interpolated, i.e., presences of a species have been indicated also for grid squares in which it might have not been recorded so far, but where it is probably present, because it is known from the surrounding units. Therefore this data set is artificially "complete" and especially suitable to test the effect of incomplete sampling on biogeographical analyses.

To simulate incomplete sampling, every presence of a species in a geographic unit given in the original data set has been deleted with a probability P (which we chose as 0.1, 0.2 and 0.3 in different simulations; 100 replications have been performed for all setups) under the side condition that every species is still present in the resulting simulated data. To compare the Kulczynski distance and the geco coefficient, we computed the Pearson correlation between the vector of dissimilarities between species in the original data set and the vector of dissimilarities between species in the simulated data set. We also carried out a non-metric MDS and a cluster analysis based on normal mixtures (see Hennig and Hausdorf, 2005, for the whole methodology) and compared the solutions from the contaminated data sets with the original solutions by means of a Procrustes-based coefficient (Peres-Neto and Jackson, 2001) and the adjusted Rand index (Hubert and Arabie, 1985).

In terms of Pearson correlations to the original data set, the geco coefficient yielded mean values larger than 0.975 for all values of P and outperformed the Kulczynski coefficient on all 300 simulated data sets. The results with respect to the MDS and the clustering pointed into the same direction. The tightest advantage for the geco coefficient was that its clusterings obtained a better Rand index than Kulczynski "only" in 78 out of 100 simulations for $P = 0.1$.

The second experiment explores the sensitivity against a change of the grid. The data set for this experiment includes the distribution of 47 weevil species in southern Africa. We used a presence/absence matrix for 2 degree latitude x 2 degree longitude grid cells as well as a presence/absence matrix for 1 degree latitude x 1 degree longitude grid cells, both given by Mast and Nyffeler (2003).

Hausdorf and Hennig (2003) analyzed the biotic element (species area cluster) composition of the weevil genus *Scobius* in southern Africa using Kulczynski distances. The results obtained with a 1 degree grid differed considerably from those obtained with a 2 degree grid. On the coarser 2 degree grid, a more clear clustering and more seemingly meaningful biotic elements have been found, though the finer grid in principle provides more precise information. Hausdorf and Hennig (2003) suggested that "If the grid used is too fine and the distribution data are not interpolated, insufficient sampling may introduce artificial noise in the data set".

If the 1 degree grid is analysed with the geco coefficient, the structures found on the 2 degree grid by geco and Kulczynski coefficients can be reproduced and even a further biotic element is found. The geco analyses on both grids are much more similar to each other (in terms of Pearson correlation, Procrustes and adjusted Rand index) than the two Kulczynski analyses.

6 Conclusion

We discussed and introduced dissimilarity measures between species distribution areas. We used some techniques that are generally applicable to the design of dissimilarity measures, namely the construction of archetypical extreme examples, the analysis of the behaviour under realistic transformations or perturbations of the data and the introduction of nonlinear monotone functions and clearly interpretable tuning constants to reflect the effective influence of some characteristics of the data.

References

GORDON, A. D. (1990): Constructing Dissimilarity Measures. *Journal of Classification, 7/2, 257-270.*

GOWER, J. C. and LEGENDRE, P. (1986): Metric and Euclidean Properties of Dissimilarity Coefficients. *Journal of Classification, 3/1, 5-48.*

HAUSDORF, B., and HENNIG, C. (2003): Biotic Element Analysis in Biogeography. *Systematic Biology, 52, 717-723.*

HENNIG, C. and HAUSDORF, B. (2005): A Robust Distance Coefficient between Distribution Areas Incorporating Geographic Distances. To appear in *Systematic Biology.*

HUBERT, L. and ARABIE, P. (1985): Comparing Partitions. *Journal of Classification, 2/2, 193-218.*

JACCARD, P. (1901): Distribution de la florine alpine dans la Bassin de Dranses et dans quelques regiones voisines. *Bulletin de la Societe Vaudoise des Sciences Naturelles, 37, 241-272.*

KERNEY, M. P., CAMERON, R. A. D., and JUNGBLUTH, J. H. (1983): *Die Landschnecken Nord- und Mitteleuropas.* Parey, Hamburg and Berlin.

KULCZYNSKI, S. (1927): Die Pflanzenassoziationen der Pieninen. *Bulletin International de l'Academie Polonaise des Sciences et des Lettres, Classe des Sciences Mathematiques et Naturelles, B, 57-203.*

MAST, A. R. and NYFELLER, R. (2003): Using a null model to recognize significant co-occurrence prior to identifying candidate areas of endemism. *Systematic Biology, 52, 271-280.*

PERES-NETO, P. R. and JACKSON, D. A. (2001): How well do multivariate data sets match? The advantages of a Procrustean superimposition approach over the Mantel test. *Oecologia, 129, 169-178.*

SHI, G. R. (1993): Multivariate data analysis in palaeoecology and palaeobiogeography-a review. *Palaeogeography, Palaeoclimatology, Palaeoecology, 105, 199-234.*

SIMPSON, G. G. (1960): Notes on the measurement of faunal resemblance. *American Journal of Science, 258-A, 300-311.*

Dissimilarities for Web Usage Mining

Fabrice Rossi[1], Francisco De Carvalho[2], Yves Lechevallier[1], and Alzennyr Da Silva[12]

[1] Projet AxIS, INRIA Rocquencourt, Domaine de Voluceau, Rocquencourt, B.P. 105, 78153 Le Chesnay cedex – France
[2] Centro de Informatica - CIn/UFPE Caixa Postal 7851, CEP 50732-970, Recife (PE) – Brasil

Abstract. The obtention of a set of homogeneous classes of pages according to the browsing patterns identified in web server log files can be very useful for the analysis of organization of the site and of its adequacy to user needs. Such a set of homogeneous classes is often obtained from a dissimilarity measure between the visited pages defined via the visits extracted from the logs. There are however many possibilities for defined such a measure. This paper presents an analysis of different dissimilarity measures based on the comparison between the semantic structure of the site identified by experts and the clustering constructed with standard algorithms applied to the dissimilarity matrices generated by the chosen measures.

1 Introduction

Maintaining a voluminous Web site is a difficult task, especially when it results from the collaboration of several authors. One of the best way to continuously improve the site consists in monitoring user activity via the analysis of the log file of the server. To go beyond simple access statistics provided by standard web log monitoring software, it is important to understand browsing behaviors. The (dis)agreement between the prior structure of the site (in terms of hyperlinks) and the actual trajectories of the users is of particular interest. In many situations, users have to follow some complex paths in the site in order to reach the pages they are looking for, mainly because they are interested in topics that appeared unrelated to the creators of the site and thus remained unlinked. On the contrary, some hyperlinks are not used frequently, for instance because they link documents that are accessed by different user groups.

One way to analyze browsing patterns is to cluster the content of the Web site (i.e., web pages) based on user visits extracted from the log files. The obtained clusters consist in pages that tend to be visited together and thus share some semantic relationship for the users. However, visits are complex objects: one visit can be, for example, the time series of requests sent by an user to the web server. The simplest way to cluster web pages on the server based on the visits is to define a dissimilarity between pages that take into account the way pages appear in the visits. The main problem with this approach is to choose a meaningful dissimilarity among many possibilities.

In this article, we propose a benchmark site to test dissimilarities. This small site (91 pages) has a very well define semantic content and a very dense hyperlink structure. By comparing prior clusters designed by experts according to the semantic content of the site to clusters produced by standard algorithms, we can assess the adequacy of different dissimilarities to the web usage mining (WUM) task described above.

2 Web usage data

2.1 From log files to visits

Web usage data are extracted from web server log files. A log file consists in a sequence of request logs. For each request received by the server, the log contains the name of the requested document, the time of the request, the IP address from where the request originates, etc. Log files are corrupted by many sources of noise: web proxies, browser caches, shared IP, etc. Different preprocessing methods, such as the ones described in Tanasa and Trousse (2004), allow to extract reliably visits from log files: a *visit* is a sequence of requests to a web server coming from an unique user, with at most 30 minutes between each request.

While the time elapsed between two requests of a visit is an important information, it is also quite noisy, mainly because the user might be disturbed while browsing or might doing several tasks at a time. In this paper, we don't take into account the exact date of a request. A visit consists therefore in a list of pages of the site in which the order of the pages is important.

2.2 Usage guided content analysis

As explained in the introduction, our goal is to cluster pages of a site by using the usage data. We have therefore to describe the pages via the visits. Let us consider the simple case of a web site with 4 pages, A, B, C and D. Let us define two visits, $v_1 = (A, B, A, C, D)$ and $v_2 = (A, B, C, B)$. The visits can be considered as variables that can be used to describe the pages (which are the individuals). A possible representation of the example data set is given by the following way:

	v_1	v_2
A	$\{1,3\}$	$\{1\}$
B	$\{2\}$	$\{2,4\}$
C	$\{4\}$	$\{3\}$
D	$\{5\}$	$\emptyset$

In this representation, the cell at row p and column v contains the set of the position of page p in navigation v. While this representation does not loose any information, compared to the raw data, it is quite difficult to use, as the variables don't have numerical values but variable size set values. Moreover, for voluminous web sites, the table is in general very sparse as most of the visits are short, regardless of the size of the web site.

2.3 Dissimilarities

Our solution consists in combining some data transformation methods with some dissimilarity measure in order to build a dissimilarity matrix for the pages. One of the simplest solutions is to map each set to a binary value, 0 for an empty set and 1 in the other case (then cell (p, v) contains 1 if and only if visit v contains at least one occurrence of page p). Many (dis)similarity measures have been defined for binary data (see e.g. Gower and Legendre (1986)). For WUM, the Jaccard dissimilarity is quite popular (see e.g. Foss, Wang, and Zaïane (2001)). It is given by

$$d_J(p_i, p_j) = \frac{|\{k|n_{ik} \neq n_{jk}\}|}{|\{k|n_{ik} \neq 0 \text{ ou } n_{jk} \neq 0\}|}, \tag{1}$$

where $n_{ik} = 1$ if and only if visit v_k contains page p_i and where $|A|$ denotes the size of set A.

For the Jaccard dissimilarity, two pages may be close even if one page appears many times in any visit whereas the other one only appears once in the considered visits. This is a direct consequence of the simple binary transformation. Using a integer mapping allows to keep more information: rather than using n_{ik}, we rely on m_{ik} defined as the number of occurrences of page p_i in visit v_k. Among the numerous dissimilarities available for integer valued data table, we retained the cosine and the tf×idf ones. Cosine is defined by

$$d_{\cos}(p_i, p_j) = 1 - \frac{\sum_{k=1}^{N} m_{ik} m_{jk}}{\sqrt{\left(\sum_{k=1}^{N} m_{ik}^2\right)\left(\sum_{k=1}^{N} m_{jk}^2\right)}}, \tag{2}$$

where N is the number of visits. The other dissimilarity is inspired by text mining: tf×idf takes into account both the relative importance of one visit for the page but also the length of the visit. A long visit goes through many pages and the information provided on each page is less specific than for a short visit. The dissimilarity is given by

$$d_{\text{tf}\times\text{idf}}(p_i, p_j) = 1 - \sum_{k=1}^{N} w_{ik} w_{jk}, \tag{3}$$

with

$$w_{ik} = \frac{m_{ik} \log \frac{P}{P_k}}{\sqrt{\sum_{l=1}^{N} m_{il}^2 \log\left(\frac{P}{P_l}\right)^2}}, \tag{4}$$

where P is the number of pages and P_k the number of distinct pages in visit v_k (see for instance Chen(1998)).

3 Comparison of dissimilarities

3.1 Benchmark Web site

Comparison between dissimilarities is conducted via a benchmark web site. We have chosen the site of the CIn, the laboratory of two of the authors. The site consists in dynamic web pages, implemented by servlets. The URLs of the pages are very long (more than one hundred characters) and not very easy to remember, as they corresponds to programmatic call to the servlets. Because of this complexity, we assume that most of the users will start browsing by the first main page of the site and then navigate thanks to the hyperlinks.

The site is quite small (91 pages) and very well organized in a tree with depth 5. Most of the content lies in the leafs of the tree (75 pages) and internal nodes have mostly a navigation and organization role. The hyperlink structure is very dense. A navigation menu appears on each page: it contains a link to the main page and to the 10 first level pages, as well as a link to the parents of the current page and to its siblings in the tree. There is sometimes up to 20 links in the menu which seems too complex in this situation.

The web log ranges from June 26th 2002 to June 26th 2003. This corresponds to 2Go of raw data from which 113 784 visits are extracted.

3.2 Reference semantic

We have classified the content of the site into 13 classes of pages, based on their content. Dissimilarity are compared by building clusters with a clustering algorithm and by comparing the obtained classes to the reference classes. As some of the classes are quite small (see Table 1), we also consider a prior clustering into 11 classes, where classes 9, 10 and 11 of the 13 classes partition are merged (they contain documents for graduate students).

Table 1. Size of the prior classes

1	2	3	4	5	6	7
Publications	Research	Partners	Undergraduate	Objectives	Presentation	Directory
8	9	10	11	12	13	
Team	Options	Archives	Graduate	News	Others	

4 Results

4.1 Partition quality assessment

To compare the dissimilarities presented in section 2.3, we produce homogeneous classes of pages, then we compare these classes with those resulting from the expert analysis on the site reference. For classification, we use a k-means like algorithm adapted to dissimilarity data (see Kaufman and Rousseeuw (1987)) and a standard hierarchical clustering based on average linkage.

To analyze the results, we use two criteria. The first algorithm works with a user specified number of classes. We compare the obtained partition with the prior partition thanks to the corrected Rand index (see Hubert and Arabie (1985)). It takes values in $[-1, 1]$ where 1 corresponds to a perfect agreement between partitions, whereas a value equal or below 0 corresponds to completely different partition.

For the hierarchical clustering, we monitor the evolution of the F measure (see van Rijsbergen (1979)) associated to each prior class with the level of the cut in the dendrogram: the F measure is the harmonic mean of the precision and the recall, i.e. respectively of the percentage of elements in the obtained class that belong to the prior class and the percentage of the prior class retrieved in the obtained class. This method allows seeing if some prior classes can be obtained thanks to the clustering algorithm without specifying an arbitrary number of classes. In a sense this analysis can reveal specific weaknesses or skills of dissimilarities by showing whether they can discover a specific class.

4.2 Dynamic clustering

The dynamic clustering algorithm requires a prior number of classes. To limit the effects of this choice, we study the partitions produced for a number of classes from 2 to 20. The results are summarized in Table 2.

Table 2. Dynamic clustering results

Dissimilarity	Rand index	Found classes	min F mesure
Jaccard	0.5698 (9 classes)	6	0.4444
Tf×idf	0.5789 (16 classes)	7	0.5
Cosinus	0.3422 (16 classes)	4	0.3

For a global analysis (corrected Rand index), we indicate the size of the partition which maximizes the criterion. It is clear that tf×idf and Jaccard give rather close results (slightly better for the first one), whereas cosine obtains very unsatisfactory results. For a detailed analysis, we search to each prior class a corresponding class (by means of the F measure) in the set of classes produced while varying the size of the partition from 2 to 20. We indicate the number of perfectly found classes and the worst F measure for the not found classes. The tf×idf measure seems to be the best one. The classes perfectly found by other dissimilarities are also obtained by tf×idf (which finds classes 3, 4, 5, 7, 8, 9 and 12). However, we can notice that the perfectly found classes are in different partitions, which explains the relatively bad Rand indices, compared to the results class by class.

4.3 Hierarchical clustering

We carry out the same analysis for the case of hierarchical classification. We vary here the number of classes by studying all the levels of possible cut in the dendrogramme. We obtain the results summarized in Table 3.

Table 3. Hierarchical clustering results

Dissimilarity	Rand index	Found classes	min F mesure
Jaccard	0.6757 (11 classes)	3	0.5
Tf×idf	0.4441 (15 classes)	3	0.4
Cosinus	0.2659 (11 classes)	5	0.4

In general, we can notice a clear domination of Jaccard and an improvement of the results for this one. The criterion of the average link used here, as well as the hierarchical structure, seems to allow a better exploitation of the Jaccard dissimilarity, whereas the results are clearly degraded for other measures. The results class by class are more difficult to analyze and seem not to depend on measure. However, the satisfactory performances of tf×idf and cosine correspond to a good approximation of the classes for very different cutting levels in the dendrogramme: it is thus not possible to obtain with these measures a good recovery of the set of classes, whereas Jaccard is overall better.

5 Discussion

Overall, the Jaccard dissimilarity appears to be the best one for recovering the prior clusters from the usage data. The tf×idf dissimilarity gives also satisfactory results, while the cosine measure fails to recover most of the prior structure. It is important however to balance the obtained results according to the way prior clusters were designed.

The organization of the CIn web site is a bit peculiar because of the generalization of navigation pages. Class 6, for instance, contains 8 pages that describe the CIn; class 5 contains 6 pages that describe the objectives of the CIn. One of the page of class 5 acts as an introductory page to the detailed presentations of the objectives, sorted into the 5 other pages. This introduces two problems: 1) this page acts as a bridge between the general description of the CIn and the detailed description of its objectives 2) there is no simple way to avoid this page and yet to access to the description of CIn's objectives. The decision to put this page in the prior class 5 has some effect on the Jaccard dissimilarity. Indeed, as there is no simple way to view other pages of class 5 without viewing the bridge page, the Jaccard dissimilarity will tend to consider that the bridge page is close to the other pages. Moreover, as there is no way to reach the bridge page without viewing the main page of the description of the CIn, the Jaccard dissimilarity will have difficulties to separate class 5 from class 6. More generally, the tree structure of the CIn's web site and the navigation (or bridge) pages are quite difficult to handle for the Jaccard dissimilarity. It appears for instance that if we cut the dendrogram constructed via this dissimilarity in order to obtain 13 classes, we face problems that seem to be directly related to the organization of the site. For instance, class 5 and class 6 are merged into one cluster, except for two pages of class 6: the first of those pages is the general presentation of the CIn (a bridge page from the main page to presentation pages) and the localization page that gives instruction to reach the CIn (this page can be accessed directly from the main page).

Tf×idf is less sensitive to this type of problem. The 13 classes obtained from the hierarchical clustering contain one class for all pages of class 5 together with one page from class 6 (a page that describes the mission of the CIn) and another class with all the remaining pages from class 6. However, tf×idf suffers from the reduction of the relevance of long visits induced by its definition. Some pages with a low number of visits tend to appear in longer visits, from people that try to get a general view of the CIn. Clustering tends therefore to produce small classes of pages unrelated to other pages, and then to merge those classes in a quite meaning less.

The case of the cosine dissimilarity is far less clear. That bad results seem to be linked the early creation (in the hierarchical clustering) of a big cluster that mix pages from the research part of CIn's site to the pages for graduate students. The dissimilarity appears to be dominated by some long visits that tend to go through all the pages of the site. The exact source of the limitations of the cosine dissimilarity are still under investigation, however.

It is clear that it would be interesting to investigate how to modify the weighting in the tf×idf to get results closer to the one of Jaccard, will keeping the correct behavior in some circumstances. It seems also important to find a way to take into account both the structure of the site and the visits, because the global organization of the visits seem to be dominated by the structure: it would therefore be interesting to emphasize "surprising" co-occurrence of pages in a visit rather than considering all co-occurrences equality.

6 Conclusion

The results presented here give interesting insight on the adequacy of three dissimilarity measures to a clustering problem related to Web Usage Mining. While they tend to support earlier results of Foss, Wang, and Zaïane (2001) that consider the Jaccard dissimilarity to be well adapted to this type of problem, they also show that the design of the benchmark and the structure of the reference web site can have strong impact on the outcome of the comparison. Further works include the comparison the chosen dissimilarities on other prior clustering of the reference web site as well as an analysis of the effect of the dissimilarities on the results of other clustering algorithm, such as an adapted version of Kohonen's Self Organized Map, as used in Rossi, El Golli, and Lechevallier (2005).

References

CELEUX, G., DIDAY, E., GOVAERT, G., LECHEVALLIER, Y. and RALAMBONDRAINY, H. (1989): *Classification Automatique des Données.* Bordas, Paris.

CHEN, C. (1998): Generalized similarity analysis and pathfinder network scaling. *Interacting with Computers*, 10:107–128.

FOSS, A., WANG, W. and ZAÏANE, O.R. (2001): A non-parametric approach to web log analysis. In *Proc. of Workshop on Web Mining in First International SIAM Conference on Data Mining (SDM2001)*, pages 41–50, Chicago, IL, April 2001.

GOWER, J. and LEGENDRE, P. (1986): Metric and euclidean properties of dissimilarity coefficients. *Journal of Classification*, 3:5–48.

HUBERT, L. and ARABIE, P. (1985): Comparing partitions. *Journal of Classification*, 2:193–218.

KAUFMAN, L. and ROUSSEEUW, P.J. (1987): Clustering by means of medoids. In Y. Dodge, editor, *Statistical Data Analysis Based on the L1-Norm and Related Methods*, pages 405–416. North-Holland, 1987.

ROSSI, F., EL GOLLI, A. and LECHEVALLIER, Y. (2005): Usage guided clustering of web pages with the median self organizing map. In *Proceedings of XIIIth European Symposium on Artificial Neural Networks (ESANN 2005)*, pages 351–356, Bruges (Belgium), April 2005.

TANASA, D. and TROUSSE, B. (2004): Advanced data preprocessing for intersites web usage mining. *IEEE Intelligent Systems*, 19(2):59–65, March-April 2004. ISSN 1094-7167.

TANASA, D. and TROUSSE, B. (2004): Data preprocessing for wum. *IEEE Potentials*, 23(3):22–25, August-September 2004.

VAN RIJSBERGEN, C.J. (1979): *Information Retrieval* (second ed.). London: Butterworths.

Properties and Performance of Shape Similarity Measures

Remco C. Veltkamp[1] and Longin Jan Latecki[2]

[1] Dept. Computing Science, Utrecht University
Padualaan 14, 3584 CH Utrecht, The Netherlands
Remco.Veltkamp@cs.uu.nl

[2] Dept. of Computer and Information Sciences, Temple University
Philadelphia, PA 19094, USA
latecki@temple.edu

Abstract. This paper gives an overview of shape dissimilarity measure properties, such as metric and robustness properties, and of retrieval performance measures. Fifteen shape similarity measures are shortly described and compared. Their retrieval results on the MPEG-7 Core Experiment CE-Shape-1 test set as reported in the literature and obtained by a reimplementation are compared and discussed.

1 Introduction

Large image databases are used in an extraordinary number of multimedia applications in fields such as entertainment, business, art, engineering, and science. Retrieving images by their content, as opposed to external features, has become an important operation. A fundamental ingredient for content-based image retrieval is the technique used for comparing images. It is known that human observers judge images as similar if they show similar objects. Therefore, similarity of objects in images is a necessary component of any useful image similarity measure. One of the predominant features that determine similarity of objects is shape similarity.

There exist a large variety of approaches to define shape similarity measures of planar shapes, some of which are listed in the references. Since an objective comparison of their qualities seems to be impossible, experimental comparison is needed. The Motion Picture Expert Group (MPEG), a working group of ISO/IEC (see `http://www.chiariglione.org/mpeg/`) has defined the MPGE-7 standard for description and search of audio and visual content. A region based and a contour based shape similarity method are part of the standard. The data set created by the MPEG-7 committee for evaluation of shape similarity measures (Bober et al. (1999), Latecki, Lakaemper and Eckhardt (2000)) offers an excellent possibility for objective experimental comparison of the existing approaches evaluated based on the retrieval rate. The shapes were restricted to simple pre-segmented shapes defined by their outer closed contours. The goal of the MPEG-7 Core Experiment CE-Shape-1 was to evaluate the performance of 2D shape descriptors under change of a view point with respect to objects, non-rigid object motion, and noise. In addition, the descriptors should be scale and rotation invariant.

2 Properties

In this section we list a number of possible properties of similarity measures. Whether or not specific properties are desirable will depend on the particular application, sometimes a property will be useful, sometimes it will be undesirable. A shape dissimilarity measure, or distance function, on a collection of shapes S is a function $d : S \times S \rightarrow \mathbb{R}$. The following conditions apply to all the shapes A, B, or C in S.

1 (Nonnegativity) $d(A, B) \geq 0$.

2 (Identity) $d(A, A) = 0$ for all shapes A.

3 (Uniqueness) $d(A, B) = 0$ implies $A = B$.

4 (Strong triangle inequality) $d(A, B) + d(A, C) \geq d(B, C)$.

Nonnegativity (1) is implied by (2) and (4). A distance function satisfying (2), (3), and (4) is called a metric. If a function satisfies only (2) and (4), then it is called a semimetric. Symmetry (see below) follows from (4). A more common formulation of the triangle inequality is the following:

5 (Triangle inequality) $d(A, B) + d(B, C) \geq d(A, C)$.

Properties (2) and (5) do not imply symmetry.

Similarity measures for partial matching, giving a small distance $d(A, B)$ if a part of A matches a part of B, in general do not obey the triangle inequality. A counterexample is the following: the distance from a man to a centaur is small, the distance from a centaur to a horse is small, but the distance from a man to a horse is large, so $d(man, centaur) + d(centaur, horse) \geq d(man, horse)$ does not hold. It therefore makes sense to formulate an even weaker form:

6 (Relaxed triangle inequality) $c(d(A, B) + d(B, C)) \geq d(A, C)$, for some constant $c \geq 1$.

7 (Symmetry) $d(A, B) = d(B, A)$.

Symmetry is not always wanted. Indeed, human perception does not always find that shape A is equally similar to B, as B is to A. In particular, a variant A of prototype B is often found more similar to B than vice versa.

8 (Invariance) d is invariant under a chosen group of transformations G if for all $g \in G$, $d(g(A), g(B)) = d(A, B)$.

For object recognition, it is often desirable that the similarity measure is invariant under affine transformations.

The following properties are about robustness, a form of continuity. They state that a small change in the shapes lead to small changes in the dissimilarity value. For shapes defined in $\mathbb{R}^2$ we can require that an arbitrary small change in shape leads to an arbitrary small in distance, but for shapes in $\mathbb{Z}^2$ (raster images), the smallest change in distance value can be some fixed value larger than zero. We therefore speak of an 'attainable $\epsilon > 0$'.

9 (Deformation robustness) For each attainable $\epsilon > 0$, there is an open set F of homeomorphisms sufficiently close to the identity, such that $d(f(A), A) < \epsilon$ for all $f \in F$.

10 (Noise robustness) For shapes in $\mathbb{R}^2$, noise is an extra region anywhere in the plane, and robustness can be defined as: for each $x \in (\mathbb{R}^2 - A)$, and each attainable $\epsilon > 0$, an open neighborhood U of x exists such that for all B, $B - U = A - U$ implies $d(A, B) < \epsilon$. When we consider contours, we interpret noise as an extra region attached to any location on the contour, and define robustness similarly.

3 Performance

First we shortly describe the settings of the MPEG-7 Core Experiment CE-Shape-1. The core experiment was divided into part A: robustness to scaling (A1) and rotation (A2), part B: performance of the similarity-based retrieval, and part C: robustness to changes caused by non-rigid motion.

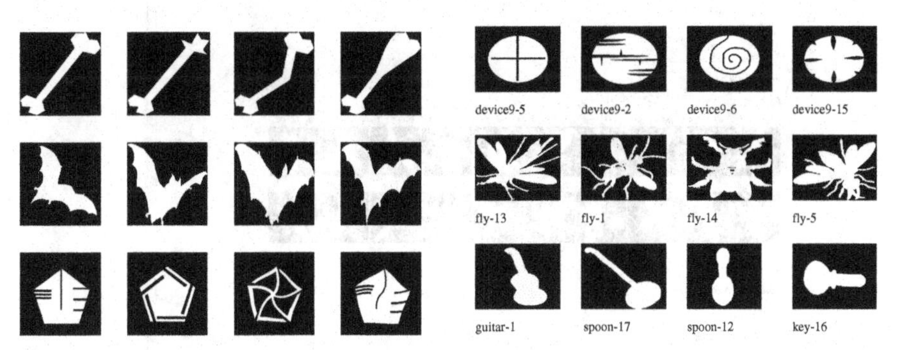

Fig. 1. Some shapes used in MPEG-7 Core Experiment CE-Shape-1 part B.

Fig. 2. The shapes with the same name prefix belong to the same class.

Part A can be regarded as a useful condition that every shape descriptor should satisfy. The main part is part B, where a set of semantically classified images with a ground truth is used. Part C can be viewed as a special case of part B. Here also the performance of the similarity-based retrieval is tested, but only the deformation due to non-rigid motion is considered. Only one query shape is used for part C.

The test set consists of 70 different classes of shapes, each class containing 20 similar objects, usually (heavily) distorted versions of a single base shape. The whole data set therefore consists of 1400 shapes. For example, each row in Fig. 1 shows four shapes from the same class.

We focus our attention on the performance evaluation of shape descriptors in experiments established in Part B of the MPEG-7 CE-Shape-1 data set (Bober et al. (1999)). Each image was used as a query, and the retrieval rate is expressed by the so called Bull's Eye Percentage (BEP): the fraction of images that belong to the same class in the top 40 matches. Since the maximum number of correct matches for a single query image is 20, the total number of correct matches is 28000.

Strong shape variations within the same classes make that no shape similarity measure achieves a 100% retrieval rate. E.g., see the third row in Fig. 1 and the first and the second rows in Fig. 2. The third row shows spoons that are more similar to shapes in different classes than to themselves.

Fig. 3. SIDESTEP interface.

To compare the performance of similarity measures, we built the framework SIDESTEP – Shape-based Image Delivery Statistics Evaluation Project, `http://give-lab.cs.uu.nl/sidestep/`. Performance measures such as the number of true/false positives, true/false negative, specificity, precision, recall, negative predicted value, relative error, k-th tier, total performance, and Bull's Eye Percentage can be evaluated for a single query, over a whole class, or over a whole collection, see Fig. 3.

4 Shape similarity measures

In this section we list several known shape similarity measures and summarize some properties and their performance in Table 1 on the MPEG-7 CE-Shape-1 part B data set. The discussion of the results follows in Section 5.

Shape context (Belongie, Malik and Puzicha (2002)) is a method that first builds a shape representation for each contour point, using statistics of other contour points 'seen' by this point in quantized angular and distance intervals.

The obtained view of a single point is represented as a 2D histogram matrix. To compute a distance between two contours, the correspondence of contour points is established that minimizes the distances of corresponding matrices.

Image edge orientation histogram (Jain and Vailaya (1996)) is built by applying an edge detector to the image, then going over all pixels that lie on an edge, and histogramming the local tangent orientation.

Hausdorff distance on region is computed in the following way. First a normalization of the orientation is done by computing the principal axes of all region pixels, and then rotating the image so that the major axis is aligned with the positive x-axis, and the minor axis with the positive y-axis. The scale is normalized by scaling the major axes all to the same length, and the y-axes proportionally. Then the Hausdorff distance between the sets A and B of region pixels is computed: the maximum of all distances of a pixel from A to B, and distances of a pixel from B to A. The Hausdorff distance has been used for shape retrieval (see for example Cohen (1995)), but we are not aware of experimental results on the MPEG-7 Core Experiment CE-Shape-1 test set reported in the literature.

Hausdorff distance on contour is computed in the same way, except that it is based on set of all contour pixels instead of region pixels.

Grid descriptor (Lu and Sajjanhar (1999)) overlays the image with a coarse grid, and assigns a '1' to a grid cell when at least 15% of the cell is covered by the object, and a '0' otherwise. The resulting binary string is then normalized for scale and rotation. Two grid descriptors are compared by counting the number of different bits.

Fourier descriptors are the normalized coefficients of the Fourier transformation, typically applied to a 'signal' derived from samples from the contour, such as the coordinates represented by complex numbers. Experiments have shown that the centroid distance function, the distance from the contour to the centroid, is a signal that works better than many others (Zhang and Lu (2003)).

Distance set correspondence (Grigorescu and Petkov (2003)) is similar to shape contexts, but consists for each contour point of a set of distances to N nearest neighbors. Thus, in contrast to shape contexts, no angular information but only local distance information is obtained. The distance between two shapes is expressed as the cost of a cheapest correspondence relation of the sets of distance sets.

Delaunay triangulation angles are used for shape retrieval in Tao and Grosky (1999) by selecting high curvature points on the contour, and making a Delaunay triangulation on these points. Then a histogram is made of the two largest interior angles of each of the triangles in the triangulation. The distance between two shapes is then simply the L_2-distances between the histograms.

Deformation effort (Sebastian, Klien and Kimia (2003)) is expressed as the minimal deformation effort needed to transform one contour into the other.

Curvature scale space (CSS) (Mokhtarian and Bober (2003)) is included in the MPEG-7 standard. First simplified contours are obtained by convolution with a Gaussian kernel. The arclength position of inflection points (x-axis) on contours on every scale (y-axis) forms so called Curvature Scale Space (CSS) curve. The positions of the maxima on the CSS curve yield the shape descriptor. These positions when projected on the simplified object contours give the positions of the mid points of the maximal convex arcs obtained during the curve evolution. The shape similarity measure between two shapes is computed by relating the positions of the maxima of the corresponding CSS curves.

Convex parts correspondence (Latecki and Lakaemper (2000)) is based on an optimal correspondence of contour parts of both compared shapes. The correspondence is restricted so that at least one of element in a corresponding pair is a maximal convex contour part. Since the correspondence is computed on contours simplified by a discrete curve evolution (Latecki and Lakaemper (1999)), the maximal convex contour parts represent visually significant shape parts. This correspondence is computed using dynamic programming.

Contour-to-centroid triangulation (Attalla and Siy (2005)) first picks the farthest point from the centroid of the shape and use it as the start point of segmenting the contour. It then divides the contour into n equal length arcs, where n can be between 10 and 75, and considers the triangles connecting the endpoints of these arcs with the centroid. It builds a shape descriptor by going clockwise over all triangles, and taking the left interior contour angle, the length of the left side to the centroid, and the ratio contour segment length to contour arc length. To match two descriptors, the triangle parameters are compared to the correspond triangle of the other descriptor, as well as to its left and right neighbor, thereby achieving some form of elastic matching.

Contour edge orientation histogram are built by going over all pixels that lie on object contours, and histogramming the local tangent orientation. It is the same as the 'image edge orientation histogram', but then restricted to pixels that lie on the object contour.

Chaincode nonlinear elastic matching (Cortelazzo et al. (1994)) represents shape in images as a hierarchy of contours, encoded as a chaincode string: characters '0' to '7' for the eight directions travelling along the contour. Two images are compared by string matching these chaincode strings. Various different string matching methods are possible, we have taken the 'nonlinear elastic matching' method.

Angular radial transform (ART) is a 2-D complex transform defined on a unit disk in polar coordinates. A number of normalized coefficients form the feature vector. The distance between two such descriptors is simply the L_1 distance. It is a region-based descriptor, taking into account all pixels describing the shape of an object in an image, making it robust to noise. It is the region-based descriptor included in the MPEG-7 standard (Salembier and Sikora (2002)).

Table 1. Performances and properties of similarity measures.

method	unique	deform	noise	BEP reported	BEP reimpl
Shape context	+	+	+	76.51	
Image edge orientation histogram	−	+	+		41
Hausdorff region	+	+	−		56
Hausdorff contour	+	+	+		53
Grid descriptor	−	+	+		61
Distance set correspondence	+	+	+	78.38	
Fourier descriptor	−	+	+		46
Delaunay triangulation angles	−	−	−		47
Deformation effort	+	+	+	78.18	
Curvature scale space	−	+	+	81.12	52
Convex parts correspondence	−	+	+	76.45	∼
Contour-to-centroid triangulation	−	−	−	84.33	79
Contour edge orientation histogram	−	+	+		41
Chaincode nonlinear elastic matching	+	+	+		56
Angular radial transform	+	+	+	70.22	53

5 Discussion

The Angular radial transform, the grid descriptor, the 'Hausdorff region', and image edge orientation histogram are region based methods, all others work only for shapes defined by a single contour. Naturally, the region based methods can also be applied to contour shapes.

Even though invariance under transformations is not always a property of the base distance, such as the Hausdorff distance, it can be easily obtained by a normalization of the shape or image, as many of the methods do.

Table 1 tabulates a number of properties and performances of the similarity measures listed in section 4. The column 'unique' indicates whether (+) or not (−) the method satisfies the uniqueness property, 'deform' indicates deformation robustness, 'noise' indicates robustness with respect to noise, 'BEP reported' lists the Bull's Eye Percentage reported in the literature, 'BEP reimpl' lists the BEP of the reimplementations (performed by master students) plugged into SIDESTEP. The symbol $\sim$ indicates that the method is of one of the authors.

Methods that are based on sampling, histogramming, or other reduction of shape information do not satisfy the uniqueness property: by throwing away information, the distance between two shapes can get zero even though they are different.

The methods that are based on angles, such as the 'Contour-to-centroid triangulation' and 'Delaunay triangulation angles' methods, are not robust to deformation and noise, because a small change in the shape can lead to a large change in the triangulation.

The Hausdorff distance on arbitrary sets is not robust to noise (an extra region anywhere in the plane), and therefore also not for regions. However, for contours, we interpret noise as an extra point attached to any contour location. As a result the Hausdorff distance on contours is robust to noise.

Fourier descriptors have been reported to perform better than CSS (Zhang and Lu (2003)), but the comparison has not been done in terms of the Bull's Eye Percentage.

It is remarkable that the 'Contour-to-triangulation' does not satisfy, theoretically, uniqueness and robustness properties, while in practice it performs so well. This is explained by the fact that the method does not satisfy the property for *all* shapes, while the performance is measured only on a limited set of shapes, where apparently the counterexamples that prevent the method from obeying the property simply don't occur.

The difference between the Bull's Eye Percentages of the method as reported in the literature and the performances of the reimplement methods is significant. Our conjecture is that this is caused by the following. Firstly, several methods are not trivial to implement, and are inherently complex. Secondly, the description in the literature is often not sufficiently detailed to allow a straightforward implementation. Thirdly, fine tuning and engineering has a large impact on the performance for a specific data set. It would be good for the scientific community if the reported test results are made reproducible and verifiable by publishing data sets and software along with the articles.

The most striking differences between the performances reported in the literature and obtained by the reimplementation are the ones that are part of the MPEG-7 standard: the Curvature Scale Space and the Angular Radial Transform. In the reimplementation of both methods we have followed closely the precise description in the ISO document (Yamada et al. (2001)), which is perhaps less tweaked towards the specific MPEG-7 Core Experiment CE-Shape-1 test set.

The time complexity of the methods often depends on the implementation choices. For example, a naive implementation of the Hausdorff distance inspects all $O(N^2)$ pairs of points, but a more efficient algorithm based on Voronoi Diagrams results in a time complexity of $O(N \log N)$, at the expense of a more complicated implementation.

Acknowledgement This research was supported by the FP6 IST projects 511572-2 PROFI and 506766 AIM@SHAPE, and by a grant NSF IIS-0534929. Thanks to Dènis de Keijzer and Geert-Jan Giezeman for their work on SIDESTEP.

References

ATTALLA, E. and SIY, P. (2005): Robust shape similarity retrieval based on contour segmentation polygonal multiresolution and elastic matching. Patt. Recogn. 38, 22292241.

BELONGIE, S., MALIK, J. and PUZICHA, J. (2002): Shape Matching and Object Recognition Using Shape Contexts. IEEE PAMI, 24(24), 509-522.

BOBER, M., KIM, J.D., KIM, H.K., KIM, Y.S., KIM, W.-Y. and MULLER, K. (1999): Summary of the results in shape descriptor core experiment. ISO/IEC JTC1/SC29/WG11/MPEG99/M4869.

COHEN, S. (1995): Measuring Point Set Similarity with the Hausdorff Distance: Theory and Applications. Ph.D thesis, Stanford University.

CORTELAZZO, G., MIAN, G.A., VEZZI, G. and ZAMPERONI, P. (1994): Trademark Shapes Description by String-Matching Techniques. Patt. Recogn. 27(8), 1005-1018.

GRIGORESCU, C. and PETKOV, N. (2003): Distance Sets for Shape Filters and Shape Recognition. IEEE Trans. Image Processing, 12(9).

JAIN, A.K. and VAILAYA, A. (1996): Image Retrieval using Color and Shape. Patt. Recogn. 29(8), 1233-1244.

LATECKI, L.J. and LAKAEMPER, R. (1999): Convexity Rule for Shape Decomposition Based on Discrete Contour Evolution. Computer Vision and Image Understanding 73, 441-454.

LATECKI, L.J. and LAKAEMPER, R. (2000): Shape Similarity Measure Based on Correspondence of Visual Parts. IEEE PAMI 22, 1185-119.

LATECKI, L.J., LAKAEMPER, R. and ECKHARDT, U. (2000): Shape descriptors for non-rigid shapes with a single closed contour. Proc. CVPR, 424-429.

LU, G. and SAJJANHAR, A. (1999): Region-based shape representation and similarity measure suitable for content-based image retrieval. Multimedia Systems, 7, 165174.

MOKHTARIAN, F. and BOBER, M. (2003): Curvature Scale Space Representation: Theory, Applications and MPEG-7 Standardization. Kluwer Academic.

SEBASTIAN, T.B., KLIEN, P. and KIMIA, B.B. (2003): On aligning curves. IEEE PAMI, 25, 116-125.

SALEMBIER, B.S.M.P. and SIKORA, T., editors (2002): Introduction to MPEG-7: Multimedia Content Description Interface. JohnWiley and Sons.

TAO, Y. and GROSKY, W.I. (1999): Delaunay trriangularion for image object indexing: a novel method for shape representation. Proc. 7th SPIE Symposium on Storage and Retrieval for Image and Video Databases, 631-642.

YAMADA, A., PICKERING, M., JEANNIN, S., CIEPLINSKI, L., OHM, J.R. and KIM, M. (2001): MPEG-7 Visual part of eXperimentation Model Version 9.0. ISO/IEC JTC1/SC29/WG11/N3914.

ZHANG, D. and LU, G. (2003): Evaluation of MPEG-7 shape descriptors against other shape descriptors. Multimedia Systems 9, 1530.

Part II

Classification and Clustering

Hierarchical Clustering for Boxplot Variables

Javier Arroyo[1], Carlos Maté[2], and Antonio Muñoz-San Roque[2]

[1] Departamento de Sistemas Informáticos, Universidad Complutense de Madrid, Profesor García-Santesmases s/n, 28040 Madrid, Spain
[2] Instituto de Investigación Tecnológica, ETSI (ICAI), Universidad Pontificia Comillas, Alberto Aguilera 25, 28015 Madrid, Spain

Abstract. Boxplots are well-known exploratory charts used to extract meaningful information from batches of data at a glance. Their strength lies in their ability to summarize data retaining the key information, which also is a desirable property of symbolic variables. In this paper, boxplots are presented as a new kind of symbolic variable. In addition, two different approaches to measure distances between boxplot variables are proposed. The usefulness of these distances is illustrated by means of a hierarchical clustering of boxplot data.

1 Introduction

Symbolic data analysis (SDA) proposes an alternative approach to deal with large and complex datasets as it allows the summary of these datasets into smaller and more manageable ones retaining the key knowledge. In a symbolic dataset, items are described by symbolic variables such as lists of (categorical or quantitative) values, intervals and distributions (Bock and Diday (2000)). However, new types of symbolic variables can be considered.

We believe that boxplots (henceforth BPs) can be proposed as a new kind of quantitative symbolic variable, which have interesting features. BPs, developed by Tukey (1977) and also called box-and-whiskers plots, are an univariate data display of a summary of an empirical distribution which, conventionally, consists of five values: the extremes, the lower and upper quartiles (hinges) and the median. BPs are an extremely useful exploratory tool, which is widely used in data analysis. The reasons are given by Benjamini (1988):

- BPs offer an excellent summary for a batch of data available at a glance.
- Good compromise among detailed description and a condensed display of summaries.
- Easy implementation and computation.
- Allowing comparison between many batches of data.
- Easy explanation to non-statisticians.

These properties also explain why BPs can be considered as an interesting type of symbolic variable. The main feature of symbolic variables is that they provide more information than classical ones, where only one single number or category is allowed as value. They also are well suited for characterizing the properties of classes obtained after an aggregation process, time-varying patterns or complex scenarios. BP variables fit in the symbolic context, as BPs represent empirical distributions in a shortened but explanatory way, reporting about their location, spread, skewness and normality.

BP variables offer an intermediate point between the simplicity of interval variables and the detailed information provided by histogram variables. Interval variables do not provide information about the central area of an empirical distribution, while BP variables do this by means of the three quartiles. In the other hand, histogram variables report in detail about the empirical distribution, but their structure is more complex: a set of consecutive intervals with associated weights is needed. The structure of the BP variable is simpler, but it is enough to report about the shape of the distribution. In addition, BP variables do not suffer from the subjectivity of histogram variables, that is, a BP is always defined by five definite components, while the components of the histogram depends on the analyst criterion.

This paper defines the concept of boxplot variable in the SDA framework,proposes two ways to measure distances between BP variables and illustrates the approach by a hierarchical clustering of BP data.

2 Definition of boxplot variable

Let Y be a variable defined for all elements of a finite set $E = \{1, ..., N\}$, Y is termed a BP variable with domain of values $\mathcal{Y}$, if Y is a mapping from E to a range $\mathcal{B}(\mathcal{Y})$ with the structure $\{m, q, Me, Q, M\}$, where $-\infty < m \leq q \leq Me \leq Q \leq M < \infty$, and m represents the minimum, q the lower quartile, Me the median, Q the upper quartile, and M the maximum.

This definition follow the style of the definitions of other symbolic variables given by Billard and Diday (2003). Equivalently, a BP variable can be considered as an special case of an interval-valued modal variable. Interval-valued modal variables are variables where each individual take as value one or more intervals, each one with a weight attached (weights usually represent frequency, probability or possibility). BP variables are a kind of modal interval-valued variable with the following restrictions: the number of intervals is always four, intervals are consecutive, the weights represent frequency or probability, and the weight of each interval is 0.25. Consequently, BP variables can also be defined by: a lower-whisker interval $\xi_1 = [m, q)$, a lower-mid-box interval $\xi_2 = [q, Me)$, an upper-mid-box interval $\xi_3 = [Me, Q)$, and an upper-whisker interval $\xi_4 = [Q, M)$, with $p_i = 0.25$, $i = 1, ..., 4$.

In a SDA context, BP variables can describe the quantitative properties of the classes obtained after an aggregation process, i.e. a BP variable summarizes the batch of values of a quantitative feature for each class. In these cases, determining the value of the elements describing each BP is needed.

Regarding the quartiles, Frigge et al (1989) show that different definitions are applied in the statistical software packages. It has been verified that the lack of standardization still prevails nowadays. Frigge et al (1989) suggested the definition proposed by Tukey (1977). Tukey's definition for the lower quartile in terms of the ordered observations $x_{(1)} \leq x_{(2)} \leq ... \leq x_{(n)}$ is

$$q = (1 - g)x_{(j)} + gx_{(j+1)}, \tag{1}$$

where $[(n+3)/2]/2 = j + g$ (note that g=0 or g=1/2).

In the case of extreme values, outliers require special attention. According to Hoaglin et al (1986), $[q - k(Q - q), Q + k(Q - q)]$ with $k = 1.5$ or $k = 2$ is a resistant rule for flagging outliers in exploratory contexts. In practice, deciding wether outliers should be removed or not depends on issues such as the problem domain, the properties of the data set, the aims of the analysis and so on.

BP variables can also describe properties of symbolic individuals where the theoretical distribution is known. In those cases, if the minimum (resp. maximum) value of the distribution is $-\infty$ (resp.$+\infty$) we recommend the use of the 0.01- or 0.05-quantile (resp. 0.99- or 0.95-quantile). Trenkler (2002) offers further information about turning theoretical distributions into boxplots.

3 Distances for BP variables

In this section, two approaches for measuring distances for BP data are considered: the first entails considering that a BP variable is a symbolic variable described by five single-valued variables and the second entails considering that a BP variable is a special case of interval-valued modal variable.

3.1 The classical approach

If a BP variable is considered as a variable compound of five sub-variables (m, q, Me, Q, M) the distance between a pair of BPs can be measured by means of a classical metric in $\mathbb{R}^5$. If a Minkowski metric with a given order $q = 1$ is considered, we obtain the Manhattan distance between a pair of BPs, BP_1 and BP_2:

$$\begin{aligned} d_{q=1}(BP_1, BP_2) = &|m_1 - m_2| + |q_1 - q_2| + |Me_1 - Me_2| + \\ &|Q_1 - Q_2| + |M_1 - M_2| . \end{aligned} \tag{2}$$

Other Minkowski metrics such as Euclidean or Chebychev or other kinds of distances can be applied. As the approach is straightforward, further details will not be given.

3.2 The Ichino-Yaguchi distance function

Ichino and Yaguchi (1994) define a generalized Minkowski metric for a multidimensional space of mixed variables (quantitative, qualitative and structural variables), which is based on the Cartesian join and meet operators.

Let $A = [\underline{A}, \overline{A}]$ and $B = [\underline{B}, \overline{B}]$ be a pair of intervals, the Cartesian meet and join operations for intervals are defined as $A \otimes B = A \cap B$ and $A \oplus B = [\min(\underline{A}, \underline{B}), \max(\overline{A}, \overline{B})]$, respectively. The Ichino-Yaguchi dissimilarity measure is defined by:

$$\phi(A, B) := |A \oplus B| - |A \otimes B| + \gamma(2|A \otimes B| - |A| - |B|), \tag{3}$$

where $|X| = \overline{X} - \underline{X}$ and $\gamma \in [0, 0.5]$ is a parameter which controls the effects of the inner-side nearness and the outer-side nearness between A and B. If $\gamma = 0.5$, the resulting dissimilarity function is:

$$\phi(A, B)_{\gamma=0.5} = |A \oplus B| - 0.5(|A| + |B|). \tag{4}$$

Ichino and Yaguchi consider that (4) is a suitable distance function. It can be seen that (4) is equivalent to:

$$\phi(A, B)_{\gamma=0.5} = \frac{|\underline{A} - \underline{B}| + |\overline{A} - \overline{B}|}{2}. \tag{5}$$

In order to measure distances between a pair of individuals, X and Y, described by p variables of different kinds, Ichino and Yaguchi (1994) proposed the generalized Minkowski metric of order q, with $q \geq 1$:

$$\delta(X, Y) = \left(\sum_{i=1}^{p} \phi(X_i, Y_i)^q \right)^{1/q}. \tag{6}$$

If BPs are considered as elements compound by four consecutive intervals, (6) can be applied to measure distances between them. Given (5) and (6) with order $q = 1$, the Ichino-Yaguchi distance function between a pair of BPs, BP_1 and BP_2, is defined as follows:

$$\begin{aligned} \delta_{q=1}(BP_1, BP_2) =& \frac{1}{2}(|m_1 - m_2| + 2\,|q_1 - q_2| + 2\,|Me_1 - Me_2| + \\ & 2\,|Q_1 - Q_2| + |M_1 - M_2|). \end{aligned} \tag{7}$$

This distance assigns more weight to the differences between quartiles than to the difference between the extremes. Therefore, it is not greatly conditioned by the behavior of the extremes, which is usually less stable than the behavior of the quartiles. Hence (7) seems suitable in contexts where the interest mainly lies in the behavior of the central area of the considered distributions.

As the BP distance in (7) is defined as the combination of the distances between the four intervals that characterizes each BP, it has to be divided into four in order to be integrated in the Ichino-Yaguchi metric for symbolic data given in (6). If not, BP variables would have quadruple weight than the other of the variables considered. The resulting BP distance is defined as follows:

$$\eta(BP_1, BP_2) = \frac{\delta(BP_1, BP_2)}{4}. \tag{8}$$

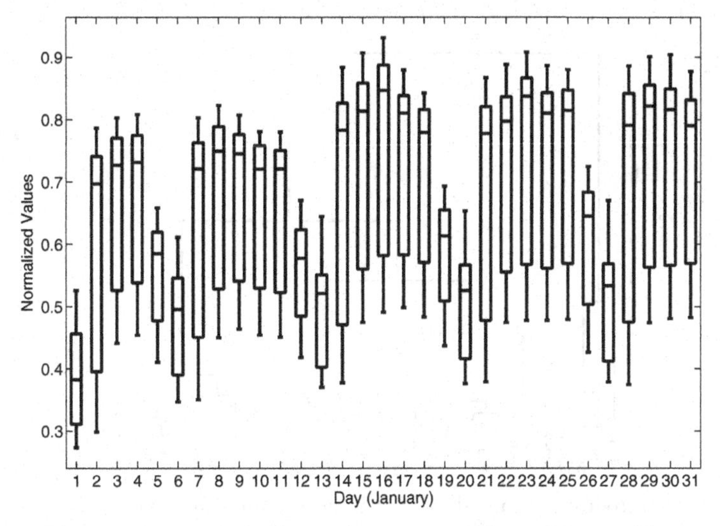

Fig. 1. Electric daily demand represented by a BP variable.

4 Hierarchical BP clustering: an example based on the daily electric demand

In order to illustrate our approach, an agglomerative clustering using the distances proposed is carried out in a real-life dataset which is available at: http://www.iit.upcomillas.es/~cmate/. This dataset describes the hourly electric demand in a Spanish distribution area during 31 days, from 1-Jan-1991 until 31-Jan-1991, i.e. 31 individuals (days) described by 24 quantitative variables (hours) which records the value of the demand (the value of the demand is normalized). The dataset is summarized by means of a BP variable which represents the empirical distribution of the 24 values of the hourly demand along each day. The resulting dataset, shown in Fig. 1, is described by one BP variable.

An agglomerative clustering algorithm with the complete linkage method is applied to the distance matrix obtained with the Ichino-Yaguchi distance for BP data. The resulting dendrogram is shown in Figure 2. It can be seen that at 0.9 the dataset is clearly partitioned into two clusters: the first one with the holidays (Sundays and New Year's Day) and the second one with the working-days. In the second cluster, a subcluster containing all the January's Saturdays and other subcluster containing Mondays can be seen. This dendrogram is quite similar to the one shown in Figure 4, which is obtained by clustering the original dataset (described by 24 classical variables) using the Manhattan distance matrix. This strong resemblance reinforces the proposed approach and the Ichino-Yaguchi distance for BP variables.

In addition, Figure 3 shows the dendrogram yielded by the clustering algorithm of the distance matrix obtained by applying the Manhattan distance

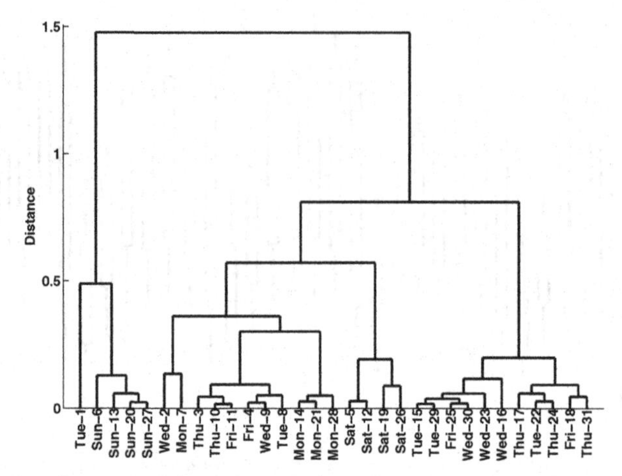

Fig. 2. Dendrogram obtained yielded by the Ichino-Yaguchi BP distance.

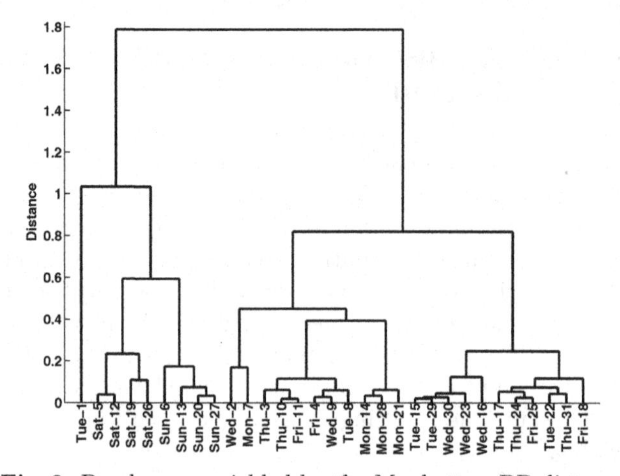

Fig. 3. Dendrogram yielded by the Manhattan BP distance.

for BP variables in the symbolic dataset. The main difference with the other dendrograms is that the subcluster of Saturdays belongs to the cluster of the Holidays. This makes sense as Saturdays are holidays for certain sectors.

Finally, the classical dataset has been summarized into a symbolic dataset described by a histogram variable representing the daily demand. The histogram variable has the same structure for all the 31 elements (days) of the set. Each histogram is composed by 10 fixed intervals partitioning the range of the variable, each interval with an attached weight representing its frequency. The distance matrix has been estimated with the quadratic-form distance proposed by Niblack et al. (1993) for image retrieval:

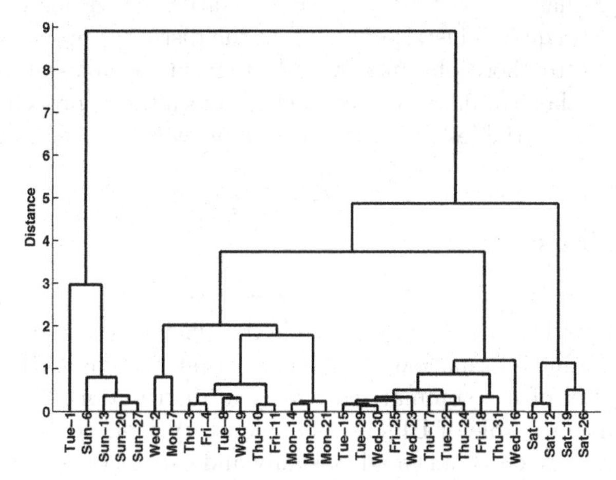

Fig. 4. Dendrogram yielded by the Manhattan distance in the classical dataset.

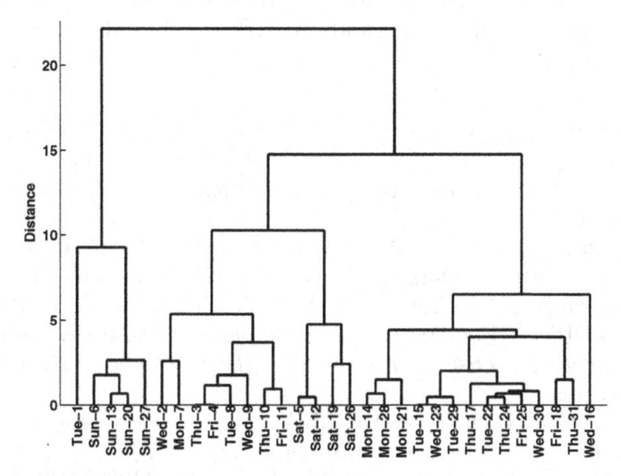

Fig. 5. Dendrogram obtained using the quadratic-form distance in the dataset described by a histogram variable.

$$d(H, K) = \sqrt{(h-k)'A(h-k)}, \tag{9}$$

where h and k are vectors representing the interval weights of the histograms H and K, respectively; matrix $A = [a_{ij}]$ denote the similarity between intervals i and j; $a_{ij} = 1 - d_{ij}/\max(d_{ij})$ where d_{ij} is the ground distance between the midpoints of intervals i and j. The dendrogram obtained with the quadratic-form distance and shown in Figure 5 generally resembles with those shown in Figures 2 and 4.

The example has shown that the summarization by means of BP variables and the subsequent clustering based on the distances proposed obtains analogous results to those obtained with histogram variables and with the clustering of the classical dataset. This fact endorses the approach proposed and shows that BP variables are a kind of variable worth considering in SDA contexts.

5 Conclusions

This paper presents BP variables as a new kind of symbolic variable with a great potential in SDA. BP variables enable the efficient summary of quantitative data providing information about the location, the spread,the skewness and the normality of the summarized empirical distribution. The proposed distances enhance the role of BPs as an exploratory tool, because, thus far, the comparison between BPs is made visually and cannot be quantified in an objective way.

Our approach to cluster BP variables is just the first step, but other clustering methods and distances for BP variables can be proposed. The adaptation of other statistical methods to BP variable also are interesting extensions to the present work.

References

BENJAMINI, Y. (1988): Opening the Box of a Boxplot. *American Statistician, 42/4, 257-262.*

BILLARD, L., and DIDAY, E. (2002): From the Statistics of Data to the Statistics of Knowledge: Symbolic Data Analysis. *Journal of the American Statistical Association, 98/462, 991-999.*

BOCK, H.H. and DIDAY, E. (2000): *Analysis of Symbolic Data: Exploratory Methods for Extracting Statistical Information ¿From Complex Data.* Springer-Verlag, Heidelberg.

FRIGGE, M., HOAGLIN, D. C., and IGLEWICZ, B. (1989): Some Implementations of the Boxplot. *American Statistician, 43/1, 50-54.*

HOAGLIN, D. C., IGLEWICZ, B., and TUKEY, J. W. (1986): Performance of Some Resistant Rules for Outlier Labeling. *Journal of the American Statistical Association, 81/396, 991-999.*

ICHINO, M., and YAGUCHI, H. (1994): Generalized Minkowski Metrics for Mixed Feature-Type Data Analysis. *IEEE Transactions on Systems, Man and Cybernetics, 24/1, 698-708.*

NIBLACK, W., BARBER, R., EQUITZ, W., FLICKNER, M.D., GLASMAN, E.H., PETKOVIC, D., YANKER, P., FALOUTSOS, C., TAUBIN, G., and HEIGHTS, Y. (1993): Querying images by content, using color, texture, and shape. *SPIE Conference on Storage and Retrieval for Image and Video Databases, 1908, 173-187.*

TRENKLER, D. (2002): Quantile-Boxplots. *Communications in Statistics: Simulation and Computation, 31/1, 1-12.*

TUKEY, J. W. (1977): *Exploratory Data Analysis.* Addison-Wesley, Reading.

Evaluation of Allocation Rules Under Some Cost Constraints

Farid Beninel[1] and Michel Grun Rehomme[2]

[1] Université de POITIERS, UMR CNRS 6086
IUT- STID, 8 rue Archimède, 79000 Niort, FRANCE
[2] Université PARIS2, ERMES, UMR CNRS 7017
92 rue d'Assas, 75006 Paris, FRANCE

Abstract. Allocation of individuals or objects to labels or classes is a central problem in statistics, particularly in supervised classification methods such as Linear and Quadratic Discriminant analysis, Logistic Discrimination, Neural Networks, Support Vector Machines, and so on. Misallocations occur when allocation class and origin class differ. These errors could result from different situations such as quality of data, definition of the explained categorical variable or choice of the learning sample. Generally, the cost is not uniform depending on the type of error and consequently the use only of the percentage of correctly classified objects is not enough informative.
In this paper we deal with the evaluation of allocation rules taking into account the error cost. We use a statistical index which generalizes the percentage of correctly classified objects.

1 Introduction

Allocation of objects to labels is a central problem in statistics. Usually, the discussed problems focus on the way to build the assignment rules, tests and validation. We are concerned here with the posterior evaluation of a given rule taking into account the error cost. Such an evaluation allows one to detect different problems including the ability of a procedure to assign objects to classes, the quality of available data or the definition of the classes.
Allocations considered here could be descriptive or inductive.The first situation consists in allocations as primary data. The second situation concerns supervised learning methods as discriminant analysis (LDA, QDA), logistic discrimination, support vector machines (SVM), decision trees, and so on.
For these two ways of allocation, the errors depend on the quality of the data. Specially for supervised classification methods, errors could result from various different causes such as the ability of these data to predict, the definition of the predicted categorical variable, the choice of the learning sample, the methodology to build the allocation rule, the time robustness of the rule and so on.
Our point, here, is the study of missallocations when the associated costs are non uniform and consequently using only the correctly classified rate is insufficient.

In the statistical literature, the hypothesis of a non uniform cost is only considered when elaborating and validating the decision rule(Breiman, 1984). Unfortunately, in real situations validated allocation rules minimizing some cost function could generate higher misclassification costs.
This paper deals with a post-learning situation *i.e.* the allocation rule is given and we have to evaluate it using a new observed sample. We, frequently, encounter such a situation when one needs confidentiality or when allocation and origin class are only realized for a sample of individuals. In insurance, for instance, the sample could be individuals subject to some risk observed over the first year from subscription.
The proposed approach consists of evaluating allocation rules, using an index which is some generalization of the correctly classified rate. To compute the significance level of an observed value of such an index, we consider the p-value associated with a null hypothesis consisting in an acceptable cost. The determination of the p-value leads to a non linear programming problem one could resolve, using available packages of numerical analysis. For the simple case of 3 cost values, we give an analytical solution.

2 Methodology

DEFINITIONS AND NOTATIONS.
Let us denote by Ω the set of individuals and by Y the associated label variable *i.e.*

$$Y : \Omega \longrightarrow G = (g_1, ..., g_q),$$
$$\omega \longrightarrow Y(\omega).$$

We denote by $C_{k,l}$ the error cost, when assigning to label g_k an individual from g_l, by $\psi(\Omega \longrightarrow G)$ the labelling rule (i.e. $\psi(\omega)$ consists of the allocation class of the individual ω and by C the cost variable (i.e. $C(\omega) = C_{k,l}$ when $\psi(\omega) = g_k$ and $Y(\omega) = g_l$).
Consider the random variable $Z(\Omega \longrightarrow [0,1])$ where $Z(\omega)$ measures the level of concordance between the allocation and the origin class of individual ω.
Given a stratified sample, the problem of interest is to infer a comparison between allocations and origin classes for all individuals of Ω.
We propose a statistical index which measures the level of concordance between functions ψ and Y, using observed data. Such an index is a linear combination of the concordance sampling variables.

ACCEPTABLE COST HYPOTHESIS.
Let us denote by $\{(\alpha_{k,l}, p_{k,l}) : k, l = 1, \ldots, q\}$ the probability distribution of the variable Z *i.e.* $(\alpha_{k,l})$ are the possible values of Z and $(p_{k,l})$ the associated probabilities.
Obviously, the cost of a decision decreases as the associated level of concordance rises. Hence, $\alpha_{k,l} \geq \alpha_{i,j}$ when $C_{k,l} \leq C_{i,j}$.

The mean of the misallocation cost is given by $E(Z) = \sum_{k,l} C_{k,l}p_{k,l}$ and for a fixed threshold $\delta \in \mathbb{R}$ the null hypothesis of acceptable cost is

$$\mathcal{H}_0(C, \delta) : \{(p_{k,l}) \in \mathbb{R}^{+,q\times q},\ \sum_{k,l} p_{k,l} = 1,\ \sum_{k,l} C_{k,l}p_{k,l} \leq \delta\}.$$

Let us denote by $Z_{h,j} \quad (h = 1,..,q \quad j = 1,..,n_h)$ the sampling variables distributed as Z.

We consider, here, the non independent case derived from sampling without replacement and with replacement when group sizes are large.

GENERALIZED correctly classified rate.

As a statistical index, to measure the concordance between functions Y, ψ given a sample $Z^{(n)} = (Z_{h,j})_{h,j}$, we propose

$$T_{\psi,Y}(Z^{(n)}, \alpha, w) = \sum_{h=1}^{q} \sum_{j=1}^{n_h} \omega_{n,h}\ Z_{h,j}.$$

Here $n = n_1 + ... + n_q$ and $w = (w_{n,h})_h$ is a weighting parameter where samplings from a same group are weighted identically. We suppose, without loss of generality ($w.o.l.g.$), positive components, $\sum_1^q n_h w_{n,h} = 1$ and consequently, $T_{\psi,Y}(Z^{(n)}, \alpha, w) \in [0, 1]$.

Note that, for equal weighting and a unique type of error, we deal with the classical correctly classified rate or $T_{\psi,Y}(Z^{(n)}, (\delta_k^l), (\frac{1}{n}))$ where δ_k^l is the Kronecker coefficient.

3 Probabilistic study

ASYMPTOTIC DISTRIBUTION.

$T_{\psi,Y}(Z^{(n)}, \alpha, w)$ is a linear combination of multinomial variables. Under the assumptions of the independence of sampling variables and convergence of $\sum_h n_h\ w_{n,h}^2$, we obtain from the Lindeberg theorem (Billingsley 1995 p.359-362)

$$\frac{T_{\psi,Y} - \mu_n}{\sigma_n} \xrightarrow{d} N(0, 1), \quad n \to \infty, \tag{1}$$

where $\mu_n(\bar{p}) = \sum_{k,l} \alpha_{k,l}\ p_{k,l}$ and $\sigma_n^2(\bar{p}) = (\sum_{k,l} \alpha_{k,l}^2\ p_{k,l} - (\sum_{k,l} \alpha_{k,l}\ p_{k,l})^2)$ $\sum_h n_h\ w_{n,h}^2$.

From a practical point of view, to use the previous result leading to the gaussian model, we have to consider carefully the atypical individuals.

THE OPTIMIZATION PROBLEM to compute p-value.
Given an observed value t of $T_{\psi,Y}$, we deal with computation of the associated *p-value.* Here, the definition corresponding to the most powerful test consists of $p-value(t) = \max_{\bar{p}}\ prob(T_{\psi,Y} \leq t/\bar{p} \in H_0(C,\delta))$.

Let $F_{\alpha,t}([0,1] \longrightarrow \mathbb{R})$ be the function such that $\frac{t-\mu_n(\bar{p})}{\sigma_n(\bar{p})} = \frac{1}{\sqrt{{}_h n_h\ w_{n,h}^2}} F_{\alpha,t}(\bar{p})$.

Using the asymptotic distribution, we obtain

$$p-value(t) = \Phi(\frac{1}{\sqrt{\sum_h n_h\ w_{n,h}^2}} \max\ F_{\alpha,t}(\bar{p})), \tag{2}$$

where Φ is the CDF of the $N(0,1)$ distribution.

The calculation of the $p-value$ leads to the following optimization problem.

$$\text{Problem(M.d)} : \begin{cases} \max F_{\alpha,t}(\bar{p}) = \frac{L(\bar{p})}{\sqrt{Q(\bar{p})}}, \\ \\ \bar{p} \in \mathcal{H}_0(C,\delta). \end{cases}$$

The order d of the problem corresponds to the number of distinct non null cost values.
Here, the constraints of $\mathcal{H}_0(C,\delta)$ are linear inequalities and as L is a linear function of the components of $\bar{p}$ and Q a quadratic one, $F_{\alpha,t}(\bar{p})$ is non linear function. To solve the general problem (M.d) one could use or adapt *non linear programming* optimization procedures.
An *ad hoc* SCILAB program called VACC(VAlidation under Cost Constraints) is available at http://www-math.univ-poitiers.fr/Stat/VACC.html.

4 Application: The case of problem (M.2)

The case of problem (M.2) is relatively frequent in real situations and has simple analytical solutions. For this case, we deal with three cost values *w.o.l.g.* $C_{k,l} = 1,\ C,\ 0$ with $1\ >\ C >\ 0$. The concordance variable Z is defined as follows.

$$Z(\omega) = \begin{cases} 1 & if\ \ C(\omega) = 0 & (p_1), \\ \alpha & if\ \ C(\omega) = C & (p_\alpha), \\ 0 & if\ \ C(\omega) = 1 & (1-p_1-p_\alpha). \end{cases}$$

and the optimization problem is

$$\text{Problem(M.2)} : \begin{cases} \max F_{\alpha,t}(\bar{p}) = F_{\alpha,t}(p_1, p_\alpha) = \frac{t - p_1 - \alpha p_\alpha}{\sqrt{p_1 + \alpha^2 p_\alpha - (p_1 + \alpha p_\alpha)^2}}, \\ p_1 \geq 0, \quad p_\alpha \geq 0, \quad p_1 + p_\alpha \leq 1, \\ C p_\alpha + (1 - p_1 - p_\alpha) \leq \delta. \end{cases}$$

We derive the solution of problem (M.2), using the following result.

Lemma 1. *For $t, \alpha \in [0,1]$ and C, δ such that $1 > C > \delta > 0$, the maximum of $F_{\alpha,t}(x,y)$ is attained for $(C-1)y - x + 1 = \delta$.*

Let us set $r = \frac{\delta - 1}{C-1}$ and $s = \frac{1}{C-1}$. For (x,y) such that $y = sx + r$ (*i.e.* the cost constraint at boundary) $F_{\alpha,t}(x,y) = G(x)$ where

$$G(x) = \frac{-(1+\alpha s)x + t - \alpha r}{\sqrt{-(1+\alpha s)^2 x^2 + (1 + \alpha^2 s - 2\alpha r - 2\alpha^2 rs)x + \alpha^2(r - r^2)}}. \tag{3}$$

We establish the following result.

Proposition 1. *Let $\Delta = \Delta(\alpha, C, t) = (1 - \alpha s)(1 - \alpha^2 s - 2t + 2\alpha st)$ and $x_0(t) = -(\alpha^3 rs - \alpha^2 st(2r-1) - 2\alpha^2 r + 2\alpha rt + \alpha r - t)/\Delta(\alpha, C, t)$. Then, the following result holds.*

$$\max F_{\alpha,t}(x,y) = \begin{cases} G(x_0(t) & if \quad x_0(t) \in [1 - \frac{\delta}{C}, 1 - \delta] \quad and \quad \Delta < 0, \\ \max(G(1 - \frac{\delta}{C}), G(1 - \delta)) & elsewhere. \end{cases}$$

On the parameters.

The parameters α, δ, C are fixed by the users of the discussed methodology and the value t is derived from data. The choice $\alpha = 0.5$ leads to the UMP test, when using the class of statistics $T_{\psi,Y}(Z^n, \alpha, w)$

The choice $w_{n,h} = \frac{1}{n}$ minimizes $\sum_h n_h w_{n,h}^2$ when $\sum_h n_h w_{n,h} = 1$. For such a choice of w, $p - value(t) = \Phi(\sqrt{n}\ maxF_{\alpha,t}(\overline{p}))$ and constitutes an upper bound for the other choices.

Example: $\alpha = C = 0.5$ and $\delta < C$

The problem, here, consists of maximizing $F_{\alpha,t}(x,y) = \frac{t - x - 0.5y}{\sqrt{x + 0.25y - (x + 0.5y)^2}}$ with the system of constraints

$$\mathcal{H}_0 = \{(x, y) : \; y > 0, \; x + y \leq 1, \; x + 0.5y \geq 1 - \delta\}.$$

We derive from the previous proposition

$$\max \; p - value(t) = \begin{cases} \frac{\sqrt{n}(t-1+\delta)}{\sqrt{0.5\delta-\delta^2}} & if \quad t > 1 - \delta, \\ \\ \frac{\sqrt{n}(t-1+\delta)}{\sqrt{\delta-\delta^2}} & \text{elsewhere.} \end{cases}$$

As an illustration, we give the following table. The considered t values relate real situations. The associated p-value is calculated for some δ values.

Table 1. Computations for $n = 50$

t *value*	δ *value*	$\sqrt{n}\ maxF_{\alpha,t}$	$max\ p - value(t)$
0.90	0.20	2.887	0.998
	0.15	1.543	0.939
	0.10	0.000	0.500
	0.05	-2.357	0.009
0.75	0.20	-1.443	0.074
	0.15	-3.086	0.001
	0.10	-5.303	0.000
	0.05	-9.428	0.000

5 Conclusion

Using the generalized correctly classified rate, the interpretation of an observed value integrates the case of non uniform error cost. As forthcoming applications and extensions, we have in mind:

- The study of the same concordance statistic in the non gaussian case. Such a case appears when n is small or when weights are unbalanced.
- The study of other statistics measuring the quality of an allocation rule, under cost constraints.
- The extension of the cost notion to take into account the structure of classes. This structure is, sometimes, given by proximity measures between classes as for the terminal nodes of a decision tree or for ordered classes.

References

ADAMS,N.M., HAND, D.J. (1999): Comparing classifiers when the misallocation costs are uncertain. *Pattergn recognition, 32, 1139-1147*

BILLINGSLEY, P. (1990): *Probability and measure.* Wiley series in probability and mathematical statistics), New York, pp.593.

BREIMAN, L., FRIEDMAN, J., OHLSEN, R., STONE,C. (1984): *Classification and regression trees.* Wadsmorth, Belmont.

GIBBONS, J.D., PRATT, J.W. (1975): p-value: interpretation and methodology. *JASA, 29/1, 20-25*

GOVAERT, G. (2003): *Analyse des données.* Lavoisier serie "traitement du signal et de l'image", Paris, pp.362.

SEBBAN, M., RABASEDA,S., BOUSSAID,O. (1996): Contribution of related geometrical graph in pattern and recognition. In: E. Diday, Y. Lechevallier and O. Optiz (Eds.): *Ordinal and symbolic data Analysis.* Springer, Berlin, 167–178.

Crisp Partitions Induced by a Fuzzy Set

Slavka Bodjanova

Department of Mathematics, Texas A&M University-Kingsville,
MSC 172, Kingsville, TX 78363, U.S.A.

Abstract. Relationship between fuzzy sets and crisp partitions defined on the same finite set of objects X is studied. Granular structure of a fuzzy set is described by rough fuzzy sets and the quality of approximation of a fuzzy set by a crisp partition is evaluated. Measure of rough dissimilarity between clusters from a crisp partition of X with respect to a fuzzy set A defined on X is introduced. Properties of this measure are explored and some applications are provided. Classification of membership grades of A into linguistic categories is discussed.

1 Introduction

The ambition of fuzzy set theory is to provide a formal setting for incomplete and gradual information as expressed by people in natural language. Membership function of a fuzzy set A defined on universal set X assigns to each element $x \in X$ value $A(x) \in [0,1]$ called the membership grade of x in A. For example, fuzzy set A representing the concept "successful" applicant for a credit card may have the following membership grades on a set of 10 applicants:

$$\begin{aligned} A = 0/x_1 + 0.2/x_2 + 0.2/x_3 + 0.2/x_4 + 0.45/x_5 + \\ + 0.5/x_6 + 0.8/x_7 + 0.8/x_8 + 1/x_9 + 1/x_{10}. \end{aligned} \tag{1}$$

There are many approaches to the construction of the membership function of a fuzzy set (Bilgic and Turksen (2000), Viertl (1996)). Because of interpretation and further applications, fuzzy sets are often approximated by "less fuzzy" (rough) fuzzy sets or even crisp sets. These approximations are derived from the level set of A,

$$\Lambda_A = \{\alpha \in [0,1] : \alpha = A(x) \text{ for some } x \in X\}. \tag{2}$$

The most rough characterization of membership grades $A(x)$, $x \in X$ is

$$\min_{x \in X} A(x) \le A(x) \le \max_{x \in X} A(x). \tag{3}$$

If $\min_{x\in X} A(x) = 0$ and $\max_{x \in X} A(x) = 1$, we do not obtain any specific information about A. By splitting X into two clusters C_1, C_2, we obtain less rough (more specific) information. For $x \in C_i, i = 1, 2$,

$$\min_{x \in C_i} A(x) \le A(x) \le \max_{x \in C_i} A(x). \tag{4}$$

Partition $C = \{C_1, C_2\}$ of X is useful for summary characterization of A by (4) only if $\max_{x \in C_1} A(x) < \min_{x \in C_2} A(x)$. Then elements $x \in C_1$ can be labeled as objects with lower membership grade in A and elements $x \in C_2$ as objects with higher membership grade in A. We say that C is a 2-category (binary) scale induced by A. The most common binary scales are created by α-cut A_α defined for all $\alpha \in (0,1]$ as follows:

$$A_\alpha = \{x \in X | A(x) \geq \alpha\}. \tag{5}$$

Then $C^\alpha = \{C_1 = X - A_\alpha, C_2 = A_\alpha\}$. The choice of $\alpha \in (0,1]$ varies from application to application and it is usually determined by a researcher familiar with the area of study. If no additional information is given, $\alpha = 0.5$ is used. In our example of credit card applicants, $C^{0.5} = \{C_1 = \{x_1, x_2, x_3, x_4, x_5\}, C_2 = \{x_6, x_7, x_8, x_9, x_{10}\}\}$. Cluster C_1 represents the subset of unsuccessful applicants (no credit card), while C_2 the subset of successful applicants (credit card will be issued). Crisp set $\tilde{A}$ with the membership function $\tilde{A}(x) = 0$ for $x \in C_1$, and $\tilde{A}(x) = 1$ for $x \in C_2$, is the crisp approximation (defuzzification) of A. The quality of approximation of A by C^α can be measured by the coefficient of accuracy or the coefficient of roughness known from the theory of rough sets (Pawlak (1982)) and rough fuzzy sets (Dubois and Prade (1990)). If the quality of approximation is high, we say that A has a reasonable 2-granule structure and the summary characterization of A by (4) is appropriate. If there is no clear distinction between the "large" and the "small" membership grades, fuzzy set A should be approximated by a crisp partition with more than two clusters. Approximation of a fuzzy set A by k-category scale is the main topic of our paper.

Partition $C = \{C_1, C_2, \ldots, C_k\}$ of X is called a scale induced by fuzzy set A, if for each cluster C_i, $i = 1, \ldots k-1$,

$$\max_{x \in C_i} A(x) < \min_{x \in C_{i+1}} A(x). \tag{6}$$

If C is a scale, then all membership grades $A(x)$ from a fuzzy set granule A/C_i can be described by a common linguistic term. Some useful terms are: $k = 2$ (small, large), $k = 3$ (small, medium, large) and $k = 5$ (very small, small, medium, large, very large). Each linguistic characterization means "with respect to the set of all different membership grades of A,"

$$\Lambda_A = \{a_1, \ldots, a_m\}, \tag{7}$$

where $a_i < a_{i+1}$, $i = 1, \ldots, m-1$. Fuzzy set A has a reasonable k-granule structure if it induces a scale $C = \{C_1, \ldots, C_k\}$ on X such that the dissimilarity between different clusters from C with respect to A is high and the roughness of approximation of A by C is low (the accuracy of approximation is high). We will introduce the measure of rough dissimilarity between crisp clusters C_r and C_s with respect to fuzzy set A . This measure evaluates how the roughness of approximation of membership grades $A(x)$, $x \in C_r \cup C_s$ decreases, when we approximate $A(x)$ instead of by the coarser cluster $C_r \cup C_s$ by two separate clusters C_r and C_s. Rough dissimilarity will be used in the search for a reasonable 2-granule structure of A, and consequently for defuzzification of A . If the quality of approximation of A by 2-cluster scale is low, we will search for a reasonable k-granule structure, $k > 2$. Finally, we will use rough dissimilarity to create a fuzzy proximity relation on the set of all elementary crisp clusters induced by A. Proximity relation will lead to the construction of a hierarchical crisp approximation of fuzzy set A.

2 Relationship between fuzzy sets and crisp partitions

Let $\mathcal{F}(X)$ denote the set of all fuzzy sets defined on X and let $\mathcal{P}(X)$ denote the set of all crisp partitions of X. Relationship between $A \in \mathcal{F}(X)$ and $C \in \mathcal{P}(X)$ can be described by some concepts from the theory of rough fuzzy sets (Dubois and Prade (1990)). In this paper we assume that X is a finite set with cardinality $|X| = n$, and $A \neq \emptyset$. Cardinality of A is given by

$$|A| = \sum_{i=1}^{n} A(x_i). \tag{8}$$

We will adjust the description of rough fuzzy sets to these assumptions.

Let $C = \{C_1, \ldots, C_k\} \in \mathcal{P}(X)$, $k \leq n$. The rough fuzzy set associated with C and A is the pair $(\underline{A}_C, \overline{A}_C)$ of fuzzy sets from $X/C \to [0,1]$ defined as follows: for all $C_i \in C$,

$$\underline{A}_C(C_i) = \min_{x \in C_i} A(x), \text{ and } \overline{A}_C(C_i) = \max_{x \in C_i} A(x). \tag{9}$$

We will use the simplified notations

$$\underline{A}_C(C_i) = \underline{A}_{C_i}, \text{ and } \overline{A}_C(C_i) = \overline{A}_{C_i}. \tag{10}$$

The accuracy of approximation of A in the approximation space (X, C) is evaluated by the coefficient of accuracy

$$\lambda_C(A) = \frac{\sum_{i=1}^{k} |C_i| \underline{A}_{C_i}}{\sum_{i=1}^{k} |C_i| \overline{A}_{C_i}}. \tag{11}$$

The roughness of approximation is measured by the coefficient of roughness

$$\kappa_C(A) = 1 - \lambda_C(A). \tag{12}$$

Rough fuzzy set $(\underline{A}_C, \overline{A}_C)$ is associated with two fuzzy sets $\underline{A}_C^*$ and $\overline{A}_C^*$ from $X \to [0,1]$ defined as follows: for $x \in X$, if $x \in C_i \in C$ then

$$\underline{A}_C^*(x) = \underline{A}_{C_i}, \; \overline{A}_C^*(x) = \overline{A}_{C_i}. \tag{13}$$

Fuzzy set $\underline{A}^*_C$ is the lower rough approximation of A by C and fuzzy set $\overline{A}^*_C$ is the upper rough approximation of A by C . The difference between the upper and the lower approximations evaluated for all $x \in X$ by fuzzy set

$$BND_{C(A)}(x) = \overline{A}^*_C(x) - \underline{A}^*_C(x), \tag{14}$$

describes the boundary region (rough region) of rough fuzzy set $(\underline{A}_C, \overline{A}_C)$. The size (cardinality) of the boundary region gives information about the roughness (uncertainty) of approximation of A by C. Obviously,

$$\sigma_C(A) = |BND_{C(A)}| = \sum_{i=1}^{k} |C_i|(\overline{A}_{C_i} - \underline{A}_{C_i}) = \sum_{i=1}^{k} \sigma_{C_i}(A), \tag{15}$$

where $\sigma_{C_i}(A)$ is the size of the boundary region of granule $(\underline{A}_{C_i}, \overline{A}_{C_i})$ of rough fuzzy set $(\underline{A}_C, \overline{A}_C)$. The smaller is the value of $\sigma_{C_i}(A)$ the more specific is characterization of elements $x \in C_i$ by (4).

Example 1 Assume fuzzy set A of the "successful" credit card applicants given by the membership function (1). Then $\Lambda_A = \{0, 0.2, 0.45, 0.5, 0.8, 1\}$ and there are five binary scales C^α induced by A. The maximal coefficient of accuracy and the minimal size of the rough region associated with C^α is obtained for $\alpha = 0.8$. We have $\sigma_{C^{0.8}}(A) = 3.8$ and $\lambda_{C^{0.8}} = 0.46$. Because 0.46 is considered low, we conclude that fuzzy set A does not have a reasonable 2-granule structure. We will explore its k-granule structure, $2 < k \leq 6$, in the next section.

3 Rough dissimilarity

The most refined scale induced by a fuzzy set $A \in \mathcal{F}(X)$ is partition $E = \{E_1, \ldots, E_m\} \in \mathcal{P}(X)$, such that for each $a_i \in \Lambda_A = \{a_1, \ldots, a_m\}$,

$$E_i = \{x \in X : A(x) = a_i\}. \tag{16}$$

We will call E_i the ith elementary cluster induced by A and E the elementary scale induced by A. Any partition $C \in \mathcal{P}(X)$ induced by A is obtained by aggregation of clusters from E. When we aggregate two different clusters $C_r, C_s \in C \in \mathcal{P}(X)$ then

$$\sigma_{C_r \cup C_s}(A) \geq \sigma_{C_r}(A) + \sigma_{C_s}(A). \tag{17}$$

Assume $C_r, C_s \in C \in \mathcal{P}(X)$ such that $\underline{A}_{C_r} = \underline{A}_{C_s} = a$ and $\overline{A}_{C_r} = \overline{A}_{C_s} = b$ and $C_r \neq C_s$. (Obviously, C is not a scale on A.) Then the rough approximation of each membership grade $A(x)$ for $x \in C_r$ is given by $a \leq A(x) \leq b$, which is the same as the rough approximation of membership grades $A(x)$ for $x \in C_s$. Therefore, for all $x \in C_r \cup C_s$ we have that $a \leq A(x) \leq b$ and

$$\sigma_{C_r \cup C_s}(A) = \sigma_{C_r}(A) + \sigma_{C_s}(A). \tag{18}$$

The larger is the difference between the size of the rough region of the aggregated granule $(\underline{A}_{C_r \cup C_s}, \overline{A}_{C_r \cup C_s})$ and the sum of sizes of the rough regions of the individual granules $(\underline{A}_{C_r}, \overline{A}_{C_r})$ and$(\underline{A}_{C_s}, \overline{A}_{C_s})$, the more dissimilar are clusters C_r, C_s with respect to the fuzzy set A.

Definition 1 *Assume $A \in \mathcal{F}(X)$ and $C = \{C_1, \ldots, C_k\} \in \mathcal{P}(X)$. The rough dissimilarity between clusters $C_r, C_s \in C$ with respect to A is evaluated by*

$$\delta_A(C_r, C_s) = \psi_{rs} - (\sigma_{C_r}(A) + \sigma_{C_s}(A)), \tag{19}$$

where

$$\psi_{rs} = (|C_r| + |C_s|)(\max\{\overline{A}_{C_r}, \overline{A}_{C_s})\} - \min\{\underline{A}_{C_r}, \underline{A}_{C_s}\}). \tag{20}$$

Note: When $r \neq s$ then $\psi_{rs} = \sigma_{C_r \cup C_s}(A)$ and

$$\delta_A(C_r, C_s) = \sigma_{C_r \cup C_s}(A) - (\sigma_{C_r}(A) + \sigma_{C_s}(A)). \tag{21}$$

Rough dissimilarity between two clusters C_r and C_s with respect to fuzzy set A evaluates how the roughness of approximation of membership grades of A increases when we approximate $A(x)$ by the coarser cluster $C_r \cup C_s$ instead of by two separate clusters C_r and C_s. The following properties of the measure of rough dissimilarity can be easily proved.

Proposition 1 *Assume $A \in \mathcal{F}(X)$ and $C \in \mathcal{P}(X)$. Then for $C_i, C_j, C_t \in C$,*

1. $0 \leq \delta_A(C_i, C_j) \leq |X|(\max_{x \in X} A(x) - \min_{x \in X} A(x)) = \sigma_X(A)$,
2. $\delta_A(C_i, C_i) = 0$,
3. $\delta_A(C_i, C_j) = \delta_A(C_j, C_i)$,
4. $\delta_A(C_i, C_j) = |X|(\max_{x \in X} A(x) - \min_{x \in X} A(x))$ *if* $C_i = \{x \in X : A(x) = \max_{x \in X} A(x)\}$, $C_j = \{x \in X : A(x) = \min_{x \in X} A(x)\}$ *and* $C_i \cup C_j = X$,
5. $\delta_A(C_i \cup C_j, C_t) \geq \max\{\delta_A(C_i, C_t), \delta_A(C_j, C_t)\}$.

The change in the accuracy of approximation of fuzzy set A by partition $C = \{C_1, \ldots, C_k\}$ and then by the coarser partition $L = \{L_1, \ldots, L_{k-1}\}$ created from C by aggregation of two different clusters is related to the rough dissimilarity of the aggregated clusters as follows:

Proposition 2 *Assume $A \in \mathcal{F}(X)$. Let $C = \{C_1, \ldots, C_k\}$ be a crisp partition of X and let $L = \{L_1, \ldots, L_{k-1}\}$ be partition of X obtained from C such that $L_1 = C_r \cup C_s$, $C_r, C_s \in C$, and $\{L_2, \ldots, L_{k-1}\} = C - \{C_r, C_s\}$. Then*

$$\lambda_L(A) = \frac{\sum_{j=1}^{k-1} |L_j| \underline{A}_{L_j}}{\sum_{j=1}^{k-1} |L_j| \overline{A}_{L_j}} = \frac{\sum_{i=1}^{k} |C_i| \underline{A}_{C_i} - \delta_1}{\sum_{i=1}^{k} |C_i| \overline{A}_{C_i} + \delta_2}, \tag{22}$$

where $\delta_1 \geq 0, \delta_2 \geq 0$ and $\delta_1 + \delta_2 = \delta_A(C_r, C_s)$.

Example 2 Let A be the fuzzy set from Example 1 and let $C^\alpha = \{C_1^\alpha, C_2^\alpha\}$ be partition induced on X by α-cut A_α, $\alpha \in \Lambda_A - \{0\}$. Because $C_1^\alpha \cup C_2^\alpha = X$, we have that

$$\delta_A(C_1^\alpha, C_2^\alpha) = \sigma_X(A) - (\sigma_{C_1^\alpha}(A) + \sigma_{C_2^\alpha}(A)) = \sigma_X(A) - \sigma_{C^\alpha}(A).$$

The best α-cut approximation of A (interpretation of A) is partition C^{α^*}, which provides the largest decrease of the size of the boundary region $\sigma_X(A)$. Partition C^{α^*} splits X into two best separated clusters (most rough dissimilar clusters). From Example 1 we know that the smallest σ_{C^α} was $\sigma_{C^{0.8}} = 3.8$ and $\sigma_X(A) = 10$. Therefore

$$\max_\alpha \delta_A(C_1^\alpha, C_2^\alpha) = \max_\alpha(\sigma_X(A) - \sigma_{C^\alpha}(A)) = \delta_A(C_1^{0.8}, C_2^{0.8}) = 6.2.$$

Rough dissimilarity $\delta_A(C_1^\alpha, C_2^\alpha)$ can be used as a criterion for the choice of the "best" α-cut approximation of A and consequently defuzzification of A.

Example 3 Let A be the fuzzy set from Example 1. We want to find partition $C \in \mathcal{P}(X)$ with a minimal number of clusters such that C is a scale induced by A and $\lambda_C(A)$ is "large", say larger than 0.7.
We start with $E = \{E_1, E_2, E_3, E_4, E_5, E_6\} \in \mathcal{P}(X)$, where $E_1 = \{x_1\}$, $E_2 = \{x_2, x_3, x_4\}$, $E_3 = \{x_5\}$, $E_4 = \{x_6\}$, $E_5 = \{x_7, x_8\}$, and $E_6 = \{x_9, x_{10}\}$. Because we are looking for a scale, we will consider only aggregation of adjacent clusters from E. Then for $i \in I_E = \{1, 2, 3, 4, 5\}$,

$$\min_{i \in I_E} \delta_A(E_i, E_{i+1}) = \min\{0.8, 1, 0.1, 0.9, 0.9\} = 0.1 = \delta_A(E_3, E_4).$$

By aggregation of clusters E_3, E_4 we obtain partition $D = \{D_1 = E_1, D_2 = E_2, D_3 = E_3 \cup E_4, D_4 = E_5, D_5 = E_6\}$, which approximates A with the coefficient of accuracy $\lambda_D(A) = 0.98 > 0.7$. Then for $i \in I_D = \{1, 2, 3, 4\}$,

$$\min_{i \in I_D} \delta_A(D_i, D_{i+1}) = \min\{0.8, 1.4, 1.3, 1.3, 0.8\} = 0.8 = \delta_A(D_1, D_2) =$$
$$= \delta_A(D_5, D_6).$$

By aggregation of clusters D_1, D_2 and then D_5, D_6 we obtain partition $T = \{T_1 = D_1 \cup D_2, T_2 = D_3, T_3 = D_5 \cup D_6\} = \{E_1 \cup E_2, E_3 \cup E_4, E_5 \cup E_6\}$, which approximates A with the coefficient of accuracy $\lambda_T(A) = 0.71 > 0.7$. Then for $i \in I_T = \{1, 2\}$,

$$\min_{i \in I_T} \delta_A(T_i, T_{i+1}) = \min\{2.1, 2.4\} = 2.1 = \delta_A(T_1, T_2).$$

By aggregation of clusters T_1, T_2 we obtain partition $Q = \{Q_1 = T_1 \cup T_2, Q_2 = T_3\}$, which approximates A with the coefficient of accuracy $\lambda_Q(A) = 0.44 < 0.7$. Note that Q is the same as the binary partition $C^{0.8}$ obtained by the 0.8-cut of A. We conclude that fuzzy set A has a reasonable 3-granule structure that can be approximated by the scale $T = \{T_1, T_2, T_3\}$ on X. Applicants from cluster T_1 are unsuccessful (no credit card), applicants from T_2 are possibly successful (they may get the credit card) and applicants from T_3 are successful (they will get the credit card).

For $C_r, C_s \in C \in \mathcal{P}(X)$, the value of $\delta_A(C_r, C_s)$ depends not only on the membership grades $A(x)$ of $x \in C_r \cup C_s$, but also on the size (cardinality) of C_r, C_s. It may happen that that $\min_{E_j \in E} \delta_A(E_i, E_j)$ is not $\delta_A(E_i, E_{i+1})$. Therefore, a scale obtained by aggregation of adjacent clusters by the method shown in Example 3 is not always the partition of X obtained by aggregation of the least rough dissimilar clusters. However, if a cluster $C_r \in C$ is obtained by aggregation of non adjacent clusters, C is not a scale. Now we will base our search for the k-granule structure of A on the matrix of all values $\delta_A(E_i, E_j)$, $(E_i, E_j) \in E \times E$. First of all we will introduce the coefficient of rough proximity between two clusters.

Definition 2 *Assume a non constant fuzzy set $A \in \mathcal{F}(X)$ and partition $C = \{C_1, \ldots, C_k\} \in \mathcal{P}(X)$. The coefficient of rough proximity between clusters $C_r, C_s \in C$ with respect to A is evaluated by*

$$\omega_A(C_r, C_s) = 1 - \frac{\delta_A(C_r, C_s)}{|X|(\max_{x \in X} A(x) - \min_{x \in X} A(x))} = 1 - \frac{\delta_A(C_r, C_s)}{\sigma_X(A)}. \quad (23)$$

If A is a constant fuzzy set then $\omega_A(C_r, C_s) = 1$.

Fuzzy relation $\omega_A : C \times C \to [0, 1]$ is reflexive and symmetric, and therefore it is a proximity relation. Transitive closure of a proximity relation is a fuzzy equivalence relation, called also a similarity relation. Each α-cut of a similarity relation is a crisp equivalence relation.

Example 4 Let A be the fuzzy set from Example 1. Elementary partition E induced by A was presented in Example 3. Values of $\omega_A(E_i, E_j)$ for all $(E_i, E_j) \in E \times E$ create symmetric matrix $\mathbf{R}$. Transitive closure $\mathbf{R}_T$ of $\mathbf{R}$ can be determined by a simple algorithm (Klir and Yuan (1995)) that consists of 3 steps:

1. $\mathbf{R}' = \mathbf{R} \cup (\mathbf{R} \circ \mathbf{R})$.
2. If $\mathbf{R}' \neq \mathbf{R}$, make $\mathbf{R} = \mathbf{R}'$ and go to Step 1.
3. $\mathbf{R}' = \mathbf{R}_T$.

We use the usual composition and union of fuzzy relations given by

$$(\mathbf{R} \circ \mathbf{R})(E_i, E_j) = \max\{\min_{E_t \in E} \{\mathbf{R}(E_i, E_t), \mathbf{R}(E_t, E_j)\}\},$$

and

$$(\mathbf{R} \cup (\mathbf{R} \circ \mathbf{R}))(E_i, E_j) = \max\{\mathbf{R}(E_i, E_j), (\mathbf{R} \circ \mathbf{R})(E_i, E_j)\}.$$

In our case the set of all different values of matrix $\mathbf{R}_T$ is

$$\Lambda_{\mathbf{R}_T} = \{0.91, 0.92, 0.94, 0.99, 1\}.$$

For each nonzero $\alpha \in \Lambda_{\mathbf{R}_T}$ we obtain partition π^α of E (and therefore of X) as follows:

$$E_i, E_j \in \pi^\alpha \text{ if } \mathbf{R}_T(E_i, E_j) \geq \alpha.$$

Then
$\pi^1 = \{E_1, E_2, E_3.E_4, E_5, E_6\}$, and $\lambda_{\pi^1}(A) = 1$,
$\pi^{0.99} = \{E_1, E_2, E_3 \cup E_4, E_5, E_6\}$, and $\lambda_{\pi^{0.99}}(A) = 0.98$,
$\pi^{0.94} = \{E_1 \cup E_2, E_3 \cup E_4, E_5, E_6\}$, and $\lambda_{\pi^{0.94}}(A) = 0.83$,
$\pi^{0.92} = \{E_1 \cup E_2, E_3 \cup E_4, E_5 \cup E_6\}$, and $\lambda_{\pi^{0.92}}(A) = 0.71$,
$\pi^{0.91} = \{E_1 \cup E_2 \cup E_3 \cup E_4 \cup E_5 \cup E_6\}$, and $\lambda_{\pi^{0.91}}(A) = 0$.

Note that partition $\pi^{0.92}$ is the same as the "best" scale $T = \{T_1, T_2, T_3\}$ found in Example 3. Fuzzy set A can be approximated by the sequence of five nested partitions π^α, $\alpha \in \Lambda_{\mathbf{R}_T}$. Their refinement relationship gives information about the hierarchical granular structure of A.

4 Conclusion

Our paper is a contribution to the ongoing study of the relationship between fuzzy sets and rough sets. Rough fuzzy sets were used to define the rough dissimilarity between clusters from a crisp partition of X with respect to a fuzzy set A defined on X. Crisp partitions of X induced by the membership grades of A were studied. Attention was given to special types of partitions, called scales, whose clusters can be labeled by terms from an ordered linguistic scale. We presented binary scales with the most rough dissimilar clusters created by α-cut of A, then k-category scales ($k > 2$) obtained by aggregation of the least rough dissimilar adjacent clusters from the elementary partition induced by A, and a hierarchical crisp approximation of A derived from the rough dissimilarities of all pairs of elementary crisp clusters induced by A. Further applications of rough dissimilarity in agglomerative and divisive hierarchical clustering procedures will be presented in our future work.

References

DUBOIS, D. and PRADE, H. (1990): Rough fuzzy sets and fuzzy rough sets. *International Journal of General Systems, 17, 191–229.*

KLIR, G.J. and YUAN, B. (1995): *Fuzzy sets and fuzzy logic: Theory and applications.* Prentice Hall, Upper Saddle River.

PAWLAK, Z. (1982): Rough sets. *International Journal of Computer and Information Sciences, 11, 341–356.*

BILGIC, T. and TURKSEN, I.B. (2000): Measurement of membership functions: Theoretical and empirical work. In: D. Dubois, H.Prade (Eds.): *Fundamentals of fuzzy sets.* Kluwer, Dordrecht.

VIERTL, R. (1996): *Statistical methods for non-precise data.* CRC Press, Boca Raton.

Empirical Comparison of a Monothetic Divisive Clustering Method with the Ward and the k-means Clustering Methods

Marie Chavent[1] and Yves Lechevallier[2]

[1] Mathématiques Appliquées de Bordeaux, UMR 5466 CNRS, Université Bordeaux1, 351 Cours de la libration, 33405 Talence Cedex, France
[2] Institut National de Recherche en Informatique et en Automatique, Domaine de Voluceau-Rocquencourt B.P.105, 78153 Le Chesnay Cedex, France

Abstract. DIVCLUS-T is a descendant hierarchical clustering method based on the same monothetic approach than classification and regression trees but from an unsupervised point of view. The aim is not to predict a continuous variable (regression) or a categorical variable (classification) but to construct a hierarchy. The dendrogram of the hierarchy is easy to interpret and can be read as decision tree. An example of this new type of dendrogram is given on a small categorical dataset. DIVCLUS-T is then compared empirically with two polythetic clustering methods: the Ward ascendant hierarchical clustering method and the k-means partitional method. The three algorithms are applied and compared on six databases of the UCI Machine Learning repository.

1 Introduction

Descendant hierarchical clustering algorithm consists in, starting from the main data set:

- choosing one cluster in the current partition,
- splitting the chosen cluster into two sub-clusters (bi-partitional algorithm).

In this paper we present a descendant hierachical clustering method called DIVCLUS-T where both steps are based on the within-cluster inertia and we compare empirically this new method with two well-known clustering methods: the Ward ascendant hierarchical clustering method and the k-means partitional method.

The input of DIVCLUS-T is a data matrix $X = (x_i^j)_{n\times p}$. The entries of this matrix are the values of a set of n objects $O = \{1,\ldots,i,\ldots,n\}$ on p variables $X^1 \ldots X^j \ldots X^p$ either all numerical or all categorical. A weight w_i is also given as input for each object i. For instance, the objects $\{1,...,i,...,n\}$ may be themselves summaries of different groups of objects. An object i is then described on the p variables by the mean values of those objects, and its weight w_i is the number of objects in this group.

The measure of heterogeneity of a cluster $C_\ell \subset O$ is the inertia defined by $I(C_\ell) = \sum_{i \in C_\ell} w_i \, d^2(x_i, g(C_\ell))$ where d is the Euclidean distance and $g(C_\ell)$ is the gravity centre of C_ℓ defined by $g(C_\ell) = \sum_{i \in C_\ell} \frac{w_i}{\sum_{k \in C_\ell} w_k} x_i$. When all the weights w_i are equal to 1, the inertia is the well-known sum of squares criterion (SSQ). For categorical data the inertia of a cluster is computed, as in multiple correspondence analysis, on the n row-profiles of the indicator matrix of X, weighted by $1/n$ (if the original weights are $w_i = 1$), and with the χ^2-distance.

A k-clusters partition P_k is a list $(C_1, \ldots, C_k)$ of subsets of O verifying $C_1 \cup \ldots \cup C_k = O$ and $C_\ell \cap C_{\ell'} = \emptyset$ for all $\ell \neq \ell'$. An adequacy measure of this partition P_k is the within-cluster inertia defined as the sum of the inertia of all its clusters: $W(P_k) = \sum_{\ell=1}^{k} I(C_\ell)$. We will see that in DIVCLUS-T the bi-partitional algorithm and the choice of the cluster to be split are based on the minimization of the within-cluster inertia.

Concerning the bi-partitional algorithm, the complete enumeration of all the $2^{n_\ell - 1} - 1$ bi-partitions of a cluster C_ℓ of n_ℓ objects is avoided by using the same monothetic approach than classification or regression trees. Breiman et al. (1984) proposed and used binary questions in a recursive partitional process, CART, in the context of discrimination and regression. Use of a stepwise optimal tree structure in a least squares regression dates back to the Automatic Interaction Detection (AID) program proposed by Morgan and Sonquist (1963). Both AID and CART distinguish the response variable and the predictor variables. The set of all possible binary questions is defined on the set of the predictor variables, and the within-variance of the bi-partitions induced by this set of binary questions are calculated on the response variable. This is a problem of regression. As for DIVCLUS-T, in the context of clustering, there are no predictor and no response variables. The binary questions are defined on the p variables of the data matrix X, and the within-cluster inertia of the bi-partitions induced by those binary questions are calculated on the same variables (possibly standardized). The bi-partitional algorithm will then select, among all the bi-partitions $(A_\ell, \bar{A}_\ell)$ of C_ℓ induced by all the possible binary questions, the one of smallest within-cluster inertia.

A binary question on a numerical variable is noted "$is \quad X^j \quad \leq \quad c \quad ?$". This binary question splits a cluster C_ℓ into two sub-clusters A_ℓ and $\bar{A}_\ell$ such that $A_\ell = \{i \in C_\ell \mid x_i^j \leq c\}$ and $\bar{A}_\ell = \{i \in C_\ell \mid x_i^j > c\}$. For a numerical variable there is an infinity of possible binary questions, but they induce at most $n_\ell - 1$ different bi-partitions. Usually the cut-values c of the binary questions associated with the bi-partitions are chosen as the middle of two consecutive values observed on this variable. For numerical data, the number of bi-partitions to be evaluated is then at most $p(n_\ell - 1)$ and the complexity of the bi-partitional algorithm is $o(n(\log(n) + p))$.

In the same way, a binary question on a categorical variable is noted "$is \quad X^j \quad \in \quad M \quad ?$" where M is a subset of categories of X^j. This binary question splits a cluster C_ℓ into two sub-clusters A_ℓ and $\bar{A}_\ell$ such that $A_\ell = \{i \in C_\ell \mid x_i^j \in M\}$ and $\bar{A}_\ell = \{i \in C_\ell \mid x_i^j \in \overline{M}\}$ where $\overline{M}$ is the complementary of M. For a categorical variable having m^j categories, the number of possible binary questions - and hence the number of bi-partitions - is at most $2^{m^j-1}-1$. For categorical data, the number of bi-partitions to be evaluated is then $\sum_{j=1}^{p}(2^{m^j-1}-1)$, and grows exponentially with the number of categories m^j.

At each stage, DIVCLUS-T chooses to split the cluster C_ℓ of the partition P_k whose splitting, according to the above bi-partitional algorithm, produces a new partition P_{k+1} with minimum within-cluster inertia. Because the inertia criterion is additive we have $W(P_k) - W(P_{k+1}) = I(C_\ell) - I(A_\ell) - I(\bar{A}_\ell)$ and because $W(P_k)$ is a constant value, minimizing $W(P_{k+1})$ is equivalent to maximize the inertia variation $I(C_\ell) - I(A_\ell) - I(\bar{A}_\ell)$. A well-known result is also that the inertia variation obtained when C_ℓ is split (or when A_ℓ and $\bar{A}_\ell$ are aggregated) can be rewritten as a weighted Euclidean distance between the gravity centres of the two sub-clusters. This quantity is also the measure of aggregation D of the Ward ascendant hierarchical clustering algorithm: $D(A_\ell, \bar{A}_\ell) = \frac{\mu(A_\ell)\mu(\bar{A}_\ell)}{\mu(A_\ell)+\mu(\bar{A}_\ell)}\, d^2(g(A_\ell), g(\bar{A}_\ell))$.

In divisive clustering, the hierarchical level index has to be linked to the criterion used to choose the cluster to be split. Obviously the set of clusters obtained with DIVCLUS-T after $K-1$ successive divisions is a binary hierarchy whose singletons are the K "leaf"-clusters. Nevertheless, because in descendant hierarchical clustering the divisions are not necessarily continued until clusters are reduced to one element (or made of identical objects), a cluster represented higher in the dendrogram of the hierarchy has to be split before the others. The hierarchical level index in DIVCLUS-T is then the application h: $h(C_\ell) = D(A_\ell, \bar{A}_\ell)$. This hierarchical level index is the same than in the Ward ascendant hierarchical clustering method.

DIVCLUS-T is hence a DIVisive CLUStering method where the output is not a classification or a regression tree but a CLUStering-Tree. Because the dendrogram can be read as a decision tree, it provides partitions into homogeneous clusters having a simple interpretation.

In Chavent (1998) a simplified version of DIVCLUS-T was presented for quantitative data. It was applied in Chavent et al. (1999) with another monothetic divisive clustering method (based on correspondence analysis) to a categorical data set of healthy human skin data. A first comparison of DIVCLUS-T with Ward and the k-means was given in this paper but only for one categorical dataset and for the 6-clusters partition. More recently DIVCLUS-T was also applied to accounting disclosure analysis (Chavent et al. (2005)). A hierarchical divisive monothetic clustering method based on the Poisson process has also been proposed in Pircon (2004). A complete presentation of DIVCLUS-T for numerical and for categorical data and an analysis of its complexity is given in Chavent et al. (2006).

Having a simple interpretation of the clusters is an advantage of the monothetic approach. By contrast, the monothetic approach should induce partitions of worst quality (according to the within-cluster inertia). The aim of this paper is then to compare the monothetic method DIVCLUS-T with two polythetic methods (WARD and the k-means). In order to point out the advantage of the monothetic approach in term of interpretability of the clusters, we first give the dendrogram of the hierarchy obtained with DIVCLUS-T on a simple categorical example. Then we will compare the quality of the partitions from 2 two 15 clusters, obtained with DIVCLUS-T, WARD and the k-means, on six datasets (3 continuous and 3 categorical) of the UCI Machine Learning repository (Hettich et al. (1998)). Because those three methods are based on the minimization of the within-cluster inertia, we will compare the proportion of the total inertia explained by those partitions.

2 A simple example

In this example the divisive method is applied to a categorical dataset where 27 races of dogs are described by 7 categorical variables (Saporta (1990)). The dendrogram of the hierarchy and the 7 first binary questions are given Figure 1.

At the first stage, the divisive clustering method constructs a bi-partition of the 27 dogs. There are 17 different binary questions and 17 bi-partitions to evaluate: two variables are binary (and induce two different bi-partitions) and the five other variables have 3 categories and induce then 5×3 different bi-partitions. The question "Is the size large?" which induces the bi-partition of smallest within-cluster inertia is then chosen. The inertia variation obtained by splitting the 15 "large" dogs is slightly smaller than the one obtained by splitting the 12 "small or medium" dogs. This latter cluster is then divided. For comparison purpose with the Ward hierarchy, the process is repeated here until getting singleton clusters or clusters of identical dogs. The Pekingese and the Chihuahua for instance have exactly the same description and can not then be divided. Finally the divisions are stopped after 25 iterations.

The Ward ascendant clustering method is defined for quantitative data. We thus applied WARD on the 12 principal components from the Multiple Factorial Analysis. The hierarchy obtained with WARD is identical to the one obtained with DIVCLUS-T. In DIVCLUS-T the inertia of categorical data is calculated as in multiple correspondence analysis on the row profiles of the indicator matrix of the data matrix X. It is well-known that this inertia is equal to the inertia performed on all the principal components from Multiple Factorial Analysis. For this reason, we checked that DIVCLUS-T and WARD give the same value for the inertia of the partitions of the hierarchy. Moreover because they use the same hierarchical level index, the dendrogram obtained with WARD is identical to the one of Figure 1 but without the binary questions.

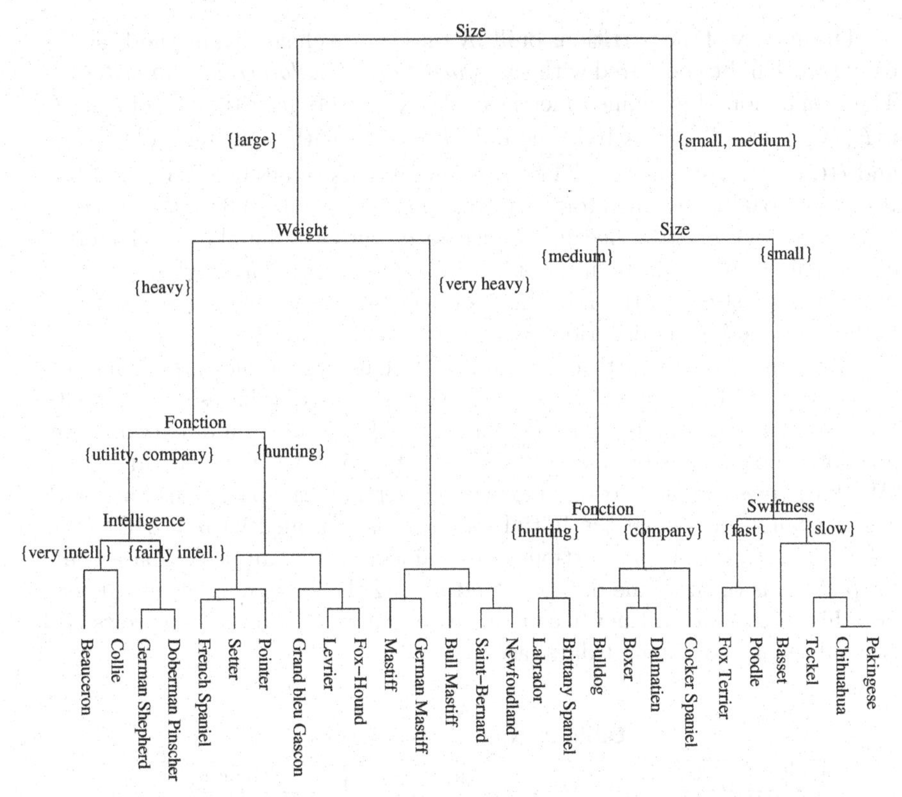

Fig. 1. DIVCLUS-T dendrogram for categorical data

3 Empirical comparison with WARD and the k-means

We have applied DIVCLUS-T, WARD and the k-means algorithms on three numerical and three categorical datasets of the UCI Machine Learning repository (Hettich et al. (1998)). For the three numerical datasets, the variables have been standardized previously. For the three categorical datasets, WARD has been applied to the principal components obtained by Multiple correspondence analysis. A short description of the six databases is given Table 1.

Table 1. Databases descriptions

Name	Type	Nb objects	Nb variables(nb categories)
Glass	numerical	214	8
Pima Indians diabete	numerical	768	8
Abalone	numerical	4177	7
Zoo	categorical	101	15(2) + 1(6)
Solar Flare	categorical	323	2(6) + 1(4) + 1(3) + 6(2)
Contraceptive Method Choice (CMC)	categorical	1473	9(4)

The quality of the partitions built by these three clustering methods on the 6 datasets can be compared with the *proportion of explained inertia criterion.* The proportion of explained inertia E of a k-clusters partition P_k of a set O is $E(P_k) = (1 - \frac{W(P_k)}{I(O)}) \times 100$, where $W(P_k)$ is the within-cluster inertia of P_k and $I(O)$ is the total inertia. This criterion takes its values between 0 and 100. It is equal to 0 for the singleton partition and it is equal to 100 for the partition reduced to one cluster. Because E decreases with the number of clusters k of the partition, it can be used only to compare partitions having the same number of clusters. In the following comments, we say that a partition P is better (for the inertia criterion) than a partition P' if $E(P) > E(P')$.

We have built the partitions with 2 to 15 clusters for the three numerical databases (see Table 2) and for the three categorical databases (see Table 3). For each database the two first columns give the proportion of explained inertia of the partitions built with DIVCLUS-T and WARD. The third column (W+km) gives the proportion of explained inertia of the partitions built with the k-means (km) when the initial partition is obtained with WARD (W). As already stated two proportions of explained inertia can be compared only for partitions of the same database and having the same number of clusters. For this reason we will never compare two values in two different rows and two values of two different databases.

Table 2. Numerical databases

	Glass			Pima			Abalone		
K	DIV	WARD	W+km	DIV	WARD	W+km	DIV	WARD	W+km
2	21.5	22.5	22.8	14.8	13.3	16.4	60.2	57.7	60.9
3	33.6	34.1	34.4	23.2	21.6	24.5	72.5	74.8	76.0
4	45.2	43.3	46.6	29.4	29.4	36.2	81.7	80.0	82.5
5	53.4	53.0	54.8	34.6	34.9	40.9	84.2	85.0	86.0
6	58.2	58.4	60.0	38.2	40.0	45.3	86.3	86.8	87.8
7	63.1	63.5	65.7	40.9	44.4	48.8	88.3	88.4	89.6
8	66.3	66.8	68.9	43.2	47.0	51.1	89.8	89.9	90.7
9	69.2	69.2	71.6	45.2	49.1	52.4	91.0	90.9	91.7
10	71.4	71.5	73.9	47.2	50.7	54.1	91.7	91.6	92.4
11	73.2	73.8	75.6	48.8	52.4	56.0	92.0	92.1	92.8
12	74.7	76.0	77.0	50.4	53.9	58.0	92.3	92.4	93.0
13	76.2	77.6	78.7	52.0	55.2	58.8	92.6	92.7	93.3
14	77.4	79.1	80.2	53.4	56.5	60.0	92.8	93.0	93.7
15	78.5	80.4	81.0	54.6	57.7	61.0	93.0	93.2	93.9

First we compare the results for the three numerical databases (Table 2). For the Glass database the partitions obtained with DIVCLUS-T are either better (for 4 clusters), worse (for 2, 3, and from 12 to 15 clusters) or equivalent (from 5 to 11 clusters). For the Pima database the partitions of DIVCLUS-T are better or equivalent until 4 clusters, and WARD takes the lead from 5 clusters on. Because DIVCLUS-T is descendant and WARD is ascendant, it is not really surprising that when the number of clusters increases WARD tends to become better than DIVCLUS-T. For the Abalone database which is bigger than the two others (4177 objects), DIVCLUS-T makes better than WARD until 4 clusters and the results are very close afterwards. A reason for having better results of DIVCLUS-T on the Abalone dataset is perhaps the greater number of objects in this database. Indeed the number of bi-partitions considered for optimization at each stage increases with the number of objects. We can then expect to have better results with bigger databases. In the third column (W+km) of the three databases, the k-means algorithm is executed on the WARD partition (taken as initial partition) and the proportion of explained inertia is then necessarily greater than the one in the second column WARD. Finally, on those three continuous databases, DIVCLUS-T seems to perform better for few clusters partitions and for bigger datasets.

Table 3. Categorical databases

	Zoo			Solar Flare			CMC		
K	DIV	WARD	W+km	DIV	WARD	W+km	DIV	WARD	W+km
2	23.7	24.7	26.2	12.7	12.6	12.7	8.4	8.2	8.5
3	38.2	40.8	41.8	23.8	22.4	23.8	14.0	13.1	14.8
4	50.1	53.7	54.9	32.8	29.3	33.1	18.9	17.3	20.5
5	55.6	60.4	61.0	38.2	35.1	38.4	23.0	21.3	24.0
6	60.9	64.3	65.1	43.0	40.0	42.7	26.3	24.9	27.7
7	65.6	67.5	68.4	47.7	45.0	47.6	28.4	28.1	29.8
8	68.9	70.6	71.3	51.6	49.8	52.1	30.3	30.7	32.7
9	71.8	73.7	73.7	54.3	53.5	54.6	32.1	33.4	35.2
10	74.7	75.9	75.9	57.0	57.1	58.3	33.8	35.5	37.7
11	76.7	77.5	77.5	59.3	60.4	61.7	35.5	37.5	40.1
12	78.4	79.1	79.1	61.3	62.9	64.4	36.9	39.4	41.5
13	80.1	80.6	80.6	63.1	65.2	65.7	38.1	41.0	42.9
14	81.3	81.8	81.8	64.5	66.2	67.7	39.2	42.0	44.2
15	82.8	82.8	82.8	65.8	68.6	69.3	40.3	43.1	44.9

With the three categorical databases (Table 3) we obtain the same kind of results. For the Solar Flare and the CMC databases, DIVCLUS-T is better than WARD until respectively 10 and 8 clusters. For the Zoo database DIVCLUS-T performs worst than WARD. This is maybe because all the variables in the Zoo database are binary and, as already stated, the quality of the results (in term of inertia) may depend on the number of categories and variables.

4 Conclusion

Imposing the monotheticity of the clusters as in DIVCLUS-T is an advantage for the interpretation of the results: the dendrogram gives a very simple interpretation of the levels of the hierarchy. This advantage has to be balanced by a relative rigidity of the clustering process. Simple simulations should be able to show easily that DIVCLUS-T is unable to find correctly clusters of specific shapes. But what are the shapes of the clusters in real datasets ? We have seen on the six databases of the UCI Machine Learning repository that the proportions of explained inertia of the partitions obtained with DIVCLUS-T are very comparable to those obtained with the Ward or the k-means algorithms, particularly for partitions with few clusters (at the top of the dendrograms). A more complete comparative study of these three clustering methods is necessary in order to better understand the influence of the number of objects, categories and variables in the quality of the results, as well as a study of their stability.

References

BREIMAN, L., FRIEDMAN, J.H., OLSHEN, R.A. and STONE, C.J. (1984): *Classification and regression Trees*, C.A:Wadsworth.

CHAVENT, M. (1998): A monothetic clustering method. *Pattern Recognition Letters, 19, 989-996.*

CHAVENT, M., GUINOT, C., LECHEVALLIER Y. and TENENHAUS, M. (1999): Méthodes divisives de classification et segmentation non supervisée: recherche d'une typologie de la peau humaine saine. *Revue Statistique Appliquée, XLVII (4), 87-99.*

CHAVENT, M., DING, Y., FU, L., STOLOWY and H., WANG, H. (2005): Disclosure and Determinants Studies: An extension Using the Divisive Clustering Method (DIV). *European Accounting Review, to publish.*

CHAVENT, M., BRIANT, O. and LECHEVALLIER, Y. (2006): *DIVCLUS-T: a new descendant hierarchical clustering method.* Internal report U-05-15, Laboratoire de Mathématiques Appliquées de Bordeaux.

HETTICH, S., BLAKE, C.L. and MERZ, C.J. (1998): *UCI Repository of machine learning databases, http://www.ics.uci.edu/ mlearn/MLRepository.html.* Irvine, CA: University of California, Department of Information and Computer Science.

MORGAN, J.N. and SONQUIST, J.A. (1963): Problems in the analysis of survey data, and proposal. *J. Aler. Statist. Assoc., 58, 415-434.*

PIRCON, J.-Y. (2004): *La classification et les processus de Poisson pour de nouvelles méthodes de partitionnement.* Phd Thesis, Facultés Universitaires Notre-Dame de la Paix, Belgium.

SAPORTA, G. (1990): *Probabilités Analyse des données et Statistique*, Editions TECHNIP.

Model Selection for the Binary Latent Class Model: A Monte Carlo Simulation

José G. Dias

Department of Quantitative Methods – UNIDE,
Higher Institute of Social Sciences and Business Studies – ISCTE,
Av. das Forças Armadas, Lisboa 1649–026, Portugal,
jose.dias@iscte.pt

Abstract. This paper addresses model selection using information criteria for binary latent class (LC) models. A Monte Carlo study sets an experimental design to compare the performance of different information criteria for this model, some compared for the first time. Furthermore, the level of separation of latent classes is controlled using a new procedure. The results show that AIC3 (Akaike information criterion with 3 as penalizing factor) has a balanced performance for binary LC models.

1 Introduction

In recent years latent class (LC) analysis has become an important technique in applied research. Let $\mathbf{y} = (\mathbf{y}_1, ..., \mathbf{y}_n)$ denote a sample of size n; J represents the number of manifest or observed variables; and y_{ij} indicates the observed value for variable j in observation i, with $i = 1, ..., n$, $j = 1, ..., J$. The latent class model with S classes for $\mathbf{y}_i = (y_{i1}, ..., y_{iJ})$ is defined by $f(\mathbf{y}_i; \boldsymbol{\varphi}) = \sum_{s=1}^{S} \pi_s f_s(\mathbf{y}_i; \boldsymbol{\theta}_s)$, where the latent class proportions π_s are positive and sum to one; $\boldsymbol{\theta}_s$ denotes the parameters of the conditional distribution of latent class s, defined by $f_s(\mathbf{y}_i; \boldsymbol{\theta}_s)$; $\boldsymbol{\pi} = (\pi_1, ..., \pi_{S-1})$, $\boldsymbol{\theta} = (\boldsymbol{\theta}_1, ..., \boldsymbol{\theta}_S)$, and $\boldsymbol{\varphi} = (\boldsymbol{\pi}, \boldsymbol{\theta})$. For binary data, Y_j has 2 categories, $y_{ij} \in \{0, 1\}$, and follows a Bernoulli distribution. From the local independence assumption – the J manifest binary variables are independent given the latent class –, $f_s(\mathbf{y}_i; \boldsymbol{\theta}_s) = \prod_{j=1}^{J} \theta_{sj}^{y_{ij}} (1 - \theta_{sj})^{1-y_{ij}}$, where θ_{sj} is the probability that observation i belonging to latent class s falls in category 1 (success) of variable j. Therefore, the binary LC model has probability mass function

$$f(\mathbf{y}_i; \boldsymbol{\varphi}) = \sum_{s=1}^{S} \pi_s \prod_{j=1}^{J} \theta_{sj}^{y_{ij}} (1 - \theta_{sj})^{1-y_{ij}}.$$

The number of free parameters in vectors $\boldsymbol{\pi}$ and $\boldsymbol{\theta}$ are $d_{\boldsymbol{\pi}} = S - 1$ and $d_{\boldsymbol{\theta}} = SJ$, respectively. The total number of free parameters is $d_{\boldsymbol{\varphi}} = d_{\boldsymbol{\pi}} + d_{\boldsymbol{\theta}}$. The likelihood and log-likelihood functions are $L(\boldsymbol{\varphi}; \mathbf{y}) = \prod_{i=1}^{n} f(\mathbf{y}_i; \boldsymbol{\varphi})$

and $\ell(\varphi; \mathbf{y}) = \log L(\varphi; \mathbf{y})$, respectively. It is straightforward to obtain the maximum likelihood (ML) estimates of φ using the EM algorithm.

The increased use of the binary LC model in applied research as result of the widespread application of batteries of dichotomous ('yes/no') variables has added pressure on making available reliable model selection tools for the number of latent classes. From a probabilistic viewpoint, the likelihood ratio test (LRT) has been used extensively as a model selection tool, because under regularity conditions, has a simple asymptotic theory (Wilks, 1938). However, these regularity conditions fail for LC models. For example, in testing the hypothesis of a single latent class against more than one, the mixing proportion under H_0 is on the boundary of the parameter space, and consequently the LRT statistic is not asymptotically chi-squared distributed. Information criteria have become popular alternatives as model selection tools. In particular, the Akaike Information Criterion (AIC) and Bayesian Information Criterion (BIC) have been widely used. Despite that, little is known about the performance of these and other information criteria for binary LC models. Most of the simulation studies have been set for finite mixtures of Gaussian distributions (McLachlan and Peel, 2000).

This paper is organized as follows. Section 2 reviews the literature on model selection criteria. Section 3 describes the design of the Monte Carlo study. Section 4 presents and discusses the results. The paper concludes with a summary of main findings, implications, and suggestions for further research.

2 Information criteria

The Akaike's information criterion (AIC) is based on the Kullback-Leibler distance between the true density and the estimated density (Akaike, 1974). AIC chooses S which minimizes

$$\mathrm{AIC} = -2\ell(\hat{\varphi}; \mathbf{y}) + 2d_{\varphi},$$

where $\hat{\varphi}$ is the ML estimate, $\ell(\hat{\varphi}; \mathbf{y})$ is the log-likelihood value at the ML estimate and d_{φ} is the number of independent parameters (Akaike, 1974). It can be a drastically negatively biased estimate of the expected Kullback-Leibler information of the fitted model (Hurvich and Tsai, 1989). Bozdogan (1993) argues that the marginal cost per free parameter, the so-called magic number 2 in AIC's equation above, is not correct for finite mixture models. Based on Wolfe (1970), he conjectures that the likelihood ratio for comparing mixture models with K and k free parameters is asymptotically distributed as a noncentral chi-square with noncentrality parameter δ and $2(K-k)$ degrees of freedom instead of the usual $K-k$ degrees of freedom as assumed in AIC. Therefore, AIC3 uses 3 as penalizing factor. The consistent AIC criterion (CAIC: Bozdogan, 1987) chooses S which minimizes

$$\mathrm{CAIC} = -2\ell(\hat{\varphi}; \mathbf{y}) + d_{\varphi}(\log n + 1).$$

The Bayesian information criterion (BIC), proposed by Schwarz (1978), utilizes the marginal likelihood $p(\mathbf{y}) = \int L(\varphi; \mathbf{y})p(\varphi)d\varphi$, which is the weighted average of the likelihood values. Using the Laplace approximation about the posterior mode ($\tilde{\varphi}$, where $L(\varphi; \mathbf{y})p(\varphi)$ is maximized), it results (Tierney and Kadane, 1986)

$$\log p(\mathbf{y}) \approx \ell(\tilde{\varphi}; \mathbf{y}) + \log p(\tilde{\varphi}) - \frac{1}{2}\log|\mathbf{H}(\tilde{\varphi}; \mathbf{y})| + \frac{d_\varphi}{2}\log(2\pi),$$

where $\mathbf{H}(\tilde{\varphi}; \mathbf{y})$ is the negative of the Hessian matrix of the log-posterior function, $\log L(\varphi; \mathbf{y})p(\varphi)$, evaluated at the modal value $\varphi = \tilde{\varphi}$. BIC assumes a proper prior, which assigns positive probability to lower dimensional subspaces of the parameter vector. For a very diffuse (almost non-informative, and consequently ignorable) prior distribution, $\mathbf{H}(\tilde{\varphi}; \mathbf{y})$ can be replaced by the observed information matrix $\mathbf{I}(\tilde{\varphi}; \mathbf{y})$. Replacing the posterior mode by the ML estimate $\hat{\varphi}$, the approximation becomes

$$\log p(\mathbf{y}) \approx \ell(\hat{\varphi}; \mathbf{y}) + \log p(\hat{\varphi}) - \frac{1}{2}\log|\mathbf{I}(\hat{\varphi}; \mathbf{y})| + \frac{d_\varphi}{2}\log(2\pi). \quad (1)$$

From the asymptotic behavior of the approximation above, the Bayesian information criterion (BIC) chooses S which minimizes

$$\mathrm{BIC} = -2\ell(\hat{\varphi}; \mathbf{y}) + d_\varphi \log n.$$

Approximation (1) can be used itself as suggested by McLachlan and Peel (2000). The resulting Laplace-empirical criterion (LEC) chooses S which minimizes

$$\mathrm{LEC} = -2\ell(\hat{\varphi}; \mathbf{y}) - 2\log p(\hat{\varphi}) + \log|\mathbf{I}(\hat{\varphi}; \mathbf{y})| - d_\varphi \log(2\pi).$$

The prior distribution $p(\varphi)$ assumes that parameters are *a priori* independent, $p(\varphi) = p(\boldsymbol{\pi})\prod_{s=1}^{S}\prod_{j=1}^{J} p(\boldsymbol{\theta}_{sj})$. The Dirichlet distribution is a natural prior for these parameters. For $\omega = (\omega_1, \omega_2, ..., \omega_k)$, it is denoted by $\mathcal{D}(\xi_1, ..., \xi_k)$ with parameters $(\xi_1, ..., \xi_k)$ and density function $p(\omega_1, \omega_2, ..., \omega_k) = \frac{\Gamma(\xi_0)}{\prod_{j=1}^{k}\Gamma(\xi_j)}\prod_{j=1}^{k}\omega_j^{\xi_j - 1}$, where $\omega_j \geq 0$ for $j = 1, ..., k$, $\sum_{j=1}^{k}\omega_j = 1$, $\Gamma(.)$ is the gamma function, and $\xi_0 = \sum_{j=1}^{k}\xi_j$. The expected value and variance of ω_j are $\mathrm{E}(\omega_j) = \xi_j/\xi_0$ and $\mathrm{Var}(\omega_j) = \xi_j(\xi_0 - \xi_j)/\left[\xi_0^2(\xi_0 + 1)\right]$, respectively. LEC-U and LEC-J criteria are defined by the uniform and Jeffreys' priors for φ, respectively [1]:

1. The uniform prior (U) corresponding to Dirichlet distributions with $\boldsymbol{\pi} \sim \mathcal{D}(1, ..., 1)$ and $\boldsymbol{\theta}_{sj} \sim \mathcal{D}(1, 1)$ is given by

$$\log p(\varphi) = \log\left[(S-1)!\right];$$

[1] Note that for binary data the Dirichlet distribution for $\boldsymbol{\theta}_{sj} = (\theta_{sj}, 1 - \theta_{sj})$ reduces to a Beta distribution. However, because of $\boldsymbol{\pi}$, we keep the general case, simplifying the expressions whenever it applies.

2. The Jeffreys' prior (J) corresponding to Dirichlet distributions with $\boldsymbol{\pi} \sim \mathcal{D}(1/2, ..., 1/2)$ and $\boldsymbol{\theta}_{sj} \sim \mathcal{D}(1/2, 1/2)$ is

$$\log p(\boldsymbol{\varphi}) = -2JS \log \Gamma\left(\frac{1}{2}\right) + \frac{1}{2}\sum_{s=1}^{S}\sum_{j=1}^{J} \log\left[\theta_{sj}(1-\theta_{sj})\right]$$
$$+ \log \Gamma\left(\frac{S}{2}\right) - S \log \Gamma\left(\frac{1}{2}\right) + \frac{1}{2}\sum_{s=1}^{S} \log \pi_s.$$

Complete data information criteria are based on data augmentation, where the observed data ($\mathbf{y}$) is expanded to a new space ($\mathbf{y}, \mathbf{z}$), which includes the missing data ($\mathbf{z}$). The missing datum (z_{is}) indicates whether latent class s has generated observation i. The expected value of z_{is} is given by

$$\alpha_{is} = \frac{\pi_s f_s(\mathbf{y}_i; \boldsymbol{\theta}_s)}{\sum_{v=1}^{S} \pi_v f_v(\mathbf{y}_i; \boldsymbol{\theta}_v)}, \tag{2}$$

and corresponds to the posterior probability that $\mathbf{y}_i$ was generated by the latent class s. The entropy of the matrix $\boldsymbol{\alpha} = (\alpha_{is})$, $i = 1, ..., n$, $s = 1, ..., S$ is defined by $\mathrm{EN}(\boldsymbol{\alpha}) = -\sum_{i=1}^{n}\sum_{s=1}^{S} \alpha_{is} \log \alpha_{is}$. For $\mathrm{EN}(\boldsymbol{\alpha}) \simeq 0$, latent classes are well separated. Note that $\boldsymbol{\alpha}$ is function of $\boldsymbol{\varphi}$ and $\mathbf{y}$. Celeux and Soromenho (1996) introduced an entropic measure for the selection of S. As $\mathrm{EN}(\boldsymbol{\alpha})$ has no upperbound, they proposed the normalized entropy criterion (NEC). NEC chooses S that minimizes

$$NEC = \frac{EN(\hat{\boldsymbol{\alpha}})}{\ell(\hat{\boldsymbol{\varphi}}; \mathbf{y}) - \ell_1(\hat{\boldsymbol{\varphi}}; \mathbf{y})},$$

where $\ell_1(\hat{\boldsymbol{\varphi}}; \mathbf{y})$ is the log-likelihood value for the one-latent-class model and $\hat{\boldsymbol{\alpha}}$ comes from (2) at the ML estimate. To overcome the impossibility of deciding between $S = 1$ and $S > 1$, Biernacki *et al.* (1999) proposed the following rule: if there is no S such that NEC< 1, then $S = 1$ has to be preferred.

3 Experimental design

The relative performance of these criteria for the binary LC model is assessed by a Monte Carlo (MC) simulation study. In our simulations all estimated LC models have non-singular estimated information matrix. The Monte Carlo experimental design controls the number of latent classes, the number of variables, the sample size, the balance of latent class sizes, and the level of separation of latent classes. The number of latent classes is set to 2 and 3, and models with one, two, three, four latent classes are estimated. The number of variables (J) was set at levels 5 and 8. From preliminary analises with $J = 5$, we concluded that datasets with a non-singular estimated information matrix for the three-latent-class LC model with sample sizes smaller than

600 are difficult to obtain. Therefore, the factor sample size (n) assumes the levels: 600, 1200, and 2400. The latent class sizes were generated using the expression $\pi_s = a^{s-1} \left(\sum_{v=1}^{S} a^{v-1} \right)^{-1}$, with $s = 1, ..., S$ and $a \geq 1$. For $a = 1$ yields equal proportions; for larger values of a, latent class sizes become more unbalanced. In our MC study, we set three levels for a: 1, 2 and 3.

Despite the importance of controlling the level of separation of latent classes in Monte Carlo studies, the approach has mostly been based on *ad hoc* procedures such as randomly generated parameters of the first latent class, and the other latent classes are obtained by adding successively a different constant in low and high level of separation of latent classes. In this paper, we apply a sampling procedure recently introduced by Dias (2004). The vector $\boldsymbol{\theta}$ is generated as:

1. Draw $\boldsymbol{\theta}_{1j}$ from the Dirichlet distribution with parameters (ϕ_1, ϕ_2), $j = 1, .., J$;
2. Draw $\boldsymbol{\theta}_{sj}$ from the Dirichlet distribution with parameters $(\delta\theta_{1j}, \delta\,(1 - \theta_{1j}))$, $j = 1, ..., J$, $s = 2, ..., S$.

This procedure assumes that parameters $\boldsymbol{\theta}$ of the LC model are sampled from a superpopulation defined by the hyperparameters δ and (ϕ_1, ϕ_2), $j = 1, ..., J$, and defines a hierarchical (Bayesian) structure. We set $(\phi_1, \phi_2) = (1, 1)$, which corresponds to the uniform distribution. For $s = 2, ..., S$, we have $\mathrm{E}(\theta_{sj}) = \theta_{1j}$ and $\mathrm{Var}(\theta_{sj}) = \theta_{1j}\,(1 - \theta_{1j})\,/\,(\delta + 1)$. With this procedure, on average, all latent classes are centered at the same parameter value generated from a uniform distribution (first latent class). The constant $\delta > 0$ controls the level of separation of the latent classes. As δ increases, the latent class separation decreases as a consequence of the decreasing of the variance. As $\delta \to \infty$, all latent classes tend to share the same parameters. Based on results in Dias (2004), three levels of δ give a good coverage of the level of separation of the latent classes. The values of δ set in this study were: 0.1 (well-separated latent classes), 1 (moderately-separated latent classes), and 5 (weakly-separated latent classes).

This MC study sets a $2^2 \times 3^3$ factorial design with 108 cells. The main performance measure used is the frequency with which each information criterion picks the correct model. For each dataset, each criterion is classified as *underfitting*, *fitting*, or *overfitting*, based on the relation between S and the estimated S by those criteria.

Special care needs to be taken before arriving at conclusions based on MC results. In this study, we performed 100 replications within each cell to obtain the frequency distribution of selecting the true model, resulting in a total of 10800 datasets. To avoid local optima, for each number of latent classes (2, 3, and 4) the EM algorithm was repeated 5 times with random starting centers, and the best solution (maximum likelihood value out of the 5 runs) and model selection results were kept. The EM algorithm ran for

1500 iterations, which was enough to ensure the convergence in all cells of the design.

Comparing our design with previous analyses by Lin and Dayton (1997) and Nadif and Govaert (1998), we extend their studies by varying the number of latent classes (both previous studies keep $S = 2$), the number of variables (they keep it fixed at 4 and 10 variables, respectively), the sample size (fixed at 960 and 200, respectively), the latent class proportions (fixed at equal and unequal levels, respectively). Moreover, Lin and Dayton (1997) analyzed AIC, BIC and CAIC and Nadif and Govaert (1998) used AIC, AIC3, BIC and NEC. We analyzed all these criteria plus the LEC for the first time.

4 Results

The key feature of the results is the overall remarkable performance of AIC and AIC3 (Table 1). While many criteria often perform satisfactory, AIC and AIC3 find the true model 69.5% and 69.1% of the times, respectively. Overall, the BIC-like criteria perform reasonably well. As in other studies, our results document the tendency of AIC to overfit. LEC-U presents the same pattern. BIC, CAIC, and LEC-J tend to choose slightly more parsimonious models. By comparing LEC-U and LEC-J results, we conclude that LEC is very sensitive to the prior setting. Finally, NEC tends to underfit in latent class modeling as has been shown by Nadif and Govaert (1998).

Table 1. Overall results

	AIC	AIC3	CAIC	BIC	LEC-U	LEC-J	NEC
Underfit	18.37	30.33	48.47	45.79	27.88	47.99	85.26
Fit	69.48	69.13	51.53	54.21	57.68	52.01	13.89
Overfit	12.15	0.54	0.00	0.00	14.44	0.00	0.85

A second objective of the study was to compare these criteria across the factors in the design (Table 2). Increasing the number of latent classes (S) reduces the performance of all information criteria. Increasing the sample size tends to improve the performance of information criteria. Increasing the number of variables (J) mostly reduces the underfitting, and improves the performance of the information criteria. For AIC, increasing the number of variables (J) is associated with overfitting. In general, the more balanced the latent class sizes are, the better is the performance of these criteria. Moreover, increasing the balance of latent class sizes tends to overfit and reduces underfitting. The level of separation of latent classes has a dramatic effect on the performance of these criteria. For example, AIC3 finds the correct model in 92.7% of the cases for well-separated latent classes, but just in 30.8% for the ill-separated case. Moreover, BIC and CAIC can be extremely conservative for ill-separated latent classes. AIC3 tends to succeed in most of the

Table 2. Results across the experimental design

Factors		AIC	AIC3	CAIC	BIC	LEC-U	LEC-J	NEC
Latent classes (S)								
	Underfit	8.30	16.70	31.67	29.20	18.02	31.52	71.78
2	Fit	79.59	82.56	68.33	70.80	74.46	68.48	26.72
	Overfit	12.11	0.74	0.00	0.00	7.52	0.00	1.50
	Underfit	28.44	43.97	65.28	62.37	37.74	64.46	98.74
3	Fit	59.37	55.70	34.72	37.63	40.89	35.54	1.06
	Overfit	12.19	0.33	0.00	0.00	21.37	0.00	0.20
Sample size (n)								
	Underfit	23.42	38.17	55.08	52.97	31.33	55.31	83.34
600	Fit	65.17	60.67	44.92	47.03	46.28	44.69	14.58
	Overfit	11.41	1.16	0.00	0.00	22.39	0.00	2.08
	Underfit	19.58	30.14	49.53	46.42	28.06	51.53	84.56
1200	Fit	70.38	69.78	50.47	53.58	58.25	48.47	14.97
	Overfit	10.04	0.08	0.00	0.00	13.69	0.00	0.47
	Underfit	12.11	22.69	40.81	37.97	24.25	37.14	87.89
2400	Fit	72.89	76.95	59.19	62.03	68.50	62.86	12.11
	Overfit	15.00	0.36	0.00	0.00	7.25	0.00	0.00
Number of variables (J)								
	Underfit	23.56	34.35	54.31	50.28	28.39	52.02	88.63
5	Fit	72.68	65.61	45.69	49.72	48.44	47.98	11.22
	Overfit	3.76	0.04	0.00	0.00	23.17	0.00	0.15
	Underfit	13.19	26.31	42.63	41.30	27.37	43.96	81.88
8	Fit	66.28	72.65	57.37	58.70	66.91	56.04	16.56
	Overfit	20.53	1.04	0.00	0.00	5.72	0.00	1.56
Proportions (a)								
	Underfit	17.50	26.36	43.81	41.94	21.58	43.56	85.42
1	Fit	68.06	72.81	56.19	58.06	59.31	56.44	12.75
	Overfit	14.44	0.83	0.00	0.00	19.11	0.00	1.83
	Underfit	15.06	29.33	46.83	44.64	28.39	47.39	84.75
2	Fit	74.42	70.06	53.17	55.36	59.03	52.61	14.69
	Overfit	10.53	0.61	0.00	0.00	12.58	0.00	0.56
	Underfit	22.56	35.31	54.78	50.78	33.67	53.03	85.61
3	Fit	65.97	64.52	45.22	49.22	54.69	46.97	14.22
	Overfit	11.47	0.17	0.00	0.00	11.64	0.00	0.17
Level of separation (δ)								
	Underfit	2.69	6.86	15.25	10.75	6.11	14.00	64.53
0.1	Fit	85.19	92.72	84.75	89.25	77.86	86.00	33.83
	Overfit	12.12	0.42	0.00	0.00	16.03	0.00	1.64
	Underfit	5.50	15.06	38.33	36.44	12.69	40.11	91.61
1	Fit	80.86	83.88	61.67	63.56	68.25	59.89	7.47
	Overfit	13.64	1.06	0.00	0.00	19.06	0.00	0.92
	Underfit	46.92	69.08	91.83	90.17	64.83	89.86	99.64
5	Fit	42.39	30.78	8.17	9.83	26.92	10.14	0.36
	Overfit	10.69	0.14	0.00	0.00	8.25	0.00	0.00

experimental conditions, presenting balanced results across different levels of separation of latent classes. Despite AIC outperforms AIC3 for some conditions, the former tends to overfit. Even for well-separated latent classes, AIC gives a high percentage of overfitting.

5 Conclusion

The paper compared the performance of information criteria for binary LC models. Because most of the information criteria are derived from asymptotics, this extensive Monte Carlo study allowed their assessment for realistic sample sizes. We have included traditional and recently proposed information criteria, some of them are compared for the first time. A large experimental design was set, controlling sample size, number of variables, number of categories, relative latent class sizes, and separation of latent classes. The level of separation of latent classes was controlled using a new procedure.

The main finding is the overall good performance of the AIC3 criterion for binary LC models. AIC has the best overall performance among all the information criteria, however it tends to overfit. Therefore, AIC3 becomes a more attractive overall information criterion.

Future research could extend our findings to other latent class models (multinomial or count data) or more general latent models (Lin and Dayton, 1997). These results suggest that the type of approximation for the marginal likelihood needed for the derivation of the LEC and BIC has to be further studied. Indeed, despite the difficulty of the ill-separated scenario, approximations other than the Laplace may improve the performance of the information criteria, in particular for discrete data models.

References

AKAIKE, H. (1974): A New Look at Statistical Model Identification, *IEEE Transactions on Automatic Control, AC-19, 716–723.*

BIERNACKI, C., CELEUX, G., and GOVAERT, G. (1999): An Improvement of the NEC Criterion for Assessing the Number of Clusters in a Mixture Model, *Pattern Recognition Letters, 20, 267–272.*

BOZDOGAN, H. (1987): Model Selection and Akaike's Information Criterion (AIC): The General Theory and Its Analytical Extensions, *Psychometrika, 52, 345–370.*

BOZDOGAN, H. (1993): Choosing the Number of Component Clusters in the Mixture-Model Using a New Informational Complexity Criterion of the Inverse-Fisher Information Matrix. In: O. Opitz, B. Lausen, and R. Klar (Eds.): *Information and Classification, Concepts, Methods and Applications.* Springer, Berlin, 40–54.

CELEUX, G., SOROMENHO, G. (1996): An Entropy Criterion for Assessing the Number of Clusters in a Mixture Model, *Journal of Classification*, 13, 195–212.

DIAS, J.G. (2004): Controlling the Level of Separation of Components in Monte Carlo Studies of Latent Class Models. In: D. Banks, L. House, F.R. McMorris, P. Arabie, and W. Gaul (Eds.): *Classification, Clustering, and Data Mining Applications.* Springer, Berlin, 77–84.

HURVICH, C.M. and TSAI, C.-L. (1989): Regression and Time Series Model Selection in Small Samples, *Biometrika*, 76, 297–307.

LIN, T.H. and DAYTON, C.M. (1997): Model Selection Information Criteria for Non-nested Latent Class Models, *Journal of Educational and Behavioral Statistics*, 22, 249–264.

MCLACHLAN, G.J. and PEEL, D. (2000): *Finite Mixture Models.* John Wiley & Sons, New York.

NADIF, M. and GOVAERT, G. (1998): Clustering for Binary Data and Mixture Models - Choice of the Model, *Applied Stochastic Models and Data Analysis*, 13, 269–278.

SCHWARZ, G. (1978): Estimating the Dimension of a Model, *Annals of Statistics*, 6, 461–464.

TIERNEY, L. and KADANE, J. (1986): Accurate Approximations for Posterior Moments and Marginal Densities, *Journal of the American Statistical Association*, 81, 82–86.

WOLFE, J.H. (1970): Pattern Clustering by Multivariate Mixture Analysis, *Multivariate Behavioral Research*, 5, 329-350.

WILKS, S.S. (1938): The Large Sample Distribution of the Likelihood Ratio for Testing Composite Hypotheses, *Annals of Mathematical Statistics*, 9, 60–62.

Finding Meaningful and Stable Clusters Using Local Cluster Analysis

Hans-Joachim Mucha

Weierstraß-Institut für Angewandte Analysis und Stochastik (WIAS),
D-10117 Berlin, Germany

Abstract. Let us consider the problem of finding clusters in a heterogeneous, high-dimensional setting. Usually a (global) cluster analysis model is applied to reach this aim. As a result, often ten or more clusters are detected in a heterogeneous data set. The idea of this paper is to perform subsequent local cluster analyses. Here the following two main questions arise. Is it possible to improve the stability of some of the clusters? Are there new clusters that are not yet detected by global clustering? The paper presents a methodology for such an iterative clustering that can be a useful tool in discovering stable and meaningful clusters. The proposed methodology is used successfully in the field of archaeometry. Here, without loss of generality, it is applied to hierarchical cluster analysis. The improvements of local cluster analysis will be illustrated by means of multivariate graphics.

1 Introduction and task

Cluster analysis models are applied frequently in order to find clusters in a heterogeneous, high-dimensional setting. As a result, often ten or more clusters are detected. It is highly recommended that the stability of the obtained clusters has to be assessed by using validation techniques (Jain and Dubes (1988), Mucha (1992), Hennig (2004)). Furthermore, if possible, the clusters should be visualized in low dimensional projections for a better understanding of the clustering results. By doing so, often one can observe that the clusters have a quite different stability. Some of them are very stable. Thus, they can be reproduced and confirmed to a high degree, for instance, by simulations based on random resampling techniques. They are both homogeneous inside and well separated from each other. Moreover, sometimes they are located far away from the main body of the data like outliers. On the other side, hidden and tight neighboring clusters are more difficult to detect and they cannot be reproduced to a high degree.

The idea of this paper is to perform local clusterings subsequent to the usual cluster analysis. For example, both global and local statistical clustering models are used recently by Priebe et al. (2004) and Schwarz and Arminger (2005). Figure 1 contains a proposal of a programming flowchart that will become more clear in sections 4 and 5. Here, the key is a general approach for the assessment of the stability of individual clusters that is applicable to every clustering algorithm (concerning hierarchical clustering, see Mucha

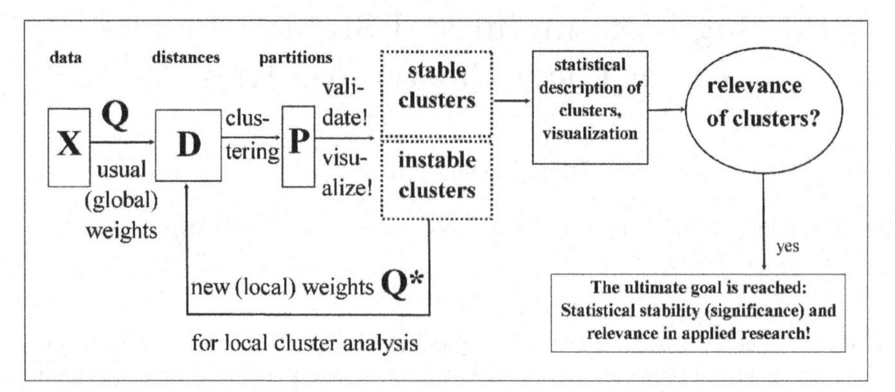

Fig. 1. Simplified graph of stages in iterative local cluster analysis methodology.

and Haimerl (2005)). Special statistical distances based on weight matrices **Q** (see section 3 for some examples) can alter dramatically by going from global to local cluster analysis. An improvement of the performance of cluster analysis can be expected by using local weight matrices $\mathbf{Q}^*$. Moreover, it can be expected that new clusters occur that are not yet detected by global clustering.

The proposed methodology is applied to hierarchical cluster analysis of more than six hundred Roman bricks and tiles. They are characterized by 19 chemical elements measured by XRF (X-Ray Fluorescence Analysis). Figure 2 shows the result of global hierarchical clustering (for details, see Mucha et al. (2005)). The bivariate non-parametric density estimation is based on the first two axes of principal components analysis (PCA). Generally, it is essential to observe that there are two main mountain chains. Figure 3 gives a more detailed view by several cuts of the bivariate density. At the right hand side, very compact clusters like "Straßburg-Königshofen" or "Groß-Krotzenburg" are characterized by dense regions, i.e. high peaks, and divided from each other by deep valleys. The mountain ridge on the left hand side could be identified with "not yet known 1", "Rheinzabern A and B", and "Worms" (from left to right). But visually, there seems to be not a sufficient possibility to distinguish the several neighboring clusters. Especially "Worms" proves almost to be only a "slope" of the cluster "Rheinzabern A and B". This potential interesting region for local cluster analysis is additionally marked by an ellipse.

2 What are meaningful and stable individual clusters?

Here we don't consider special properties like compactness and isolation (Jain and Dubes (1988)). What are stable clusters from a general statistical point of view? These clusters can be confirmed and reproduced to a high degree.

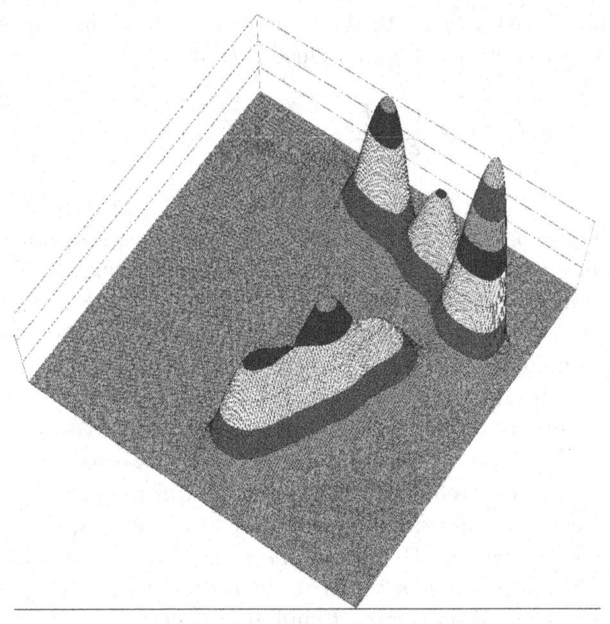

Fig. 2. Continuous visualization of cluster analysis results based on PCA.

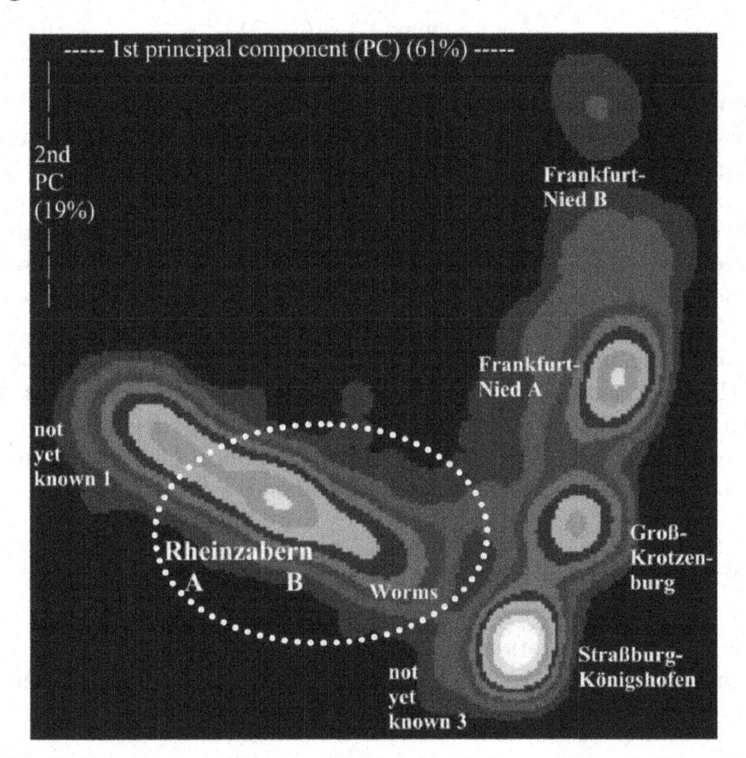

Fig. 3. Several cuts of the bivariate density that was shown in Figure 2.

To define stability with respect to the individual clusters, measures of correspondence between a cluster $\mathcal{E}$ and a cluster $\mathcal{F}$ like

$$\gamma_0(\mathcal{E},\mathcal{F}) = \frac{|\mathcal{E}\cap\mathcal{F}|}{|\mathcal{E}\cup\mathcal{F}|}, \quad \gamma(\mathcal{E},\mathcal{F}) = \frac{|\mathcal{E}\cap\mathcal{F}|}{|\mathcal{E}|} \tag{1}$$

have to be defined. ($\mathcal{E}$ and $\mathcal{F}$ are nonempty subsets of some finite set.) Hennig (2004) suggests the Jaccard coefficient γ_0. This measure is symmetric and it attains its minimum 0 only for disjoint sets and its maximum 1 only for equal ones. The asymmetric measure γ assesses the rate of recovery of subset $\mathcal{E}$ by the subset $\mathcal{F}$. It attains its minimum 0 only for disjoint sets and its maximum 1 only if $\mathcal{E}\subseteq\mathcal{F}$ holds. Obviously, it is necessary $\gamma_0 \leq \gamma$.

Now suppose, a clustering of a set of entities $\mathcal{C} = \{1,...,i,...,I\}$ into a collection of K subsets $\{\mathcal{C}_1,...,\mathcal{C}_k,...,\mathcal{C}_K\}$ of $\mathcal{C}$ has to be investigated. Let $\mathcal{C}_k$ be one individual cluster whose stability has to be assessed. To investigate the stability, validation techniques based on random resampling are recommended. Let's consider one simulation step: Clustering of a random drawn sample of the set of entities $\mathcal{C}$ into a collection of K clusters $\{\mathcal{F}_1,...,\mathcal{F}_K\}$ in the same way as the whole set $\mathcal{C}$. The definition of stability of cluster $\mathcal{C}_k$ using measure γ is based on the most similar cluster

$$\gamma_k^* = \max_{\mathcal{F}_i\in\{\mathcal{F}_1,...,\mathcal{F}_K\}} \gamma(\mathcal{C}_k,\mathcal{F}_i).$$

By repeating resampling and clustering many times, the stability of the cluster $\mathcal{C}_k$ can be assessed, for instance, by computing the median of the corresponding values of γ_k^*. Let us denote such an estimate $\hat{\gamma}_k^*$. It is difficult to fix an appropriate threshold to consider a cluster as stable (see the section below). To support the decision about stable regions, the clusters can often be visualized in low dimensional projections by applying methods like discriminant analysis, principal components analysis, and multidimensional scaling (e.g., see Figure 3). The simulation itself is computationally expensive.

What are meaningful or relevant clusters (from the researcher's point of view)? Here external information and experience of experts from application area can help to reconfirm the relevance of clusters that are stable from the statistical point of view. The ultimate aim of an application has to be finding stable and meaningful clusters in the data at hand.

3 Clustering based on statistical distances

Let a sample of I observations in R^J be given and denote by $\mathbf{X} = (x_{ij})$ the corresponding data matrix consisting of I rows and J variables. Then the generalized L_2-distance between observations (rows) x_i and x_l is

$$d_Q^2(\mathbf{x}_i,\mathbf{x}_l) = (\mathbf{x}_i-\mathbf{x}_l)^T\mathbf{Q}(\mathbf{x}_i-\mathbf{x}_l) \tag{2}$$

with a positive definite weight matrix $\mathbf{Q}$, which is the inverse covariance matrix usually. The special L_2-case $\mathbf{Q} = \mathbf{I}_J$ (same weights, that is, all the variables are measured in the same scale) will not be considered here because these distances remain unchanged by going from global to local cluster analysis. (Thus, it makes usually no sense to switch to local clustering steps.) Some interesting special weight matrices for the proposed local clustering will be given below. A well-known special distance measure of (2) is the Mahalanobis distance, where

$$\mathbf{Q} = \overline{\mathbf{S}}^{-1} \quad \text{with } \overline{\mathbf{S}} = \frac{1}{I-K} \sum_{k=1}^{K} \sum_{i \in \mathcal{C}_k} (\mathbf{x}_i - \overline{\mathbf{x}}_k)(\mathbf{x}_i - \overline{\mathbf{x}}_k)^T,$$

the pooled covariance matrix. Remember, K denotes the number of clusters. Furthermore, $\overline{\mathbf{x}}_k$ is the usual maximum likelihood estimate of expectation vector in cluster $\mathcal{C}_k$. Another special statistical distance of (2) is the squared weighted Euclidean distance, where $\mathbf{Q}$ is diagonal:

$$\mathbf{Q} = (\text{diag}(\mathbf{S}))^{-1} \quad \text{with } \mathbf{S} = \frac{1}{I-1} \sum_{i} (\mathbf{x}_i - \overline{\mathbf{x}})(\mathbf{x}_i - \overline{\mathbf{x}})^T, \tag{3}$$

the usual covariance matrix. Here $\overline{\mathbf{x}} = (1/I)\mathbf{X}^T\mathbf{1}$ is the vector of total means. In order to preserve the natural degree of variation the use of special weights

$$\mathbf{Q} = (\text{diag}(\overline{x}_1, \overline{x}_2, ..., \overline{x}_J))^{-2} \tag{4}$$

has been recommended (Underhill and Peisach (1985)). Otherwise one can use adaptive weights like the diagonal elements proportional to the inverse pooled within-cluster variances: $\mathbf{Q} = (\text{diag}(\overline{\mathbf{S}}))^{-1}$. These adaptive weights can be estimated in an iterative manner (Mucha (1992)). Diagonal weights like (3) or (4) are important for simple model-based Gaussian clustering when the variables are measured in different scales. As a consequence of this weighting scheme the variables become comparable one with each other. Concerning model-based clustering the paper of Banfield and Raftery (1993) is a good reference for a further reading on this topic. For instance, in the simplest case the well-known sum-of-squares criterion

$$V_K = \sum_{k=1}^{K} \text{tr}(\mathbf{W}_k) = \sum_{k=1}^{K} \sum_{i \in C_k} d_Q^2(\mathbf{x}_i, \overline{\mathbf{x}}_k) \tag{5}$$

has to be minimized. Here $\mathbf{W}_k = \sum_{i \in \mathcal{C}_k} (\mathbf{x}_i - \overline{\mathbf{x}}_k)(\mathbf{x}_i - \overline{\mathbf{x}}_k)^T$ is the sample cross-product matrix for the k-th cluster $\mathcal{C}_k$.

4 Local cluster analysis methodology

Mucha and Haimerl (2005) recommended an automatic validation of hierarchical clustering based on resampling techniques that can be considered as

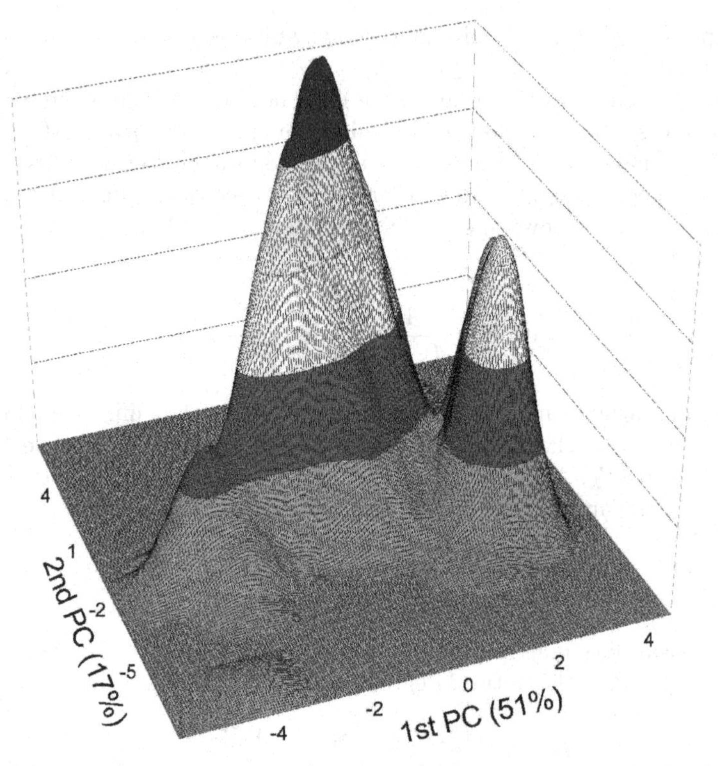

Fig. 4. Bivariate density estimation based on the first two axes of the local PCA.

a three level assessment of stability. The first and most general level is decision making about the appropriate number of clusters. The decision is based on such well-known measures of correspondence between partitions like the Rand index, the adjusted Rand index, and the index of Fowlkes and Mallows (Hubert and Arabie (1985)). Second, the stability of each individual cluster is assessed based on measures of similarity between subsets, e.g., the symmetric Jaccard measure γ_0 or the asymmetric measure of rate of recovery γ. It should be mentioned that it makes sense to investigate the (often quite different) specific stability of clusters. This is the basis of the methodology for global and successive local clustering that is presented in Figure 1. In the third and most detailed level of validation, the reliability of the cluster membership of each individual observation can be assessed.

The general approach of such a three level build-in validation is based on a contingency table that is obtained by crossing two partitions (Mucha (2004)). One of them, i.e. that one that has to be validated, is fixed during the simulations whereas the other ones are the result of clustering other samples that are drawn randomly from the data at hand $\mathbf{X}$. As a result of several hundred simulations one gets a corresponding set of estimates of the measures of (2). Working with a appropriate threshold on the average or median of the set of measures one can decide to consider clusters as stable or instable. The instable ones are set aside for further local analysis.

Hierarchical cluster analysis is in some sense more general than partitioning methods because the resultant hierarchy can be considered as a sequence of nested partitions. Thus, cluster validation based on comparing partitions is a more complex task here.

5 Application of local cluster analysis

The application of local cluster analysis in archeaometry is based on the result of hierarchical clustering that was already shown in Figure 2 and 3 (for details, see Mucha et al. (2005). The hierarchical Ward method minimizes (5) (Ward (1963)). It is applied for both the global and the local cluster analyses. The special weights (4) are used in order to handle the quite different scales of the variables. Some of the underlying estimates of these weights are heavily affected by going from global to local cluster analysis.

The thresholds $\hat{\gamma}_k^* > 0.99$ and $\hat{\gamma}_k^* > 0.95$ are used to consider a cluster as very stable and stable, respectively. Thus, "Groß-Krotzenburg" and "Straßburg-Königshofen" are the most stable clusters with a median $\hat{\gamma}_k^* = 1.0$ and an average rate of recovery of 0.999 and 0.995, respectively (Figure 3). On the other side, the most instable clusters are "Rheinzabern A and B" with ($\hat{\gamma}_k^* = 0.806$) followed by "Frankfurt-Nied B" ($\hat{\gamma}_k^* = 0.929$) and "Worms" ($\hat{\gamma}_k^* = 0.966$).

To demonstrate local clustering, the two (former) global clusters "Rheinzabern A and B" and "Worms" were selected for further analysis. Figure 4 shows a stable two cluster solution of local cluster analysis. The bivariate density is figured out using the first two components of the local PCA of the covariance matrix. Obviously, the local clusters are zoned much better from each other than the global ones. The corresponding medians of rates of recovery are $\hat{\gamma}_k^* = 0.9896$ (local "Rheinzabern A and B") and $\hat{\gamma}_k^* = 1.0$ (local "Worms", i.e., the smaller but more compact mountain at the right hand side), respectively. These medians were improved considerably. All the results here are obtained with respect to 250 simulations. Taking into account external information and experience of experts from the application area, the meaningfulness of local clustering can be confirmed (Mucha et al. (2005)).

6 Conclusion

Usually, subsequent cluster analysis requires the selection of clusters for local use. This can be done by general validation of results of cluster analysis based on contingency tables and by looking at multivariate graphics. The stability of results of cluster analysis based on statistical distances can be improved by subsequent local clustering with weight matrices that are based on local statistics. In applications, the stabilized local clusters (from a statistical point of view) can become often also more meaningful ones from the experts point of view.

References

BANFIELD, J.D. and RAFTERY, A.E. (1993): Model-Based Gaussian and non-Gaussian Clustering. *Biometrics, 49, 803–821.*

HENNIG, C. (2004): A General Robustness and Stability Theory for Cluster Analysis. *Preprint, 7,*, Universität Hamburg.

HUBERT, L.J. and ARABIE, P. (1985): Comparing Partitions. *Journal of Classification, 2, 193–218.*

JAIN, A. K. and DUBES, R. C. (1988): *Algorithms for Clustering Data.* Prentice Hall, Englewood.

MUCHA, H.-J. (1992): *Clusteranalyse mit Mikrocomputern.* Akademie Verlag, Berlin.

MUCHA, H.-J. (2004): Automatic Validation of Hierarchical Clustering. In: J. Antoch (Ed.): *Proceedings in Computational Statistics, COMPSTAT 2004, 16th Symposium.* Physica-Verlag, Heidelberg, 1535–1542.

MUCHA, H.-J., BARTEL, H.-G., and DOLATA, J. (2005): Model-based Cluster Analysis of Roman Bricks and Tiles from Worms and Rheinzabern. In: C. Weihs and W. Gaul, W. (Eds.): *Classification - The Ubiquitous Challenge*, Springer, Heidelberg, 317–324.

MUCHA, H.-J. and HAIMERL, E. (2005): Automatic Validation of Hierarchical Cluster Analysis with Application in Dialectometry. In: C. Weihs and W. Gaul, W. (Eds.): *Classification - The Ubiquitous Challenge*, Springer, Heidelberg, 513–520.

PRIEBE, C. E., MARCHETTE, D. J., PARK, Y., WEGMAN, E. J., SOLKA, J. L., SOCOLINSKY, D. A., KARAKOS, D., CHURCH, K. W., GUGLIELMI, R., COIFMAN, R. R., LIN, D., HEALY, D. M., JACOBS, M. Q., and TSAO, A. (2004): Iterative Denoising for Cross-Corpus Discovery. In: J. Antoch (Ed.): *Proceedings in Computational Statistics, COMPSTAT 2004, 16th Symposium.* Physica-Verlag, Heidelberg, 381–392.

SCHWARZ, A. and ARMINGER, G. (2005): Credit Scoring Using Global and Local Statistical Models. In: C. Weihs and W. Gaul, W. (Eds.): *Classification - The Ubiquitous Challenge*, Springer, Heidelberg, 442–449.

UNDERHILL, L.G. and PEISACH, M. (1985): Correspondence analysis and its application in multielement trace analysis. *J. Trace and microprobe techniques 3 (1 and 2), 41–65.*

WARD, J.H. (1963): Hierarchical Grouping Methods to Optimise an Objective Function. *JASA, 58, 235–244.*

Comparing Optimal Individual and Collective Assessment Procedures

Hans J. Vos[1], Ruth Ben-Yashar[2], and Shmuel Nitzan[2]

[1] Department of Research Methodology, Measurement and Data Analysis, University of Twente, P.O. Box 217, 7500 AE Enschede, the Netherlands
[2] Department of Economics, Bar-Ilan University, 52900 Ramat-Gan, Israel

Abstract. This paper focuses on the comparison between the optimal cutoff points set on single and multiple tests in predictor-based assessment, that is, assessing applicants as either suitable or unsuitable for a job. Our main result specifies the condition that determines the number of predictor tests, the collective assessment rule (aggregation procedure of predictor tests' recommendations) and the function relating the tests' assessment skills to the predictor cutoff points.

1 Introduction

The existing psychological and educational literature discusses how cutoff points can be determined, while there is only one psychological or educational test or one measure which weighs the scores on a number of psychological or educational tests as a composite score, or, for many tests, how the cutoff point on each predictor test can be determined separately. However, no results are reported how in case of a multiple test composed of several tests the cutoff points on each separate test and the collective assessment rule (i.e., aggregation procedure of predictor tests' recommendations) can be determined dependently. For example, take a predictor-based assessment system in which the collective assessment rule is that an applicant must pass $(n+1)/2$ out of n predictor tests for being selected, then one must decide on a cutoff point for each separate predictor test. Therefore, the goal of this paper is to present a model that takes into account the dependence between the cutoff points on a number of predictor tests composing a multiple test and its aggregation process to come to a collective assessment in terms of rejecting or admitting an applicant for a job in industrial/organizational (I/O) psychology. Doing so, Bayesian decision theory will be used as a conceptual framework (e.g., Lehmann, 1959). In other words, the predictor cutoffs and the collective assessment rule will be optimized simultaneously by maximizing the multiple test's common expected utility. It should be emphased that in the remainder with test is meant a psychological test, and thus, no hypothesis test.

The model advocated here has been applied earlier successfully by Ben-Yashar and Nitzan (2001) to economics where organizations face the comparable problem of deciding on approval or rejection of investment projects. A team of n decision makers has to decide which ones of a set of projects are to be accepted so as to maximize the team's common expected utility.

2 The model

In the field of personnel selection, it often occurs that an applicant is assessed as being either accepted or rejected for a job based on a multiple test composed of several predictor tests, i.e., a battery of n ($n \geq 1$) performance measures such as psychological tests, role-plays, and work sample tasks. It is assumed that the true state of an applicant regarding the current job performance (usually a supervisory performance rating) is unknown and can be assessed as either suitable ($s = 1$) or unsuitable ($s = -1$). An applicant is assessed as suitable if his or her performance is at least equal to a pre-established cutoff point (performance level) on the criterion variable(s) represented by the current job performance. Furthermore, based on applicant's performance on predictor test i ($1 \leq i \leq n$), it is decided if an applicant is qualified as being passed ($a_i = 1$) or failed ($a_i = -1$) on predictor test i. The predictor tests i will usually differ in their outcomes regarding passing or failing of applicants.

The true state of an applicant, however, is unknown on each of the n predictor tests. Instead, an applicant receives a test score x_i (i.e., a performance rating) on each predictor test i which depends on applicant's performance in a certain skill area. The pass-fail decision a_i is now made by setting a cutoff point on each test score x_i in the form of a threshold R_i (i.e., predictor cutoff) such that

$$x_i \geq R_i \rightarrow a_i = 1$$
$$x_i < R_i \rightarrow a_i = -1$$

The test score x_i is drawn from a distribution function represented by the density $f_1(x_i)$ for suitable and $f_2(x_i)$ for unsuitable applicants. Therefore, the conditional probabilities p_i^1 and p_i^2 that a predictor test i makes a correct pass-fail decision under the two possible states of nature (the *assessment skills* of each predictor test) are:

$$p_i^1 = \Pr\{a_i = 1 \mid s = 1\} = \int_{R_i}^{\infty} f_1(x_i)dx_i$$

$$p_i^2 = \Pr\{a_i = -1 \mid s = -1\} = \int_{-\infty}^{R_i} f_2(x_i)dx_i,$$

where $(1 - p_i^1)$ and $(1 - p_i^2)$ can be interpreted as Type I and Type II error probabilities (i.e., probabilities of making incorrect fail and pass decisions) of each predictor test i. Assessment skills of predictor tests are assumed to be endogeneous variables that depend on the cutoff points to be set.

The vector $a = (a_1, ..., a_n)$ is referred to as the *assessment profile* of a set of n predictor tests for an individual applicant, where $a_i = 1$ or $a_i = -1$ denotes if the applicant is either passed or failed on predictor test i ($1 \leq i \leq n$). The collective assessment, acceptance (1) or rejection (-1) of an applicant, is then determined by means of a collective assessment rule g that transforms the profile of assessments of n predictor tests into a collective assessment. g is referred to as the *structure* of the collective assessment process and assigns 1 or -1 (acceptance or rejection of an applicant) to any assessment profile a in $\Omega = \{1, -1\}^n$. That is, g: $\Omega \rightarrow \{1, -1\}$.

To formally define the objective function (i.e., the multiple test's common expected utility), we need to present the conditional probabilities of reaching a correct collective assessment, given the structure g. Let us therefore partition the set Ω of all assessment profiles into $A(g/1)$ and $A(g/-1)$, where $A(g/1) = \{a \in \Omega \mid g(a) = 1\}$ and $A(g/-1) = \{a \in \Omega \mid g(a) = -1\}$, and where $g(a)$ is the collective assessment rule for an assessment profile a. For a given structure g, the collective assessment accepts a suitable applicant and rejects an unsuitable applicant with probability $\varphi(g/1)$ and $\varphi(g/-1)$, respectively, where $\varphi(g/1) = \Pr\{a \in A(g/1) \mid s = 1\}$ and $\varphi(g/-1) = \Pr\{a \in A(g/-1) \mid s = -1\}$. Note that for a single test i, $\varphi(g/1)$ and $\varphi(g/-1)$ are equal to respectively p_i^1 and p_i^2 .

3 Necessary conditions for optimal cutoff points

For a multiple test, our goal is to derive the collective assessment rule g and cutoff point R_i $(1 \leq i \leq n)$ on predictor test i $(1 \leq i \leq n)$ dependently that maximize the multiple test's common expected utility. Therefore, the following problem is faced:

$$\text{Max}_{R_i, g} \quad \alpha U(1/1)\varphi(g/1) + \alpha U(-1/1)[1 - \varphi(g/1)] + \\ (1-\alpha)U(-1/-1)\,\varphi(g/-1) + (1-\alpha)U(1/-1)\,[1 - \varphi(g/-1)], \quad (1)$$

where $U(1/1)$, $U(1/-1)$, $U(-1/-1)$ and $U(-1/1)$ are the (economic) utilities corresponding to the four possible assessment outcomes on each predictor test, that is, correct passing (true positive), incorrect passing (false positive), correct failing (true negative), and incorrect failing (false negative). Furthermore, α and $(1-\alpha)$ denote the a priori probabilities that an applicant is assessed as either suitable (1) or unsuitable (-1). Since $[\alpha U(-1/1) + (1-\alpha)U(1/-1)]$ does not depend on R_i, the above maximization problem can be reduced to the following form:

$$\text{Max}_{R_i, g} \quad \alpha U(1/1)\varphi(g/1) - \alpha U(-1/1)\varphi(g/1) + \\ (1-\alpha)U(-1/-1)\,\varphi(g/-1) - (1-\alpha)U(1/-1)\,\varphi(g/-1). \quad (2)$$

Note that the optimal assessment method for a multiple test consists of a collective assessment rule g and a vector of optimal predictor cutoff values.

4 Qualified majority rule (QMR)

Quite often the collective assessment rule g is given and not necessarily optimal. However, it might still be possible to improve the predictor-based assessment process by controlling its optimal cutoff point R_i^* on each predictor test i ($1 \leq i \leq n$). Suppose now that a qualified majority rule (QMR) is employed, which is defined as follows:

$$g = \begin{cases} -1 \text{ for } N(-1) \geq kn \\ 1 \ \text{ otherwise,} \end{cases} \tag{3}$$

where $N(-1)$ is the number of predictor tests failed by the applicant, n is the number of predictor tests, and k ($1/n \leq k \leq 1$ and kn is an integer) is the minimal proportion of predictor tests failed by the applicant necessary for the collective assessment to be -1 (rejection of applicant). The parameter k represents the collective assessment rule g, or the structure of the assessment process. For instance, a simple majority rule $k = \frac{n+1}{2n}$ implies that an applicant is rejected if $N(-1) \geq \frac{n+1}{2}$ and accepted otherwise. Let $U(1) = [U(1/1) - U(-1/1)]$ and $U(-1) = [U(-1/-1) - U(1/-1)]$ denote the positive net utility corresponding to respectively the correct pass and correct fail decision, it then follows that we face the following problem:

$$\text{Max}_{\ R_i} \ \alpha U(1)\varphi(k/1) + (1-\alpha)U(-1)\varphi(k/-1). \tag{4}$$

Given the structure k of the collective assessment process and the number n of predictor tests, the optimal cutoff point R_i^* on predictor test i ($1 \leq i \leq n$) of a multiple test is determined by the following necessary condition:

$$\frac{dp_i^1}{dR_i} = -Z\frac{dp_i^2}{dR_i}W_i, \tag{5}$$

where

$$Z = \frac{(1-\alpha)U(-1)}{\alpha U(1)} \tag{6}$$

$$W_i = \frac{\frac{\partial\varphi(g/-1)}{\partial p_i^2}}{\frac{\partial\varphi(g/1)}{\partial p_i^1}} = \left(\frac{p_i^2}{1-p_i^1}\right)^{kn-1}\left(\frac{1-p_i^2}{p_i^1}\right)^{n-kn} \tag{7}$$

The proof of the above assertion is given in Ben-Yashar and Nitzan (2001).

In a single test i, it obviously holds that n, and thus k, is equal to 1 implying that $W_i = 1$. It follows then immediately from the above equation that the optimal cutoff point R_i^+ on predictor test i ($1 \leq i \leq n$) in this case is determined by the following necessary condition:

$$\frac{dp_i^1}{dR_i} = -Z\frac{dp_i^2}{dR_i}. \tag{8}$$

5 Relationship between optimal cutoff points for single and multiple tests

The optimal cutoff points for single and multiple tests in predictor-based assessment are usually different. Whether or not the cutoff points for single tests are stricter than the cutoff points for multiple tests depends on the characteristics of the assessment process: the preferred assessment skills of the predictor tests, the number of predictor tests and the collective assessment rule. Our main result specifies the condition that determines the relationship between the optimal cutoff points R_i^+ and R_i^* for single and multiple tests in predictor-based assessment.

Theorem:

$$R_i^* \gtreqless R_i^+ \iff W_i \gtreqless 1 \iff k \gtreqless \lambda_i \tag{9}$$

where

$$\lambda_i = \frac{1}{n} + \frac{n-1}{n} \frac{\ln \frac{p_i^1}{1-p_i^2}}{\beta_i^1 + \beta_i^2}, \tag{10}$$

n is the fixed size of the number of predictor tests, $\beta_i^1 = \ln \frac{p_i^1}{1-p_i^1}$ and $\beta_i^2 = \ln \frac{p_i^2}{1-p_i^2}$. The parameter λ_i $(1 \leq i \leq n)$ can be interpreted as the bias/asymmetry of the tests' assessment skills. For the proof of this theorem, we refer to Ben-Yashar and Nitzan (2001).

6 Predictor-based assessment using the Assessment Center method: An illustration

To illustrate the theorem for comparing the optimal cutoff points R_i^+ and R_i^* set on single and multiple tests, the Assessment Center (AC) method is given as an empirical example. In a typical Assessment Center the candidates applying for a job participate in a variety of exercises (e.g., leadership, sensitivity, delegation, etc.) that enable them to demonstrate a particular (interpersonal) skill, knowledge, ability, or competence. The performance rating on each exercise is done by observers (called assessors). Comparing these ratings with a pre-established cutoff point, it is decided whether or not an applicant's performance on each specific exercise is satisfactorily enough to be passed. Then the assessors combine the pass-fail decisions on all the exercises and reach a collective assessment for each applicant, that is, either accept or reject the applicant for the job.

In the current example, data of candidates applying for trainee positions were available for a large company. The performance on each of the 15 exercises (i.e., the predictor tests i) of the Assessment Center (i.e., the multiple test) was rated by a team of two carefully trained assessors on a 100-point scale running from 0 to 100. So, i was running from 1 to 15 and each predictor score x_i was running from 0 to 100.

Since the company did not have any prior information of the applicants, the a priori probabilities a and $(1-\alpha)$ of assessing an applicant's true state (i.e., current job behavior) as respectively suitable ($s = 1$) or unsuitable ($s = -1$) were set equal. Hence, $\alpha = (1-\alpha) = 0.5$.

Furthermore, using the lottery method described in Luce and Raiffa (1957), the positive net utility corresponding to a correct pass decision (i.e., $U(1)$) was perceived by the company from an economic perspective twice as large as the positive net utility corresponding to a correct fail decision (i.e., $U(-1)$). Hence, since the utility ratio $U(1)/U(-1) = 2$ and $\alpha = (1-\alpha) = 0.5$, it follows that $Z = 0.5$.

In order to calculate the optimal cutoff point R_i^* on each single exercise i $(1 \leq i \leq 15)$ of the AC, given the collective assessment rule k and number n of exercises, we finally still need to specify p_i^1 and p_i^2 as functions of R_i. It was assumed that the test score distributions $f_1(x_i)$ and $f_2(x_i)$ for exercise i $(1 \leq i \leq 15)$ in the suitable and unsuitable group of applicants followed a normal distribution with mean μ_i^1 and μ_i^2 (with μ_i^2 lower than μ_i^1) and standard deviation σ_i^1 and σ_i^2, respectively. Based on a group of 127 candidates (69 accepted and 58 rejected) who all applied for actual trainee positions in the past, it will first be described how it was determined if an applicant was assessed as either suitable ($s = 1$) or unsuitable ($s = -1$).

First, depending on applicant's performance, for each applicant (both accepted and rejected ones) a test score x_i $(0 \leq x_i \leq 100)$ was assigned to each exercise i $(1 \leq i \leq 15)$ by the team of two assessors. Henceforth, the predictor score on exercise i will be denoted as X_i. Next, for each selected applicant a criterion score y_i (i.e., applicant's supervisor rating of current job performance concerning exercise i on a 100-point scale) was determined on the criterion variable Y_i $(1 \leq 15)$. Current job performance will be denoted as the composite criterion variable Y. For the group of selected applicants the following statistics could now be computed for each exercise i $(1 \leq i \leq 15)$: the means μ_{X_i} and μ_{Y_i}, the standard deviations σ_{X_i} and σ_{Y_i}, and the correlation $\rho_{X_i Y_i}$ between X_i and Y_i. Using these statistics, we then computed for each rejected applicant the predicted criterion score $\widehat{y}_i$ (i.e., job behavior on exercise i if the applicant would have been selected) as a linear regression estimate on applicant's predictor score x_i:

$$\widehat{y}_i = \mu_{Y_i} + \rho_{X_i Y_i}(\sigma_{Y_i} \,/\, \sigma_{X_i})(x_i - \mu_{X_i}). \tag{11}$$

Next, for each applicant (both accepted and rejected ones), a composite criterion score y on Y was calculated by taking his or her average criterion score over all 15 exercises. Finally, each applicant was assessed as either suitable ($s = 1$) or unsuitable ($s = -1$) by examining if applicant's composite criterion score y was above or below a pre-established cutoff point $y_c = 55$ on the criterion variable Y. The mean and standard deviation of $f_1(x_i)$ and $f_2(x_i)$ could now be estimated straightforward for each exercise i $(1 \leq i \leq 15)$.

The comparison of the optimal cutoff points R_i^+ and R_i^* set on single and multiple tests by using the theorem will be illustrated for the 9th exercise of leadership (i.e., $i = 9$). The parameters of $f_1(x_9)$ and $f_2(x_9)$ were estimated as follows: $\mu_9^1 = 74.12$, $\mu_9^2 = 50.68$, $\sigma_9^1 = 10.79$, and $\sigma_9^2 = 11.66$. The assumption of normality for $f_1(x_9)$ and $f_2(x_9)$ was tested using a Kolmogorov-Smirnov goodness-of-fit test. It turned out that the p-values were respectively 0.289 and 0.254, showing a satisfactory fit against the data $(\alpha = 0.05)$.

Then, using the customary notation $\Phi(\mu, \sigma)$ for the normal distribution with mean μ and standard deviation σ, the cumulative density is $\Phi(\mu_9^1, \sigma_9^1)$ for the suitable and $\Phi(\mu_9^2, \sigma_9^2)$ for the unsuitable applicants on Exercise 9. It then follows that $p_9^1 = 1 - \Phi((R_9 - \mu_9^1) / \sigma_9^1)$ (where $\Phi((R_9 - \mu_9^1) / \sigma_9^1)$ now represents the lower tail probability of the standard normal distribution evaluated at the cutoff point R_9), whereas $p_9^2 = 1 - \Phi((R_9 - \mu_9^2) / \sigma_9^2)$.

Relation between R_9^* and R_9^+ for given values of k and n

R_9^+ was computed by inserting $\frac{dp_9^1}{dR_9} = -\Phi((R_9 - \mu_9^1) / \sigma_9^1)$, $\frac{dp_9^2}{dR_9} = \Phi((R_9 - \mu_9^2) / \sigma_9^2)$, and $Z = 0.5$ into (8) resulting in $R_9^+ = 58.77$. R_9^+ was computed numerically using a root finding procedure from the software package Mathematica.

In order to investigate the influence of more and less lenient assessment rules on the optimal predictor cutoff, R_9^* was computed for $k = 3/15$, $k = 8/15$, and $k = 13/15$. Inserting first $k = 3/15$ and $n = 15$ into W_9, and next W_9 and $Z = 0.5$ into (5), and using again the root finding procedure from Mathematica, resulted in $R_9^* = 51.04$, $W_9 = 0.219$, $\lambda_9 = 0.224$, $p_9^1 = 0.984$, and $p_9^2 = 0.512$. So, verifying the theorem for $k = 3/15 = 0.2$, results in:

$$R_9^* = 51.04 < R_9^+ = 58.77 \iff W_9 = 0.219 < 1 \iff k = 0.2 < \lambda_9 = 0.224.$$

As can be seen from the above result, $R_9^* < R_9^+$ implying that a more tolerant cutoff point is set on Exercise 9 of the multiple test composed of 15 exercises relative to the cutoff point set on the single Exercise 9. This result can be accounted for that the collective rule $k = 3/15$ is much less lenient toward qualifying applicants as accepted than the simple majority rule since $kn = 3 < 8$ (i.e., (15+1)/2).

Next, for $k = 8/15 = 0.533$ (i.e., the simple majority rule), we obtained the following results: $R_9^* = 62.43$, $W_9 = 1.995$, $\lambda_9 = 0.520$, $p_9^1 = 0.861$, and $p_9^2 = 0.843$. According to the theorem, a somewhat stricter cutoff point R_9^* is now set on Exercise 9 of the multiple test composed of 15 exercises relative to the cutoff point R_9^+ set on the single Exercise 9. This makes sense since the simple majority rule is more lenient toward qualifying applicants as accepted than the collective rule $k = 3/15$.

Finally, for $k = 13/15 = 0.867$, we obtained the following results: $R_9^* = 73.36$, $W_9 = 14.31$, $\lambda_9 = 0.819$, $p_9^1 = 0.528$, and $p_9^2 = 0.974$. As can be verified from the theorem (i.e., $W_9 >> 1$), a much stricter cutoff point R_9^* is now set on Exercise 9 of the multiple test composed of 15 exercises relative to the cutoff point R_9^+ set on the single Exercise 9. This is because the collective rule $k = 13/15$ is much more lenient toward qualifying applicants as accepted than the simple majority rule.

7 Concluding remarks

Although the field of personnel selection is a useful application of the proposed model for dichtomous assessment procedures, it should be emphasized that the model advocated in this paper has a larger scope of application. As mentioned already in the Introduction, the proposed collective aggregation procedure can be applied to many binary decisions determined by teams of n ($n \geq 2$) decision makers or test batteries. Optimal collective aggregation procedures in organizations that face such uncertain dichotomous choices as deciding on approval or rejection of investment projects have also been extensively studied in Pete et al (1993).

For reasons of mathematical tractability, it has been assumed that suitability and unsuitability for a job consists of two "natural classes" that can be characterized by different normal distributions. However, it seems more realistic to assume that suitability and unsuitability in reality are rather defined by a more or less arbitrary cutoff on a continuous scale.

References

BEN-YASHAR, R. and NITZAN, S. (2001): Investment Criteria in Single and Multi-Member Economic Organizations. *Public Choice, 109, 1-13.*

LEHMANN, E.L. (2000): *Testing Statistical Hypotheses.* Macmillan, New York.

LUCE, R.D. and RAIFFA, H. (1957): *Games and Decisions.* Wiley, New York.

PETE, A., PATTIPATI, K.R. and KLEINMAN, D.L. (1993): Optimal Team and Individual Decision Rules in Uncertain Dichotomous Situations. *Public Choice, 75, 205-230.*

Part III

Network and Graph Analysis

Some Open Problem Sets for Generalized Blockmodeling

Patrick Doreian

Department of Sociology, University of Pittsburgh,
2603 WWPH, Pittsburgh, PA 15260, USA

Abstract. This paper provides an introduction to the blockmodeling problem of how to cluster networks, based solely on the structural information contained in the relational ties, and a brief overview of generalized blockmodeling as an approach for solving this problem. Following a formal statement of the core of generalized blockmodeling, a listing of the advantages of adopting this approach to partitioning networks is provided. These advantages, together with some of the disadvantages of this approach, in its current state, form the basis for proposing some open problem sets for generalized blockmodeling. Providing solutions to these problem sets will transform generalized blockmodeling into an even more powerful approach for clustering networks of relations.

1 Introduction

Hummon and Carley (1993) identified blockmodeling as a particularly important area of social network analysis. Blockmodeling was founded on the concept of structural equivalence, a concept introduced by Lorrain and White (1971). In essence, the goal of blockmodeling is to partition the actors (vertices) of social network simultaneously with their relational ties. Based on some conception of equivalence, the actors are clustered into *positions* and the relational ties are clustered into *blocks*. If there are k positions in a blockmodel, there will be k^2 blocks, of which k will be diagonal blocks.

Two vertices are structurally equivalent if they are connected to the network in an identical fashion. This approach caught fire with the implementation of practical programs for identifying blockmodels based on structural equivalence, notably CONCOR (Breiger et al. (1975)) and STRUCTURE (Burt (1976)). A generalization of structural equivalence is regular equivalence (Sailer (1978), White and Reitz (1983)) where vertices are regularly equivalent if they are equivalently connected to equivalent others. (Formal definitions of these equivalences can be found in the original sources and in Doreian *et al.* (2005: Chapter 6).)

As a simple example consider the sociogram (network) shown in Figure 1 and on the left in Table 1 where the vertices are in an arbitrary order and labeled. The example is taken from Doreian *et al.* (2005). In general, the essential structure of a network is not obvious when networks are large and/or complex. Blockmodeling attempts to discern and represent network structure

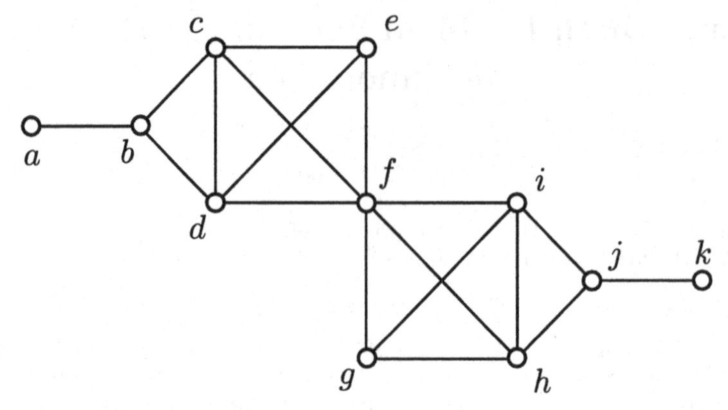

Fig. 1. An Artificial Network

Table 1. A Sociomatrix and Blockmodel for the Artificial Network

Sociomatrix	a	b	c	d	e	f	g	h	i	j	k
a	0	1	0	0	0	0	0	0	0	0	0
b	1	0	1	1	0	0	0	0	0	0	0
c	0	1	0	1	1	1	0	0	0	0	0
d	0	1	1	0	1	1	0	0	0	0	0
e	0	0	1	1	1	1	0	0	0	0	0
f	0	0	1	1	1	0	1	1	1	0	0
g	0	0	0	0	0	1	0	1	1	0	0
h	0	0	0	0	0	1	1	0	1	1	0
i	0	0	0	0	0	1	1	1	0	1	0
j	0	0	0	0	0	0	0	1	1	0	1
k	0	0	0	0	0	0	0	0	0	1	0

Blockmodel	a	k	b	j	f	c	d	h	i	e	g
a	0	0	1	0	0	0	0	0	0	0	0
k	0	0	0	1	0	0	0	0	0	0	0
b	1	0	0	0	0	1	1	0	0	0	0
j	0	1	0	0	0	0	0	1	1	0	0
f	0	0	0	0	0	1	1	1	1	1	1
c	0	0	1	0	1	0	1	0	0	1	0
d	0	0	1	0	1	1	0	0	0	1	0
h	0	0	0	1	1	0	0	0	1	0	1
i	0	0	0	1	1	0	0	1	0	0	1
e	0	0	0	0	1	1	1	0	0	0	0
g	0	0	0	0	1	0	0	1	1	0	0

and does so through appealing to some form of equivalence. In this case, specifying regular equivalence leads to the blockmodel shown on the right in Table 1 where the vertices of the network have been permuted into the order shown on the right in Table 1 and a coherent partition imposed. There are 5 positions and 25 blocks. The blocks in this table are either all null (*null blocks*) or take the form where each row and column of the block contain a 1 (*regular blocks*). If we label the regular blocks with 1 and the null blocks with 0, we get the *image matrix* shown in Table 2. This has a much simpler structure that captures the essence of Figure 1 where C_1 is {a, k}, C_2 is {b, j}, C_3 is {f}, C_4 is {c, d, h, i}, and C_5 is {e, g}.

Borgatti and Everett (1989) proved that every binary network has a class of regular equivalences that form a lattice. The network shown in Figure 1 has 21 exact regular equivalence partitions. Both Borgatti and Everett's result, and the use of generalized blockmodeling, make it clear that a network can have a (potentially large) number of optimal partitions rather than having a single blockmodel. This feature raises some interesting substantive and empirical tasks of assessing the set of such partitions in most empirical contexts when multiple optimal partitions are located.

Table 2. The Image of the Artificial Network

	C_1	C_2	C_3	C_4	C_5
C_1	0	1	0	0	0
C_2	1	0	0	1	0
C_3	0	0	0	1	1
C_4	0	1	1	1	1
C_5	0	0	1	1	0

There have been two approaches to the blockmodeling problem: indirect methods and direct methods. Indirect methods are characterized by transforming one (or more) network(s) into some (dis)similarity matrix and then clustering this matrix by some clustering procedure. Direct methods skip the transformation(s) and work with the data directly. Because of the increased combinatorial burden of considering all of the partitions of a network as the size of the network increases, examining all possible partitions is not possible for large networks. An alternative is the adoption of a (local) optimization procedure and this has been called an optimizational approach to blockmodeling by Doreian *et al.* (2005) as one distinctive feature of 'generalized blockmodeling'.

2 Generalized blockmodeling

The key step in moving from conventional blockmodeling (where an indirect approach is used) to generalized blockmodeling is the translation of an equivalence type into a set of *permitted block types*. For structural equivalence there are two: null blocks and complete blocks. Similarly, there are two permitted block types for regular equivalence: null and regular blocks (Batagelj *et al.* (1992a)). Specifying new types of equivalence is done through the specification of new permitted block types. See Doreian *et al.* (1994, 2005) for details of some new block types and new blockmodel types.

We denote a network as $N = (V, R)$ where V is the set of vertices of the network and R is a social relation. The nature of a blockmodel partition of a network is characterized by:

- A partition of vertices: $C = \{C_1, C_2, ..., C_k\}$
- With k clusters, C partitions also the relation R into k^2 blocks where

$$R(C_i, C_j) = R \cap C_i \times C_j$$

- Each block is defined in terms of units belonging to clusters C_i and C_j and consists of all arcs from units in cluster C_i to units in cluster C_j. If $i = j$, the block $R(C_i, C_i)$ is called a *diagonal block*.

With a set of empirical blocks and a corresponding set of permitted blocks, it is straight forward to think of comparing the empirical blocks with the corresponding permitted blocks on a block by block basis. Clearly, if an empirical block and an ideal permitted block are the same, there will no inconsistencies between the two. A criterion function is specified that does two things:

- For each pair of empirical and corresponding ideal blocks, it captures the difference between the blocks in an empirical partition and the blocks in a permitted (ideal) partition.
- The block by block differences (inconsistencies) are combined (added) into and overall criterion function as a measure of fit.

In general terms, for a network, $N = (V, R)$:

- We let Θ denote the set of all relations of a selected type (e.g. structural or regular equivalence);
- A criterion function $P(C)$ is defined to measure the difference between an empirical partition and a partition with permitted block types;
- This criterion function has the following the properties:

  P1. $P(C) \geq 0$
  P2. $P(C) = 0 \Leftrightarrow \sim \in \Theta$,

- And the criterion functions are defined to reflect the equivalence type.

More specifically, for a clustering $C = \{C_1, C_2, \ldots, C_k\}$

- Let $\mathcal{B}(C_u, C_v)$ denote the set of all ideal blocks corresponding to an empirical block $R(C_u, C_v)$.
- The global inconsistency of clustering C with an ideal clustering can be expressed as $P(C) = \sum_{C_u, C_v \in C} \min_{B \in \mathcal{B}(C_u, C_v)} d(R(C_u, C_v), B)$
- where $d(R(C_u, C_v), B)$ measures the difference (number of inconsistencies) between the block $R(C_u, C_v)$ and the ideal block B.
- The function d has to be compatible with the selected type of equivalence (Batagelj *et al.*, 1992b).

In practice, a local optimization procedure has been used and implemented in pajek[1] (Batagelj and Mrvar, 1998) for generalized blockmodeling. First, an initial clustering with a specified number (k) of clusters is created randomly. The neighborhood of the clustering is determined by two transformations: moving a vertex from one cluster to another or by interchanging a pair of vertices from two different clusters. If the criterion function diminishes

[1] This program is updated frequently to implement new procedures and to eliminate bugs. Version 1.12 is dated February 12, 2006.

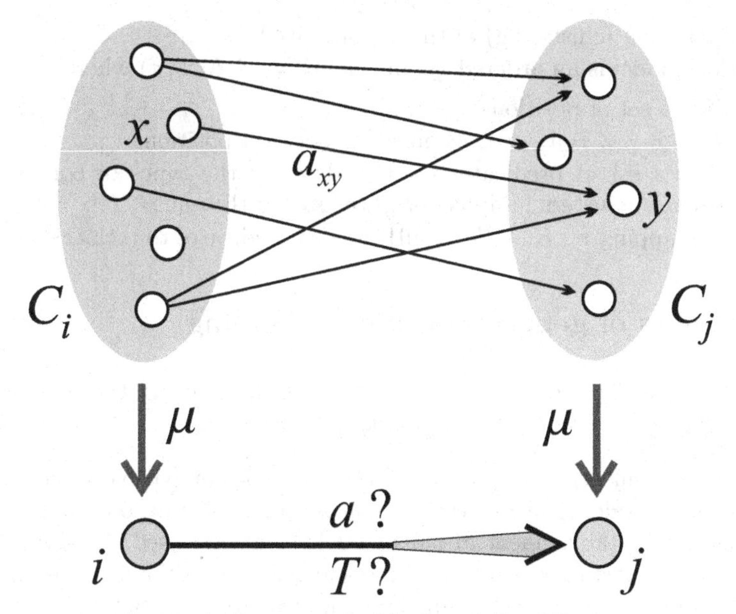

Fig. 2. A Picture of the Blockmodeling Scheme

under a transformation, the algorithm moves to it and repeats the transformation steps until the criterion function gets no smaller. This partition (or a set of such partitions) is (are) stored and the procedure is repeated many times. Only the best fitting partitions (those with the minimized value of the criterion function) are kept for further exploration as blockmodels of the network. There can be, and often are, multiple equally well fitting partitions (blockmodels) of a network.

Pictorially, the process is visualized in Figure 2. The vertices (units) clustered together in C_i are mapped under μ to the position i and the vertices (units) clustered into C_j are mapped under μ to the position j. The values of a set of ties, $\{a_{xy}\}$, between vertices in C_i and the vertices in $\{C_j\}$ are summarized in some fashion by a value, denoted by a, and assigned to a block type (predicate) T. More formally, the generalized blockmodeling scheme is specified by:

- A network with units V
- A relation $R \subseteq V \times V$.
- Let Z be a set of positions (or images of clusters of units)
- Let $\mu : V \to Z$ denote a mapping which maps each unit to its position. The cluster of units $C(t)$ with the same position $t \in Z$ is

$$C(t) = \mu^{-1}(t) = \{x \in V : \mu(x) = t\}$$

Therefore

$$C(\mu) = \{C(t) : t \in Z\}$$

is a partition (clustering) of the set of units V.

- A blockmodel is an ordered quadruple $M = (Z, K, T, \pi)$ where:
  - Z is a set of positions;
  - $K \subseteq Z \times Z$ is a set of *connections* between positions;
  - T is a set of predicates used to describe the types of connections between clusters in a network; we assume that nul $\in T$;
  - A mapping $\pi : K \to T \setminus \{\text{nul}\}$ assigns predicates to connections.

3 Benefits of generalized blockmodeling

There are considerable advantages that follow from using the generalized blockmodeling framework. They include:

- The use of an *explicit* criterion function that is compatible with a given type of equivalence means that a built in measure of fit, one that is integral to the establishment of blockmodels, is always used. This stands in stark contrast with conventional modeling which, in the main, is characterized by the absence of compelling and coherent measures of fit.
- Thus far, in comparisons with conventional blockmodeling, with attention restricted to structural and regular equivalence, partitions established by using generalized blockmodeling frequently out-perform partitions established by the conventional approach and never perform less well.
- Within conventional blockmodeling, the prevailing equivalence types are structural and regular equivalence. By focusing on the permitted block types, it is possible to expand their number and, with the expanded set of permitted block types, we can specify new types of blockmodels. See Doreian *et al.* (2005).
- The potentially unlimited set of new block types permit the inclusion of substantively driven blockmodels. Structural balance theory, for which Heider (1946) is credited with the first formal statement, is included naturally (Doreian and Mrvar (1996)). The specification of a balance theoretic blockmodel is one with positive blocks (having only positive or null ties) on the main diagonal of the image and negative blocks (having only negative or null ties) off the main diagonal are specified. The ranked clusters models of Davis and Leinhardt (1972) is included through the use of symmetric blocks on the diagonal of the image matrix and the specification of an acyclic block structure (Doreian *et al.* (2000). Routines for establishing both of these models are implemented in pajek (Batagelj and Mrvar (1998)).
- Conventional blockmodeling has been used primarily in an inductive way when an equivalence type is specified and the analyst accepts what is delineated as a result of the clustering procedure. Yet, we often know more about the expected block structure and can include this knowledge into the specification of blockmodels.

  - Not only can we specify the block types, we can specify also the location of some of them in the blockmodel. The ranked clusters model has this feature where symmetric (or null) blocks are on the diagonal of the image and null blocks are specified above the main diagonal (or below it, but not both above and below the main diagonal). In the extreme, the location of every block type can be specified. The structural balance model exemplifies this feature.
  - Constraints can be included in the specification of a blockmodel by specifying which vertices must be clustered together, or specifying pairs of vertices that should never be clustered together. In the extreme, the location of all vertices in their positions can be specified.
  - Penalties can be imposed on specific block types so that inconsistencies with these block types are penalized severely. If null blocks are particularly important, inconsistencies in such blocks can be emphasized by specifying a heavy penalty. The result, typically, is that empirical null blocks are identified without inconsistencies with a corresponding ideal null block.
- Together, the pre-specification of the location of block types in a blockmodel, the use of constraints, and invoking of penalties permit the deductive use of blockmodeling. This feature is displayed in Table 3 where a '?' means 'unspecified'.

Table 3. Inductive and Deductive Approaches to Blockmodeling

		Clustering	Blockmodel
Inductive		?	?
	Pre-specification	?	given
Deductive	Constraints	given	?
	Constrained Pre-specification	given	given

In conventional blockmodeling, with only the specification of an equivalence type, neither the clustering nor the blockmodel form are specified. With generalized blockmodeling, one or both of these features can be specified. Further discussion of this topic is presented in Doreian *et al.* (2005: Chapter 7).

Examples of the use of generalized blockmodeling are found throughout Doreian *et al.* (2005). These include partitions of the following: social relations in work settings; signed relations over time in a pseudo-fraternity; signed relations in a monastery; classroom networks; networks among the members of Little League baseball teams; search and rescue interorganizational networks; political actor networks; discussion networks for a student government; baboon grooming networks; marriage ties among noble families; and journal citation networks.

4 Some generalized blockmodeling open problem sets

The open problems that I propose we consider stem from both potential weaknesses of generalized blockmodeling and its strengths listed above.

1. We do not have a clear idea about the sensitivity of generalized blockmodeling to particular data features. Even with all network members represented, there can be missing data for the ties. In addition, some relational ties may be present in the data that are not veridical. One problem set, then, is exploring the sensitivity of generalized blockmodeling, in general, and the use of specific block types in blockmodels, in particular, to the presence of missing and inaccurate data. This can be done with simulated data where processes generating 'missing' data are built into the simulation. It also can be done with real data sets where some ties are edited (i.e. corrupted) to examine the impacts on blockmodels established when using the real data and the corrupted data.
2. A second open problem set for generalized blockmodeling is the examination of *boundary problems*. One example of network boundaries occurs with blockmodels of journal citation networks where two types of problem can occur. Even within specific fields, some journals and citation pairs are missed. This could, perhaps, be included in the previous problem set. More important, when the focus is on a specific discipline, is the exclusion of adjacent relevant disciplines. These other disciplines are the environment for a particular discipline. Omitting them creates a boundary and the omission of related fields can dramatically influence the established partitions. Laumann *et al.* (1989) point to the crucially important 'boundary specification problem'. Solutions, for example, Doreian and Woodard (1994), have been proposed as ways of appropriately delineating the boundary of a network. Often however, the boundary that has been specified still omits other relevant units. Pursuing this problem set will give us some idea of the vulnerability of generalized blockmodeling to the boundary imposed on an empirical network and a sense of the implications for the validity of the blockmodels that are established.
3. A third problem set stems from coming at the boundary problem from a different direction. This is to seek representations of the wider environment of a network so that the validity of blockmodel that has been established is preserved even though the details of the connections to the rest of the networked world are not included in the data. An example of this line of thinking is found in Doreian (1993) in the context of establishing measures of standing for a set of national scientific communities with and without the rest of the world being considered.
4. One obvious response to the boundary problem is to include more vertices and study larger networks within the generalized blockmodeling framework. Unfortunately, generalized blockmodeling is computationally burdensome and this imposes a practical network size constraint for generalized blockmodeling. One advantage of the conventional blockmodeling

approach is that it can be used for much larger networks than is the case for generalized blockmodeling[2]. This fourth set of open problems amounts to creating more efficient algorithms for establishing generalized blockmodels. It may be useful, as a part of this effort, to find ways in which both conventional and generalized blockmodeling methods can be used in conjunction. This suggests the creation of (dis)similarity measures that are compatible with, and sensitive to, the new types of blocks and new blockmodel types for use in clustering procedures via indirect methods. Such partitions could then be optimized via generalized blockmodeling methods.

5. Thinking about missing data, inaccurate data, and boundary problems points to a more general issue. For most social networks for which we gather data, our information will be incomplete. Even if board interlock data are complete for a set of organizations, such ties are only part of the broader structure of interorganizational relations, much of which will be unobservable for most network observers. Or, as a variant of this more general problem, consider having data on email exchanges between members of an organization. Do these data tell us anything about the structure of the organization? If we have intercepted exchanges between units of an enemy in a battle field, can these network data help identify the spatial structure and locations of the enemy units? The general problem is one of *identifying structure from incomplete network information.* Given that one of the goals of blockmodeling is to delineate the underlying structure of a network, can generalized blockmodeling be used to tackle this broad generic problem of identifying structure from incomplete information?
6. Most of the examples considered by Doreian *et al.* (2005) concern binary networks. The obvious extension is to consider valued networks in a principled fashion. A promising start is found in Žiberna (2005).
7. It is clear that, within the generalized blockmodeling framework, the criterion function for a particular problem is minimized for a specified blockmodel[3]. And if the number of inconsistencies is zero or very small it seems reasonable to accept that the established blockmodel fits the network data well. When the blockmodel is specified truly on substantive grounds this might be enough. However, if a specified blockmodel fits with a 'larger' number of inconsistencies, can we say that the blockmodel fits? Answering this question is complicated by the fact that different block types and types of blockmodels have criterion functions that differ in the

[2] The drop down menu for operations in pajek uses an asterisk (*) to mark operations where network size is a serious problem. This is a clear signal that there is limit to what can be done within this framework as far a network size is concerned.

[3] Because a local optimization procedure is used in generalized blockmodeling, there is always the possibility that an optimized partition is missed. One partial solution is to use many repetitions of the procedure. Even so, the general point stands.

'magnitude' of the criterion function. As a result, the magnitudes of the criterion functions cannot be compared across different types of blockmodels. It is always possible to over-fit a blockmodel in ways that are idiosyncratically responsive to specific features of a specific data by using a set of *ad hoc* of block types. Such a blockmodel is likely to be worthless. This raises the issue of evaluating *statistically* whether a selected blockmodel type fits a body of network data - *based on the structural data alone.* This last point is important. Network equivalence is a purely structural concept.

8. Conventional blockmodeling deals with the issue of multiple relations in a straightforward but simplistic fashion. The relations are simply 'stacked' and measures of (dis)similarity are computed for the vectors made up of rows and/or columns of the relations. If the relations are essentially the same, this approach *may* have value. But if the relations are qualitatively different and have different structures, then the stacking strategy is likely to obliterate those difference. Instead, we need an approach that permits different relations to have different structures with, potentially, different block type structures. Batagelj *et al.* (2006) use a simple example for an organization where one relation is the organizational hierarchy representing 'subordinate to' (which is consistent with regular equivalence and column-regular equivalence) and an advice seeking relation that conforms exactly to a ranked clusters model. Each relation has the same clustering of actors but the blockmodel types for each relation differ. This feature is not recovered by stacking the relations in the conventional blockmodeling approach. This amounts to seeking a general blockmodeling strategy for partitioning three dimensional network data. Put differently, this eighth problem set calls for methods for partitioning of a relational box in the sense of Winship and Mandel (1983) - but without the inclusion of products of relations. Baker (1986) suggested an approach to three dimensional blockmodeling within conventional blockmodeling.
9. For a long time, network analysis had a static character where the disciplinary focus was restricted to delineating network structure. Many tools were created to describe network structure and this 'static phase' may have been necessary for the development of a structural perspective. However, most networks are generated over time by network processes and, recently, more attention has been given to the study of temporal network processes. Doreian and Stokman (1997) outlined an agenda for studying the *evolution* of social networks. There is, indeed, a new focus on studying networks over time at the level of the observed network. The final problem set for generalized blockmodeling stems from the question: Can generalized blockmodeling be used to study the evolution of the fundamental (or underlying) structure of the network? The question arises naturally. If generalized blockmodeling is an appropriate strategy to delineate the underlying structure of a network and if it is correct to think of networks evolving, surely, it is the underlying structure that evolves with the ob-

served network providing the relevant data for studying evolution of the fundamental network. This can be viewed in terms of a relational box and, while this has the same logical structure of the multiple networks relational box, it seems to call for distinctive methods. It may be useful to couple this problem set with the statistical evaluation problem set by specifying generative models for the evolution of fundamental network structure in the form of generalized blockmodel images.

5 Summary

I have provided an overview of generalized blockmodeling and some if its advantages. However, there are some limitations to the approach in its current form. Together the advantages and limitations provide the foundations for proposing some open problems sets. Just as generalized blockmodeling has transformed blockmodeling, the successful pursuit of these open problems sets will transform generalized blockmodeling.

References

BAKER, W. E. (1986): Three-dimensional blockmodeling. *Journal of Mathematical Sociology,12, 191-223.*

BATAGELJ, V., DOREIAN, P. and FERLIGOJ, A. (2006): Three dimensional blockmodeling. International Sunbelt XXVI Social Network Conference, Vancouver, Canada, April 25-30.

BATAGELJ, V., DOREIAN, P. and FERLIGOJ, A. (1992a): An optimization approach to regular equivalence. *Social Networks, 14, 121-135.*

BATAGELJ, V., FERLIGOJ, A. and DOREIAN, P. (1992b): Direct and indirect methods for structural equivalence. *Social Networks, 14, 63-90.*

BATAGELJ, V. and MRVAR, A. (1998): Pajek – Program for large network analysis. *Connections, 21(2), 47-57.* See also http://vlado.fmf.uni-lj.si/pub/networks/pajek for documentation and the most recent version of this program.

BORGATTI, S. P. and EVERETT, M. (1989) The class of regular equivalences: Algebraic structure and computation. *Social Networks, 11, 65-88.*

BREIGER, R. L., BOORMAN, S. A. and ARABIE, P. (1975): An algorithm for clustering relational data with applications to social network analysis and comparison to multidimensional scaling. *Journal of Mathematical Psychology, 12, 328-383.*

BURT, R. S. (1976): Positions in networks. *Social Forces, 55, 93-122.*

DAVIS J. A. and LEINHARDT, S. (1972): The structure of positive interpersonal relations in small groups. In: J. Berger, M. Zelditch Jr and B. Anderson (Eds) *Sociological Theories in Progress, Volume 2.* Houghton Mifflin, Boston, 218-251.

DOREIAN, P.(1993): A measure of standing for citation networks within a wider environment. *Information Processing and Management, 30/1, 21-31.*

DOREIAN, P., BATAGELJ, V. and FERLIGOJ, A. (2005): *Generalized Blockmodeling.* University of Cambridge, New York.

DOREIAN, P., BATAGELJ, V. and FERLIGOJ, A. (2000): Symmetric-Acyclic Decompositions of Networks. *Journal of Classification, 17/1, 3-28.*

DOREIAN, P., BATAGELJ, V. and FERLIGOJ, A. (1994): Partitioning networks based on generalized concepts of equivalence. *Journal of Mathematical Sociology,19, 1-27.*

DOREIAN, P. and MRVAR A. (1996): A partitioning approach to structural balance. *Social Networks, 18, 149-168.*

DOREIAN, P. and STOKMAN, F. N. (Eds.) (1997): *Social Network Evolution* Gordon and Breach, New York.

DOREIAN, P. and WOODARD, K. L. (1994): Defining and locating cores and boundaries of social networks. *Social Networks, 16, 267-293.*

HEIDER, F. (1946): Attitudes and cognitive organization *Journal of Psychology, 21, 107-112.*

HUMMON, N. P. and CARLEY, K (1993) Social networks as normal science *Social Networks, 15, 71-106.*

LAUMANN E. O., MARSDEN, P., V. and PRENSKY, D. (1983): The boundary specification problem in network analysis. In: R. S. Burt and M. J. Minor (Eds) *Applied Network Analysis: A Methodological Introduction* Sage, Beverly Hills, 18-34.

LORRAIN, F. and WHITE, H. C. (1971): Structural equivalence of individuals in social networks. *Journal of Mathematical Sociology, 1, 49-80.*

SAILER, L. D. (1978): Structural equivalence: Meaning and definition, computation and application. *Social Networks, 1, 73-90.*

WHITE, D. R. and REITZ, K. P. (1983): Graph and semigroup homomorphisms on networks of relations. *Social Networks, 5, 193-234*

WINSHIP, C. and MANDEL, M (1983): Roles and positions: A critique and extension of the blockmodeling approach. In S. Leinhardt (Ed.) *Sociological Methodology 1983-4* Jossey-Bass, San Francisco, 314-344.

ŽIBERNA, A. (2005): Generalized blockmodeling of valued networks. University of Ljubljana, Slovenia.

Spectral Clustering and Multidimensional Scaling: A Unified View

François Bavaud

Section d'Informatique et de Méthodes Mathématiques
Faculté des Lettres, Université de Lausanne
CH-1015 Lausanne, Switzerland

Abstract. Spectral clustering is a procedure aimed at partitionning a weighted graph into minimally interacting components. The resulting eigen-structure is determined by a reversible Markov chain, or equivalently by a symmetric transition matrix F. On the other hand, multidimensional scaling procedures (and factorial correspondence analysis in particular) consist in the spectral decomposition of a kernel matrix K. This paper shows how F and K can be related to each other through a linear or even non-linear transformation leaving the eigen-vectors invariant. As illustrated by examples, this circumstance permits to define a transition matrix from a similarity matrix between n objects, to define Euclidean distances between the vertices of a weighted graph, and to elucidate the "flow-induced" nature of spatial auto-covariances.

1 Introduction and main results

Scalar products between features define similarities between objects, and reversible Markov chains define weighted graphs describing a stationary flow. It is natural to expect flows and similarities to be related: somehow, the exchange of flows between objects should enhance their similarity, and transitions should preferentially occur between similar states.

This paper formalizes the above intuition by demonstrating in a general framework that the symmetric matrices K and F possess an identical eigenstructure, where K (kernel, equation (2)) is a measure of similarity, and F (symmetrized transition. equation (5)) is a measure of flows. Diagonalizing K yields principal components analysis (PCA) as well as mutidimensional scaling (MDS), while diagonalizing F yields spectral clustering. By theorems 1, 2 and 3 below, eigenvectors of K and F coincide and their eigenvalues are simply related in a linear or non-linear way.

Eigenstructure-based methods constitute the very foundation of classical multivariate analysis (PCA, MDS, and correspondence analysis). In the last decade, those methods have been very extensively studied in the machine learning community (see e.g. Shawe-Taylor and Cristianini 2004, and references therein), in relationship to manifold learning and spectral clustering (Bengio et al. 2004). The general "$K - F$ connection" described here hence formalizes a theme whose various instances have already been encountered and addressed in the classical setup (see section 2.2) or in the kernel setup, at least implicitly. The relative generality of the present approach (weighted objects, weighted variables, weighted graphs) might provide some guidance for defining the appropriate objects (kernels, scalar products, similarities or affinities, etc.). Also, the same formalism permits to characterize a broad family of *separable auto-covariances*, relevant in spatial statistics.

Multi-dimensional scaling (MDS) in a nutshell: consider n objects described by p features. Data consist of $\Phi = (\varphi_{ij})$ where φ_{ij} is the value of the j-th feature on the i-th object. Let $\rho_j > 0$ denote the weight of feature j, with $\sum_{j=1}^{p} \rho_j = 1$, and define the diagonal matrix $R := \text{diag}(\rho)$. Also, let $\pi_i > 0$ denote the weight of object i, with $\sum_{i=1}^{n} \pi_i = 1$, and define $\Pi := \text{diag}(\pi)$. Also, define

$$B_{ii'} := \sum_j \rho_j \varphi_{ij} \varphi_{i'j} \qquad D_{ii'} := B_{ii} + B_{i'i'} - 2B_{ii'} = \sum_j \rho_j (\varphi_{ij} - \varphi_{i'j})^2 \quad (1)$$

The scalar product $B_{ii'}$ constitutes a measure a *similarity* between objects i and i', while the squared Euclidean distance $D_{ii'}$ is a measure of their *dissimilarity*. Classical MDS consists in obtaining distance-reproducing coordinates such that the (total, weighted) *dispersion* $\Delta := \frac{1}{2} \sum_{ii'} \pi_i \pi_{i'} D_{ii'}$ is optimally represented in a low-dimensional space. To that effect, the coordinate $x_{i\alpha}$ of object i on factor α is obtained from the spectral decomposition of the *kernel* $K = (K_{ii'})$ with $K_{ii'} := \sqrt{\pi_i \pi_{i'}} B_{ii'}$ as follows:

$$K := \sqrt{\Pi} B \sqrt{\Pi} = U \Gamma U' \qquad U = (u_{i\alpha}) \qquad \Gamma = \text{diag}(\gamma) \qquad x_{i\alpha} := \frac{\sqrt{\gamma_\alpha}}{\sqrt{\pi_i}} u_{i\alpha} \quad (2)$$

where U is orthogonal and contains the eigenvectors of K, and Γ is diagonal and contains the eigenvalues $\{\gamma_\alpha\}$ of K. Features are *centred* if $\sum_i \pi_i \varphi_{ij} = 0$. In that case, the symmetric, positive semi-definite (p.s.d) matrices B and K obey $B\pi = 0$ and $K\sqrt{\pi} = 0$, and will be referred to as a *proper similarity matrix*, respectively *proper kernel matrix*. By construction

$$D_{ii'} = \sum_{\alpha \geq 2} (x_{i\alpha} - x_{i'\alpha})^2 \qquad \Delta = \sum_{\alpha \geq 2} \gamma_\alpha \quad (3)$$

where $\gamma_1 = 0$ is the trivial eigenvalue associated with $u_1 = \sqrt{\pi}$.

Spectral clustering in a nutshell: consider the $(n \times n)$ normalised, symmetric *exchange* matrix $E = (e_{ii'})$ where $e_{ii'} = e_{i'i} \geq 0$, $e_{i\bullet} := \sum_{i'} e_{ii'} > 0$, and $\sum_{ii'} e_{ii'} = 1$. By construction, $w_{ii'} := e_{ii'}/e_{i\bullet}$ is the transition matrix of a reversible Markov chain with stationary distribution $\pi_i := e_{i\bullet}$. In a weighted graph framework, $e_{ii'}$ constitutes the weight of the undirected edge (ii'), measuring the proportion of units (people, goods, matter, news...) circulating in (ii'), and π_i is the the weight of the object (vertex) i.

The *minimal normalized cut* problem consists in partitioning the vertices into two disjoints sets A and A^c as little interacting as possible, in the sense that

$$h := \min_A \frac{e(A, A^c)}{\min(\pi(A), \pi(A^c))} \quad (\text{with } e(A, A^c) := \sum_{i\in A, i'\in A^c} e_{ii'} \ , \ \pi(A) := \sum_{i\in A} \pi_i) \tag{4}$$

where the minimum value h is the *Cheeger's constant* of the weighted graph.

The eigenvalues of $W = (w_{ii'})$ are real and satisfy $1 = \lambda_1 \geq \lambda_2 \geq \ldots \lambda_n \geq -1$, with $\lambda_2 < 1$ iff the chain is irreducible and $\lambda_n > -1$ iff the chain is not of period two (bipartite graphs). The same eigenvalues appear in the spectral decomposition of the *symmetrized transition matrix* $F = (f_{ii'})$ defined as $f_{ii'} = e_{ii'}/\sqrt{\pi_i \pi_{i'}}$:

$$F := \Pi^{-\frac{1}{2}} E \Pi^{-\frac{1}{2}} = U \Lambda U' \qquad U = (u_{i\alpha}) \qquad \Lambda = \text{diag}(\lambda) \tag{5}$$

where U is orthogonal and Λ diagonal. By construction, $F\sqrt{\pi} = \sqrt{\pi}$. A symmetric, non-negative matrix F with eigenvalues in $[-1, 1]$ and $F\sqrt{\pi} = \sqrt{\pi}$ will be refereed to as a *proper* symmetrized transition matrix.

In its simplest version, *spectral clustering* (see e.g. Ng et al. (2002); Verma and Meila (2003)) consists in partitioning the graph into two disjoints subsets $A(u) := \{i | u_{i2} \leq u\}$ and $A^c(u) := \{i | u_{i2} > u\}$, where u_{i2} is the second eigenvector and u a threshold, chosen as $u = 0$, or as the value u making $\sum_{i\in A(u)} u_{i2}^2 \cong \sum_{i\in A^c(u)} u_{i2}^2$, or the value minimising $h(u) := e(A(u), A^c(u))/\min(\pi(A(u)), \pi(A^c(u)))$. Minimal normalized cut and spectral clustering are related by the Cheeger inequalities (see e.g. Diaconis and Strook (1991); Chung (1997))

$$2h \geq 1 - \lambda_2 \geq 1 - \sqrt{1 - h^2} \tag{6}$$

where the *spectral gap* $1 - \lambda_2$ controls the speed of the convergence of the Markov dynamics towards equilibrium.

Theorem 1. *(**F** → **K**). Let E be an $(n \times n)$ exchange matrix with associated symmetrized transition matrix $F = U\Lambda U'$ and vertex weight π. Then any $(n \times n)$ matrix $K = (K_{ii'})$ of the form*

$$K := (a - b)F + (a + b)I - 2a\sqrt{\pi}\sqrt{\pi}' \tag{7}$$

constitutes, for $a, b \geq 0$, a centred proper kernel with spectral decomposition $F = U\Gamma U'$ with eigenvalues $\gamma_\alpha = (a - b)\lambda_\alpha + (a + b) - 2a\,\delta_{\alpha 1}$.

Proof : the eigenvectors u_α of I and $\sqrt{\pi}\sqrt{\pi}'$ are identical to those of F, with associated eigenvalues $\mu_\alpha \equiv 1$ and $\mu_\alpha = \delta_{\alpha 1}$ respectively. In particular, $K\sqrt{\pi} = [(a - b) + (a + b) - 2a]\sqrt{\pi} = 0$, making K centred. It remains to show the positive-definiteness of K, that is $\gamma_\alpha \geq 0$. Actually, $\gamma_1 = 0$ and, for $\alpha \geq 2$, $\gamma_\alpha = a(1 + \lambda_\alpha) + b(1 - \lambda_\alpha) \geq 0$ since $-1 < \lambda_\alpha < 1$. □

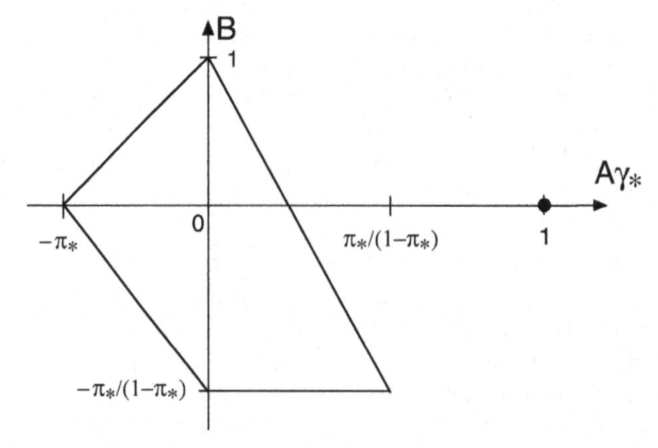

Fig. 1. domain of possible values $A\gamma_*$ and B insuring the existence of a proper symmetrized transition F from a kernel K by (8). Although allowing for non-trivial values $A, B \neq 0$, the domain is not optimal, and degenerates into $A = 0$ and $B \in [0, 1]$ for $n \to \infty$ in view of $\pi_* \to 0$. The point $(1, 0)$ depicts the values corresponding to the FCA example of section (2.2).

Theorem 2. *(**K** → **F**). Let K be an $(n \times n)$ centred kernel with trivial eigenvector $\sqrt{\pi}$. Then any $(n \times n)$ matrix $F = (f_{ii'})$ of the form*

$$F = AK + BI + (1 - B)\sqrt{\pi}\sqrt{\pi}' \tag{8}$$

constitutes, for $A \in [-\frac{\pi_*}{\gamma_*}, \frac{\pi_*}{(1-\pi_*)\gamma_*}]$ and $B \in [-\frac{\pi_*+\min(A,0)\gamma_*}{1-\pi_*}, \frac{\pi_*-|A|\gamma_*}{\pi_*}]$ (where $\gamma_* := \max_\alpha \gamma_\alpha$ and $\pi_* := \min_i \pi_i$), a non-negative symmetrized transition matrix with associated stationary distribution π (see figure 1).

Proof : treating separately the cases $A \geq 0$ and $A \leq 0$, and using (in view of the positive-definite nature of K) $\max_i K_{ii} \leq \gamma_*$, $\min_i K_{ii} \geq 0$, $\max_{i \neq i'} K_{ii'} \leq \gamma_*$ and $\min_{i \neq i'} K_{ii'} \geq -\gamma_*$ as well as $\min_{i,i'} \sqrt{\pi_i \pi_{i'}} = \pi_*$ demonstrates that F as defined in (8) obeys $\min_{i \neq i'} f_{ii'} \geq 0$ and $\min_i f_{ii} \geq 0$. Thus $e_{ii'} := \sqrt{\pi_i \pi_{i'}} f_{ii'}$ is symmetric, non-negative, and satisfies in addition $e_{i\bullet} = \pi_i$ in view of $K\sqrt{\pi} = 0$. □

The coefficients (A, B) of theorem 2 are related to the coefficients (a, b) of theorem 1 by $A = 1/(a - b)$ and $B = (b + a)/(b - a)$, respectively $a = (1 - B)/2A$ and $b = -(1 + B)/2A$. The maximum eigenvalue $\gamma_* := \max_\alpha \gamma_\alpha > 0$ of K is $\gamma_* = a(1 + \lambda_2) + b(1 - \lambda_2) = (\lambda_2 - B)/A$ for $a > b$ (i.e. $A > 0$), and $\gamma_* = a(1 + \lambda_n) + b(1 - \lambda_n) = (\lambda_n - B)/A$ for $a < b$ (i.e. $A < 0$).

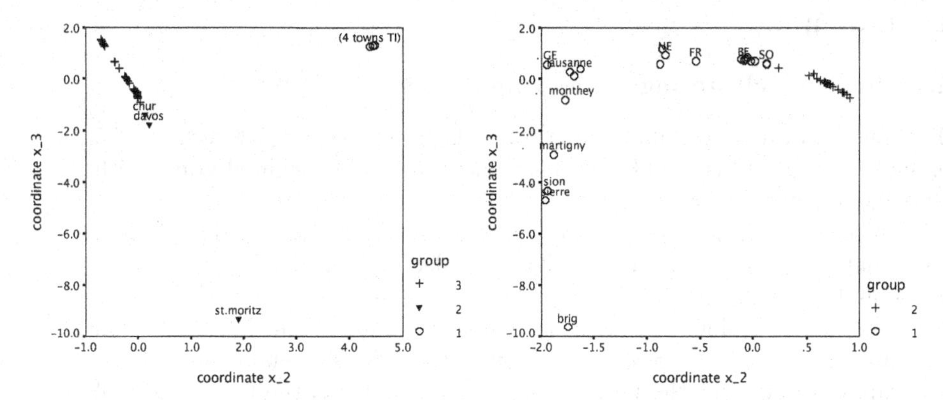

Fig. 2. Two-dimensional factorial towns configuration $x_{i\alpha}$ for $\alpha = 2, 3$ for the initial network ($n = 55$, left) and, for the largest sub-network obtained after four minimal normalized cuts ($n = 48$, right).

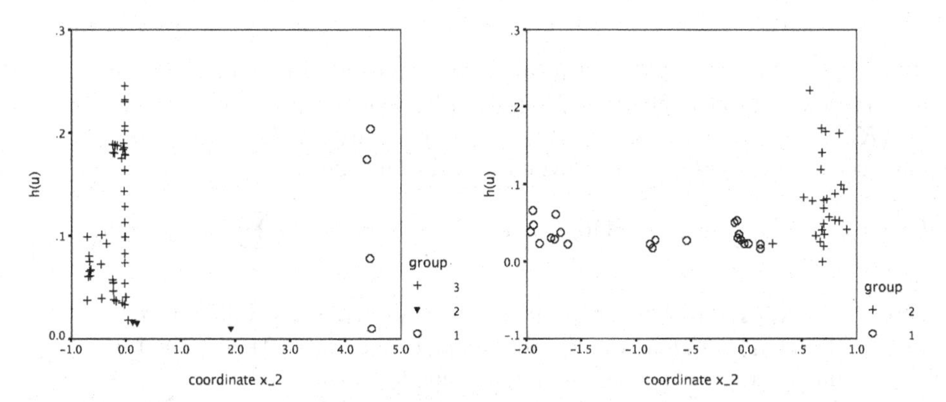

Fig. 3. Determining the minimal normalized cuts $\min_u h(u)$ along the "second-eigenvalue path" with discrete values $u_i = \sqrt{\pi_i}\, x_{i2}$. Left: 55 towns, from which Ticino (4 towns, 1st iteration) and Graubünden (3 towns, 2nd and 3rd iteration) are removed. Right: the resulting 48 towns, split into (VS-VD-GE) and (NE-FR-JU-BE-SO) for the first group, and the rest of the German-speaking towns for the second group.

2 Examples

2.1 Spectral clustering: Swiss commuters

The number of daily commuters n_{ij} from place i to place j (between $n = 55$ extended Swiss towns) yields (after symmetrization) a weighted graph with associated transition matrix F.

Eigenvalues are $\lambda_1 = 1 > \lambda_2 = .9947 > \ldots > \lambda_{55} = .5116$. Factor coordinates $x_{i\alpha}$ (figure 2) define "flow-revealed" distances $D_{ii'}$. In view of theorem 1 (and in view of the arbitrariness of $\gamma(\lambda)$, and of the closeness between the eigenvalues λ_α) the coordinates are simply defined (see equation (2)) as $x_{i\alpha} = u_{i\alpha}/\sqrt{\pi_i} = u_{i\alpha}/u_{i1}$. They are obviously reminiscent of the geographical map, but the precise mechanism producing the factor maps of figure 2 remains to be elucidated. The spectral clustering determination of the threshold u minimizing $h(u)$ (section 1) is illustrated in figure 3.

2.2 Correspondence analysis: educational levels in the region of Lausanne

Let $N = (n_{ij})$ be a $(n \times m)$ contingency table counting the number of individuals belonging to category i of X and j of Y. The "natural" kernel matrix $K = (K_{ii'})$ and transition matrix $W = (w_{ii'})$ associated with factorial correspondence analysis (FCA) are (Bavaud and Xanthos 2005)

$$K_{ii'} = \sqrt{\pi_i}\sqrt{\pi_{i'}} \sum_j \rho_j (q_{ij} - 1)(q_{i'j} - 1) \qquad\qquad w_{ii'} := \pi_{i'} \sum_j \rho_j q_{ij} q_{i'j} \quad (9)$$

where $\pi_i = n_{i\bullet}/n_{\bullet\bullet}$ are the row profiles, $\rho_j = n_{\bullet j}/n_{\bullet\bullet}$ the columns profiles, and $q_{ij} = (n_{ij}\, n_{\bullet\bullet})/(n_{i\bullet} n_{\bullet j})$ are the *independence quotients*, that is the ratio of the counts by their expected value under independence.

Coordinates $x_{i\alpha}$ (2) obtained from the spectral decomposition of K are the usual objects' coordinates in FCA (for $\alpha \geq 2$), with associated χ-square dissimilarities $D_{ii'}$ and χ-square inertia $\Delta = \mathtt{chi}^2/n_{\bullet\bullet}$ (Bavaud 2004). On the other hand, $w_{ii'}$ is the conditional probability of drawing an object of category i' starting with an object of category i and "transiting" over all possible modalities j of Y. The resulting Markov chain on n states is reversible with stationary distribution π, exchange matrix $e_{ii'} = e_{i'i} = \pi_i w_{ii'}$ and symmetrized transition matrix $f_{ii'} = \sqrt{\pi_i}\sqrt{\pi_{i'}} \sum_j \rho_j q_{ij} q_{i'j}$.

Here K and F are related as $K = F - \sqrt{\pi}\sqrt{\pi}'$, with values $A = 1$ and $B = 0$ (respectively $a = -b = 1/2$) and $\gamma_* = 1$ in theorems 2 and 1. The corresponding value lie outside the non-optimal domain of figure 1.

Data[1] give the number of achieved educational levels i (8 categories) among 169′836 inhabitants living in commune j (among $p = 12$ communes around Lausanne, Switzerland). Eigenvalues are $\gamma_1 = 0$ and $1 > \gamma_2 = \lambda_2 =$

[1] F.Micheloud, private communication

$.023 > \ldots \lambda_8 = .000026$ with inertia $\Delta = .031$. While significantly non-zero ($n_{\bullet\bullet}\Delta >> \chi^2_{.99}[77]$), those low values are close to the perfect mobility case (section 4), that is regional educational disparities are small in relative terms. Figure 4 depicts the factor configuration ($\alpha = 2.3$) with coordinates (2) as well as dual regional coordinates. The biplot confirms the existence of the well-attested West-East educational gradient of the region.

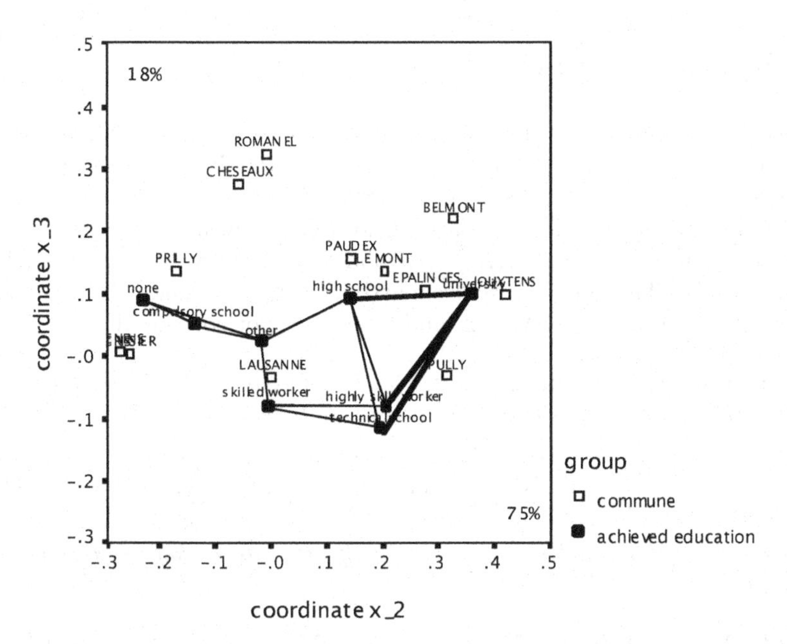

Fig. 4. Biplot: FCA rows and columns objects' coordinates. The symmetric quantity $s_{ii'} := w_{ii'}/\pi_{i'}$ is a size-independent measure of similarity with average 1 (Bavaud and Xanthos 2005), defining strong ($s \geq 1.05$), weak ($1.05 > s \geq 1$) or no ($s < 1$) links between distinct education levels.

3 Non-linear transformations

Theorem 3. *i) Let K be a proper kernel. Then K^r (for $r = 1, 2, \ldots$) and $h(K) := \sum_{r\geq 1} h_r K^r$ (where $h_r \geq 0$ and $\sum_{r\geq 1} h_r = 1$) are proper kernels.*

ii) Let F be a proper symmetrized transition. Then F^r (for $r = 0, 1, 2, \ldots$), $f(F) := \sum_{r\geq 1} f_r F^r$ (where $f_r \geq 0$ and $\sum_{r\geq 1} f_r = 1$) and $cf(F) + (1-c)I$ (where $0 < c \leq 1$) are proper symmetrized transitions.

iii) K and F can be put in non-linear correspondence by

$$h(K) = (a - \tilde{b})f(F) + (a + \tilde{b})I - 2a\sqrt{\pi}\sqrt{\pi}' \qquad a, \tilde{b} \geq 0 \qquad (10)$$

Proof : i) and ii) are immediate. Part iii) follows from theorem (1) and definition $\tilde{b} := (1-c)a + cb$. □

Since $h(U\Gamma U') = Uh(\Gamma)U'$ and $f(U\Lambda U') = Uf(\Lambda)U'$, theorem 3 exhibits a broad class of MDS - spectral clustering correspondences (see the examples of section 4), differing by their eigenvalues spectrum but sharing the same eigenvectors, in particular u_1 and hence the weights vector $\pi = u_1^2$.

4 Separable auto-covariances

The present formalism turns out to be relevant in spatial statistics, where spatial autocorrelation is defined by a covariance matrix between the objects (= regions).

To that extent, consider a spatial field $\{X_i\}_{i=1}^n$ measured on n regions, with common expectation $E(X_i) = \mu$ and associated weights $\{\pi_i\}_{i=1}^n$. Let $\bar{X} := \sum_i \pi_i X_i$. The auto-covariance matrix $\Sigma = (\sigma_{ii'})$ is said to be *separable* if, for any i, the variables $X_i - \bar{X}$ and $\bar{X} - \mu$ are not correlated.

Theorem 4. *Σ is separable iff $\Sigma\pi = \sigma^2\mathbf{1}$, where $\sigma^2 = E((\bar{X}-\mu)^2)$ and $\mathbf{1}$ is the unit vector. In this case, the $(n \times n)$ matrices*

$$K := \frac{1}{\sigma^2}\sqrt{\Pi}\Sigma\sqrt{\Pi} - \sqrt{\pi}\sqrt{\pi}' \qquad\qquad B = \frac{1}{\sigma^2}\Sigma - J \tag{11}$$

(where $J := \mathbf{1}\mathbf{1}'$ is the unit matrix) constitute a proper kernel, respectively dissimilarity.

Proof : $\Sigma\pi = \sigma^2\mathbf{1}$ iff $\sigma^2 = \sum_{i'} \pi_{i'}[E((X_i-\mu)(X_{i'}-\bar{X})) + E((X_i-\mu)(\bar{X}-\mu))] = E((X_i-\bar{X})(\bar{X}-\mu)) + E((\bar{X}-\mu)(\bar{X}-\mu))$ iff $E((X_i-\bar{X})(\bar{X}-\mu)) = 0$ and $E((\bar{X}-\mu)^2) = \sigma^2$. □

Under separability, equations (1) and (11) show the *variogram* of Geostatistics to constitute a squared Euclidean distance since $\mathrm{Var}(X_i - X_{i'}) = \sigma^2 D_{ii'}$. Observe that Σ or B as related by (11) yield (up to σ^2) the same distances. Together, theorem 3 (with $h(x) = x$) and theorem 4 imply the following

Theorem 5. *Let $f(F)$ the function defined in theorem 3 and $a, \tilde{b} \geq 0$. Then the $(n \times n)$ matrix*

$$\frac{1}{\sigma^2}\Sigma := (a-\tilde{b})\Pi^{-\frac{1}{2}}f(F)\Pi^{-\frac{1}{2}} + (a+\tilde{b})\Pi^{-1} + (1-2a)J \tag{12}$$

constitutes a separable auto-covariance.

Theorem 5 defines a broad class of "flow-induced" spatial models, among which (deriving the relations between parameters is elementary):

- the auto-regressive model $\Sigma = \sigma^2(1-\rho)(I-\rho W)^{-1}\Pi^{-1}$
- *equi-correlated* covariances $\sigma^{-2}\Sigma^2 = \tilde{a}\Pi^{-1} + \tilde{c}J$, with associated geostatistical distances $D_{ii'} = \tilde{a}(1/\pi_i + 1/\pi_{i'})$ for $i \neq i'$. This occurs under contrasted limit flows, namely (A) *perfect mobility flows* $w_{ii'} = \pi_{i'}$ (yielding $f(F) = F = \sqrt{\pi}\sqrt{\pi}'$) and (B) *frozen flows* $w_{ii'} = \delta_{ii'}$ (yielding $f(F) = F = I$).

Irrespectively of the function f, any auto-covariance Σ defined in theorem 5 must be separable, a testable fact for a given empirical Σ. Also, the factorial configuration of the set of vertices in a weighted graph or of states in a reversible chain can be obtained by MDS on the associated geostatistical distances $D_{ii'}$. As demonstrated by theorem 3, all those configurations are identical up to dilatations of the factorial axes; in particular, the low-dimensional plot $\alpha = 2, 3$ is invariant up to dilatations, provided f is increasing.

References

BAVAUD, F. (2004): Generalized factor analyses for contingency tables. In: D.Banks et al. (Eds.): *Classification, Clustering and Data Mining Applications.* Springer, Berlin, 597-606.

BAVAUD, F. and XANTHOS, A. (2005): Markov associativities. *Journal of Quantitative Linguistics, 12, 123-137.*

BENGIO, Y., DELALLEAU, O., LE ROUX, N., PAIEMENT, J.-F. and OUIMET, M. (2004): Learning eigenfunctions links spectral embedding and kernel PCA. *Neural Computation, 16, 2197-2219.*

CHUNG, F. (1997): *Spectral graph theory.* CBMS Regional Conference Series in Mathematics 92. American Mathematical Society. Providence.

DIACONIS, P. and STROOK, D. (1991): Geometric bounds for eigenvalues of Markov chains. *Ann. Appl. Probab., 1, 36-61.*

NG, A., JORDAN, M. and WEISS, Y. (2002): On spectral clustering: Analysis and an algorithm. In T. G. Dietterich et al. (Eds.): Advances in Neural Information Processing Systems 14. MIT Press, 2002.

SHAWE-TAYLOR, J. and CRISTIANINI, N. (2004): Kernel Methods for Pattern Analysis. Cambridge University Press.

VERMA, D. and MEILA, M. (2003): A comparison of spectral clustering algorithms. UW CSE Technical report 03-05-01.

Analyzing the Structure of U.S. Patents Network

Vladimir Batagelj[1], Nataša Kejžar[2], Simona Korenjak-Černe[3], and Matjaž Zaveršnik[1]

[1] Department of Mathematics, FMF, University of Ljubljana,
Jadranska 19, SI-1000 Ljubljana, Slovenia
[2] Faculty of Social Sciences, University of Ljubljana,
Kardeljeva pl. 5, SI-1000 Ljubljana, Slovenia
[3] Faculty of Economics, EF, University of Ljubljana
Kardeljeva pl. 17, SI-1000 Ljubljana, Slovenia

Abstract. The U.S. patents network is a network of almost 3.8 millions patents (network vertices) from the year 1963 to 1999 (Hall et al. (2001)) and more than 16.5 millions citations (network arcs). It is an example of a very large citation network.

We analyzed the U.S. patents network with the tools of network analysis in order to get insight into the structure of the network as an initial step to the study of innovations and technical changes based on patents citation network data.

In our approach the SPC (Search Path Count) weights, proposed by Hummon and Doreian (1989), for vertices and arcs are calculated first. Based on these weights vertex and line islands (Batagelj and Zaveršnik (2004)) are determined to identify the main themes of U.S. patents network. All analyses were done with `Pajek` – a program for analysis and visualization of large networks. As a result of the analysis the obtained main U.S. patents topics are presented.

1 Introduction

Patents are a very good source of data for studying the innovation development and technical change because each patent contains information on innovation, inventors, technical area, assignee etc. Patent data also include citations to previous patents and to scientific literature, which offer the possibility to study linkages between inventions and inventors. On the other hand we have to be aware of the limitations when using such datasets, since not all inventions are patented, the patent data are not entirely computerized, and that it is hard to handle very large datasets.

The database on U.S. patents (Hall et al. (2001)) was developed between 1975 and 1999. It includes U.S. patents granted between January 1963 and December 1999. It counts 2,923,922 patents with text descriptions and other 850,846 patents represented with scanned pictures, altogether 3,774,768 patents. There are 16,522,438 citations between them. Since it is a legal duty for the assignee to disclose the existing knowledge, a citation represents previously existing knowledge contained in the patent.

The idea of using patent data for economic research originated from Schmookler (1966), Scherer (1982), and Griliches (1984). Hall et al. (2001) included more information about patents in the analyses and also demonstrated the usefulness of citations.

The idea of our work was to look at the patents data as a large network. In the network patents are represented by vertices. Two patents (vertices) are linked with a directed link, an *arc*, when one cites the other one. We used the SPC method to obtain the weights of patents and their citations. Weight of a particular patent or particular citation can be interpreted as a relative importance of that patent or that citation in the network. We used weights to determine islands – groups of 'closely related' vertices.

Hall, Jaffe, and Trajtenberg aggregated more than 400 USPTO (United States Patent and Trademark Office) patent classes into 36 2-digit technological subcategories, and these are further aggregated into 6 main categories: Chemical, Computers and Communications, Drugs and Medical, Electrical and Electronics, Mechanical, and Others. We examined the constructed variable of technological subcategory and checked the titles of patents in order to confirm our hypothesis, that islands determine specific theme of patents.

2 Search path count method

Let us denote a network by $N = (V, L)$, where V is a set of *vertices* and L is a set of *arcs*. The arc (v, u) goes from vertex $v \in V$ to vertex $u \in V$ iff the patent represented by v cites the patent represented by u. This network is a *citation network*. Citation networks are usually (almost) *acyclic*. The cycles, if they exist, are short. Network can be converted to acyclic one by using different transformations – for example, by simply shrinking the cycles. Hummon and Doreian proposed in 1989 three arc weights to operationalize the importance of arcs in citation networks: (1) node pair projection count method, (2) search path link count method, and (3) search path node pair method.

Batagelj (1991, 2003) showed that the use of SPC (Search Path Count) method computes efficiently, in time $O(|L|)$, the last two (2) and (3) of Hummon and Doreian's weights. The SPC method assumes that the network is acyclic. In an acyclic network there is at least one *entry* – a vertex of indegree 0, and at least one *exit* – a vertex of outdegree 0. Let us denote with I and O the sets of all entries and all exits, respectively. The SPC algorithm assigns to each vertex $v \in V$ as its value the number of different I-O-paths passing through the vertex v; and similarly, to each arc $(v, u) \in L$ as its weight the number of different I-O-paths passing through the arc (v, u). These counts are usually normalized by dividing them by the number of all I-O-paths.

We calculated normalized weights of edges and vertices for the U.S. patents network using the SPC method in `Pajek`. The number of all paths through the network is 1,297,400,940,682. We multiplied the weights with one million since the normalized values of most of the weights were very small.

3 Determining islands

The following table

size	1 & 2	3 & 4	5 & 6	7 & 8	9 & 10	11 & 12	13 & 14	15 & 16	19	3,764,117
number	2830	583	276	72	35	12	6	2	1	1

shows the (not fully detailed) distribution of the size of *weak components*. A weak component is a subnetwork of vertices that are connected when disregarding the arcs direction. There exist several small weak components and one huge one (3,764,117 vertices). This implies that most of the patents are somehow connected to almost all other patents. Patents in small weak components might be the early ones (granted before the year 1975), which would be densely connected if the citation data were available or there might exist patents that started a very specific topic which is indeed just locally connected.

Based on the calculated weights more informative connected subnetworks can be also determined. For this purpose we used line and vertex islands. Islands (Batagelj and Zaveršnik (2004), Zaveršnik (2003)) are connected subnetworks (groups of vertices) that locally dominate according to the values of vertices or lines.

Let $N = (V, L, p)$ be a network with vertex property $p : V \to \mathbb{R}$. Nonempty subset of vertices $C \subseteq V$ is called a *vertex island* of network N if the corresponding induced subnetwork is connected and the weights of the neighboring vertices $N(C)$ are smaller or equal to the weights of vertices from C

$$\max_{u \in N(C)} p(u) \leq \min_{v \in C} p(v).$$

The line islands are defined similarly. Let $N = (V, L, w)$ be a network with line weight $w : L \to \mathbb{R}$. Nonempty subset of vertices $C \subseteq V$ is called a *line island* of network N if there exists a spanning tree T in the corresponding induced subnetwork, such that the lowest line of T has larger or equal weight than the largest weight of lines from C to the neighboring vertices

$$\max_{(u,v) \in L, u \notin C, v \in C} w(u, v) \leq \min_{e \in L(T)} w(u, v).$$

Let us look at values $p(v)$ of vertices as *heights of vertices*. The network can be seen as some kind of a *landscape*, where the vertex with the largest value is the highest peak. Eliminating the vertices (and the corresponding lines) with height lower than t, we obtain a group of internally connected subnetworks – islands called a *vertex cut* at *cutting level* t. Unfortunately this does not give a satisfying result. We are usually interested in subnetworks with specific number of vertices – not smaller than k and not larger than K – trying to embrace single theme clusters. To identify such islands we have to determine vertex cuts at all possible levels and select only those islands of the selected size. Batagelj and Zaveršnik (2004) developed an efficient algorithm for determining such islands. It is implemented in Pajek.

We determined vertex islands of sizes $[1, 300]$. When determining line islands of sizes $[2, 300]$ we excluded all 'weak, submerged' lines (lines in the line island with weights lower than the largest value of the line linking island to the rest of network). We obtained 24,921 vertex islands on 36,494 vertices and 169,140 line islands on 424,191 vertices.

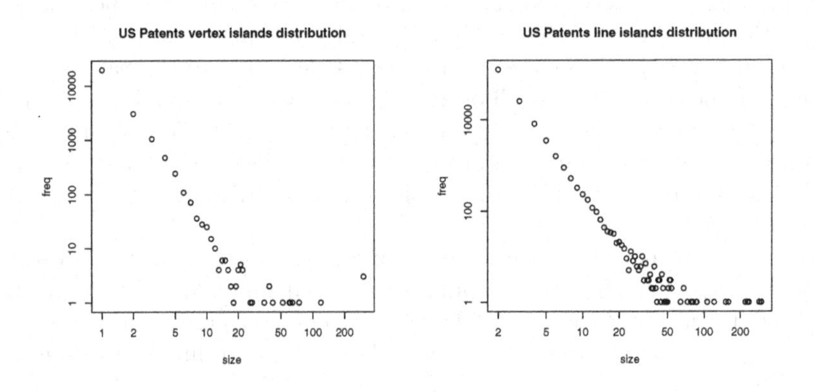

Fig. 1. Distributions of the islands based on their size.

Figure 1 shows the size distribution of islands (vertex and line islands respectively). The x axis represents the size of islands and the y axis represents the frequency of islands. It can be seen that the relation in the log-log scale is almost linear. With the increase of the size of islands, their frequency decreases with power-law.

4 Line islands

In this section some of the most interesting line islands will be presented. We chose them with respect to their size and to the size of the weights on the lines. We scanned through all the islands that have at least 21 vertices and islands that are smaller but with the minimal line weight 10. There are 231 such islands. We were interested in the technological subcategory, title of the patent and grant year for each vertex (each patent) of the island.

Titles of the patents were obtained from the website of The United States Patent and Trademark Office. We automatized their extraction using statistical system R and its package XML.

We found out that the patents in an individual island are dealing with the same (or very similar) topic which is very likely in the same technological subcategory. We noticed that islands with smaller weights are mainly of categories Others (category code 6) and Chemical (code 1). With increase in the minimal island weight, categories first change in favor of Drugs and Medical category (code 3), then Electrical and Electronics (code 4) and last to Computers and Communications (code 2). Interestingly Mechanical category was not noticed throughout the scan.

The largest island is surprisingly mostly of category Chemical. It has exactly 300 vertices and its minimal weight is the highest (3332.08). The patents are mostly from the category code 19 (Miscellaneous-Chemical). 254 vertices or approximately 84.7% are from this category. The second largest category code is 14 (Organic Compounds), which counts 27 vertices or 9%. When examining the titles of the patents we found out that this island is about *liquid crystals*, that could be used for computer displays. This somehow connects the theme to the category Computers and Communications and makes the theme in the largest island less surprising.

The second largest island (298 vertices) is even more homogenous in topic than the first one. Its theme is about *manufacturing transistors and semiconductor devices*, which is classified in category code 46 (Semiconductor Devices). There are 270 (approx. 90.6 %) vertices in even more specific classification group code 438 (USPTO 1999 classification class Semiconductor device manufacturing process).

We also observed small size islands with large minimal weights. The topic of 5 islands within 31 islands with the largest minimal weights deals with the *internal combustion engine for vehicles and its fuel evaporation system.* This very likely implies that there is a huge island (more than 300 vertices) about this theme, but due to our maximum island size restriction (300 vertices in the island) there are only its peaks (*subislands*) captured in our result. We verified this hypothesis by determining line islands of a larger size. When calculating islands of size $[2, 1000]$ there were 2 islands of 1000 vertices with the largest minimal weights. The theme of one of them is about internal combustion engines (for details see Kejžar 2005). It contains the small islands captured with the initial calculation of the line islands. This shows that this theme is much broader than most of other themes and hence it was not possible to embrace it completely with an island of maximum 300 vertices.

Figure 2 shows an island with 229 vertices and the seventh largest minimal weight, that has 3 strong theme branches. Through the years patents were granted (the oldest patents are at the bottom of the network) these 3 different topics became connected. In the Figure 3 the title of every patent in the first branch is shown separately. We can see that the topic of the longest branch is about *television market research with video on demand.* The second branch is about the *identity verification apparatus*, and the third about the *computer security system.* The three branches are thematically not far away, so the

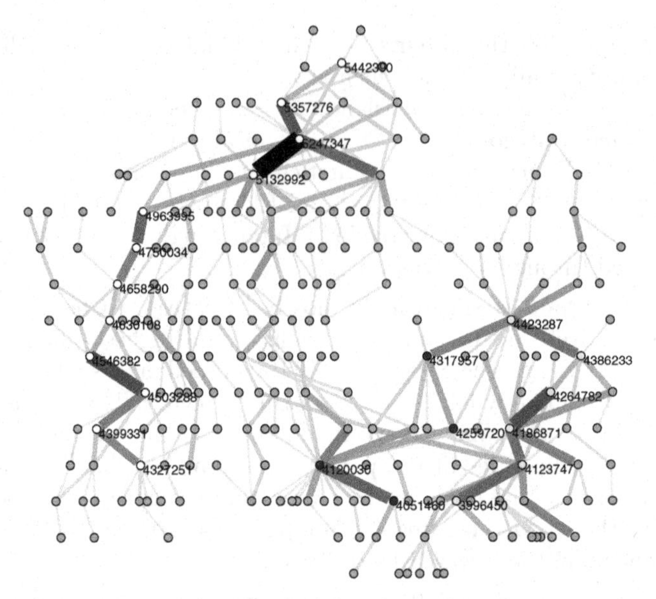

Fig. 2. An island with 3 theme branches.

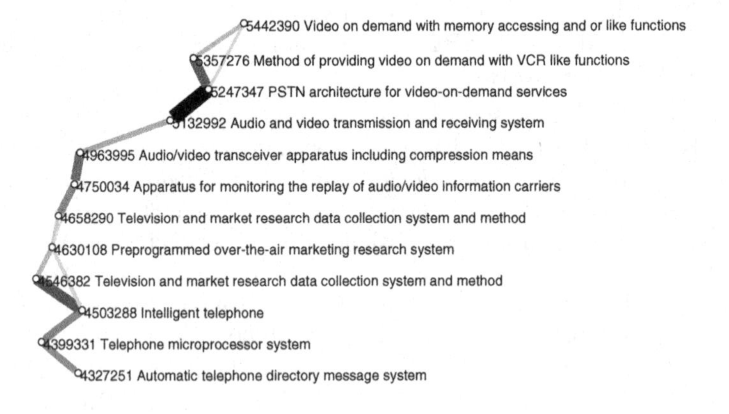

Fig. 3. First branch of the island.

findings of a patent that mainly belongs to one of them use (or are used) also in the other two branches. This shows in the island the connections among branches that are strong enough to merge the three branches together in one (a bit more diverse) island.

5 Vertex islands

Some of the most interesting vertex islands were obtained by restricting the minimal vertex island's size to 10, or the minimal vertex island's size to 5 and weights larger than 10. There are 123 such islands on 2,971 vertices.

Three of them have size 300. The one of them with the largest weights includes patents mostly from the category Chemical. The main theme in this island is the *liquid crystals* which is also the main theme in the main line island.

The next largest vertex island beside the largest three is the island on 119 patents. It is very homogenous – all patents belong to the Electrical and Electronics subcategory Semiconductor Devices (code 46), and all patents are classified into USPTO Patent Class 438.

Large and powerful vertex islands show a very similar structure of themes as in line islands. This is not surprising since weights (from the SPC method) on lines and neighboring vertices are highly correlated. It can be noticed that significantly less vertex islands than line islands are obtained when the same size range is considered.

There are also some small vertex islands with very high weights. Some of them are presented in the Table 1. The meaning of the codes for technical subcategories can be obtained from Hall, Jaffe and Trajtenberg article about the patents data.

Table 1. Some of the small vertex islands with the highest weights (minw > 200)

Island No.	Size	Weights minw maxw	Subcategory	Theme
24900	5	1018.58 1846.06	53	controlling an ignition timing for an internal combustion engine
24878	5	632.32 1039.10	46 19	fabricating monocrystalline semiconductor layer on insulating layer by laser crystallization
24874	8	590.82 1043.82	24	multiprocessor cache coherence system
24811	10	357.48 901.61	22 21, 49	area navigation system including a map display unit
24806	10	343.46 562.06	22	programmable sequence generator for in-circuit digital testing
24797	10	322.46 966.49	53	valve timing control system for engine
24796	10	318.69 1818.92	24 12, 19	track transverse detection signal generating circuit

6 Conclusion

An approach to determine main themes in large citation networks is presented, which can be viewed as a kind of network clustering. A very large network of U.S. patents was used as an example. We used the SPC (Search Path Count) method to get vertex and line weights. Vertices and lines with higher weights represent more important patents and citations in the network. We used them to determine vertex and line islands of the network. Islands are non overlapping connected subsets of vertices. Due to the citation links between the vertices, vertices have similar characteristics (similar topics in our case). Therefore islands can be viewed as thematic clusters.

The characteristics of patents in more than 300 islands were examined. The islands that were examined were selected by their size and minimal line or vertex weight. The results confirmed the hypothesis that an island consists of vertices with similar features (in our case themes). Due to the limited space in this paper we could only present the most important and the most interesting vertex and line islands. There are some differences between the vertex and the line islands, but the main themes and the main islands remain roughly the same.

References

ALBERT, R. and BARABÁSI, A.L. (2002): Statistical Mechanics of Complex Networks. *Reviews of Modern Physics, 74, 47* `http://arxiv.org/abs/cond-mat/0106096`

BATAGELJ, V. (2003): Efficient Algorithms for Citation Network Analysis. `http://arxiv.org/abs/cs.DL/0309023`

BATAGELJ, V. and FERLIGOJ, A.(2003): Analiza omrežij. (Lectures on Network analysis.): `http://vlado.fmf.uni-lj.si/vlado/podstat/AO.htm`

BATAGELJ, V. and MRVAR, A.: `Pajek`. Home page: `http://vlado.fmf.uni-lj.si/pub/networks/pajek/`

BATAGELJ, V. and MRVAR, A. (2003): Pajek – Analysis and Visualization of Large Networks. In: Jünger, M., Mutzel, P., (Eds.): *Graph Drawing Software.* Springer, Berlin, 77-103.

BATAGELJ, V. and ZAVERŠNIK, M.: Islands – identifying themes in large networks. Presented at Sunbelt XXIV Conference, Portorož, May 2004.

HUMMON, N.P. and DOREIAN, P. (1989): Connectivity in a Citation Network: The Development of DNA Theory. *Social Networks, 11, 39-63.*

HALL, B.H., JAFFE, A.B. and TRAJTENBERG, M. (2001): The NBER U.S. Patent Citations Data File. NBER Working Paper 8498. `http://www.nber.org/patents/`

KEJŽAR, N. (2005): Analysis of U.S. Patents Network: Development of Patents over Time. *Metodološki zvezki, 2, 2,195-208.* `http://mrvar.fdv.uni-lj.si/pub/mz/mz2.1/kejzar.pdf`

ZAVERŠNIK, M. (2003): Razčlembe omrežij. (Network decompositions). PhD. Thesis, FMF, University of Ljubljana.

The United States Patent and Trademark Office. `http://patft.uspto.gov/netahtml/srchnum.htm`

The R Project for Statistical Computing. Home page: `http://www.r-project.org/`

Identifying and Classifying Social Groups: A Machine Learning Approach

Matteo Roffilli and Alessandro Lomi

University of Bologna (Italy)

Abstract. The identification of social groups remains one of the main analytical themes in the analysis of social networks and, in more general terms, in the study of social organization. Traditional network approaches to group identification encounter a variety of problems when the data to be analyzed involve two-mode networks, i.e., relations between two distinct sets of objects with no reflexive relation allowed within each set. In this paper we propose a relatively novel approach to the recognition and identification of social groups in data generated by network-based processes in the context of two-mode networks. Our approach is based on a family of learning algorithms called Support Vector Machines (SVM). The analytical framework provided by SVM provides a flexible statistical environment to solve classification tasks, and to reframe regression and density estimation problems. We explore the relative merits of our approach to the analysis of social networks in the context of the well known "Southern women" (SW) data set collected by Davis Gardner and Gardner. We compare our results with those that have been produced by different analytical approaches. We show that our method, which acts as a data-independent preprocessing step, is able to reduce the complexity of the clustering problem enabling the application of simpler configurations of common algorithms.

1 Introduction

A variety of phenomena of interest to students of organization and social networks involve the collection and analysis of two-mode network data that give rise to rectangular arrays recording the association between column elements - for example, individuals - and row elements - for example, events. Statistical methods for the simultaneous representation of the elements of a two mode networks - such as correspondence analysis - have been used for many years to analyze simple two-way and multi-way tables containing some measure of correspondence between the rows and columns (Greenacre (1984)). In practice however, correspondence analysis is not well suited to the analysis of discrete network data and its results are frequently difficult to interpret in the context of social networks (Borgatti and Everett(1997)). A more direct and theoretically inspired approach to the analysis of two-mode networks is based on the observation that they embed information on

[1] The present research is part of a broader project on "Models for representing organizational knowledge." The project receives generous financial support from MIUR (The Italian Ministry of University and Scientific Research) through the FIRB research funding scheme (grant code number RBNE03A9A7_005).

two one-mode networks: a person-by-person network and an event-by-event network (Breiger (1974)). These networks can then be analyzed separately to explore dual patterns of association of events through persons, and of persons through events, an insight that is also central to the interpretation of lattice models (Freeman (2003); Freeman and White (1994); Pattison and Breiger (2002)). Until recently, the connection between the duality revealed by patterns of association between actors and events and core network-based notions of role, positions and role-structures had not been clearly articulated (White, Boorman and Breiger (1976); Boorman and White (1976)). Doreian, Batagelj and Ferligoj (2004) moved a number important steps in this direction by extending generalized blockmodeling techniques (Doreian, Batagelj and Ferligoj (2005)) to the analysis of two-mode network data. The unique advantages of the approach proposed by Doreian, Batagelj and Ferligoj (2004) reside in its ability to analyze two-mode data directly and on its reliance on an explicit optimization approach. One potential drawback of this approach is the need of an a priori definition of the number of blocks that are being sought and their type. Optimization-based blockmodeling techniques try to find the best value of the parameters of a prefixed model by minimizing a loss measure which depends on given data. Although this leads to results that are clearly interpretable, we note that this approach to blockmodeling is perhaps better suited for pattern recognition tasks - i.e. finding expected patterns inside data - rather than for pattern discovery tasks - i.e. the ability to discover unknown and unpredictable patterns. Because the discovery of social groups is one of the defining problems of modern social network analysis (Freeman (2003)), we feel that there is at least some value in thinking about approaches that allow pattern discovery. To improve on purely exploratory approaches to blockmodeling, we think that search for structural patterns in social network data should be conducted as much as possible in the context of a rigorous statistical framework. In this paper we elaborate on these considerations and present a novel approach to the identification of groups in two-mode network data. The approach is based on the theory of Support Vector Machines, a family of learning algorithms based on recent advancements in Statistical Learning Theory (Vapnik (1995)). It shares a number of features with the approach suggested by Doreian, Batagelj and Ferligoj (2004). For example, both approaches are based explicitly on optimization procedures, and they both adopt a penalty cost concept. Finally, both approaches can be applied directly to the analysis of two-mode network data and extended to the analysis of one mode-networks (provided that one is willing to interpret the column of an adjacency matrix as features of the rows) We reframe the problem of identifying social groups as a problem of finding classes and illustrate the performance of the method that we propose on the well known "Southern Women" data set collected in the 1930s by ethnographers Allison Davis, Elizabeth Stubbs, Davis, Burleigh B. Gardner, Mary R. Gardner and J. G. St. Clair Drake (Henceforth DGG). They collected data on social stratification in Natchez, Mississippi and produced a comparative study of social class in black and in white society (Davis et al (1941)). The part of this work that is of direct relevance to our study concerns the data that the original research team collected on the participation of a small group of 18 women in 14 distinct social events. This data has since become a sort of test bank for exploring the performance of various analytical approaches to the problem of group identification in social networks (Freeman (2003)). As Freeman recently noted, the Southern Women data set is useful because is it contains a relatively simple structural pattern. In Freeman's own words (): "According to DGG, the women seemed to organize themselves into two more or less distinct groups. Moreover, they reported that the positions-core and peripheral of the members of these groups could also be determined in terms of the ways in which different women had been involved in group activities." Hence a careful reanalysis of this data holds promise to yield results that may be compared with those obtained by more conventional scaling or network-analytic methods. A second advantage of using the Southern Women (SW) data set to illustrate our approach to group identification is that the results can be tested against the intuition that the members of the original research team have developed based on their extensive ethnographic work. We use the well known SW data set to establish the validity of our modeling approach and to test the consistency of its results with those produced by more established network analytical methods. In the next section we provide a non technical introduction to SVM followed by a brief outline of its recent history. In section 2 we introduce the problem of learning in the context of a binary classification problem and we discuss the problem of novelty detection - an issue that relates directly to the problem of mutual constitution of space and classes that we have mentioned in this introduction. In section 3 we illustrate the value of our methodological proposal in a reanalysis of the Southern Women data set. We conclude the paper with a short discussion of the main findings and with an outline of some directions for future research.

2 Support Vector Machines

The Support Vector Machine is a new powerful learning algorithm based on recent advances in Statistical Learning Theory (SLT) also known as Vapnik-Chervonenkis theory (Vapnik (1995)). SVM offers versatile tools to solve classification, regression and density estimation problems. Given a binary (e.g. {0,1} or {*positive,negative*}) labeling y_i of l objects, the task of finding a way to separate them into the two classes is called *learning*. Exploiting the learned model, the algorithm is able to predict the labels of unseen and unlabeled objects. A common way in the machine learning community to represent an object makes use of a collection of n real-valued characteristics, called *features*, that exploit particular properties of interest. In doing so, each object $\vec{x}$ behaves as a point in an n-dimensional space, called *input space*. In this scenario, the aim of the learning phase of SVM is to place a hyperplane in such way that all positive objects will be placed on its positive side and all negative objects on the negative one. In the testing phase, an unknown object $\vec{x}$ will be classified by checking on what side of the hyperplane it is located. From a mathematical point of view, the quest for such hyperplane is formulated as a Quadratic Programming (QP) problem with convex constraints. The dual formulation with Lagrangian multipliers α_i is:

$$\begin{aligned} &\text{Maximize}_{\vec{\alpha}} \sum_{i=1}^{l} \alpha_i - \frac{1}{2} \sum_{i,j=1}^{l} \alpha_i \alpha_j y_i y_j K(x_i, x_j) \\ &\text{subject to} : 0 \leq \alpha_i \leq C, \qquad i = 1, \ldots, l \\ &\sum_{i=1}^{l} \alpha_i y_i = 0. \end{aligned} \tag{1}$$

where K, called *kernel*, is a measure of similarity between two vectors in a highly nonlinear space called *feature space*, which substitutes the input space, while the parameter C controls the maximal quantity of admissible errors. This optimization problem does not have local maxima achieving independence from the starting point of the optimization. One of the most used kernel function, which substitutes the original dot product, is the Gaussian kernel defined as:

$$\text{Gaussian} \qquad \mathrm{K}(\vec{\mathrm{x_i}}, \vec{\mathrm{x_j}}) = \exp\left(-\frac{\| \vec{\mathrm{x_i}} - \vec{\mathrm{x_j}} \|^2}{2\sigma^2}\right) \tag{2}$$

Novelty detection

Let's imagine now that the labels y_i of the objects are not provided. In this case the presented framework of binary classification can not be applied since we do not have two classes to separate but only one class (e.g. without loss of generality we can label each object as positive). This special kind of problem is called *one-class classification* or alternatively *novelty detection*. The new goal is to find some subsets of the input space in which there is a high probability to find an object. The subset is expressed as a function which takes positive values in the area with high probability and negative ones in the other case. When a new object becomes available, we can predict if it is drawn from the same distribution or it is novel by computing the value of the estimated function in the point where it is located. In this sense we are always dealing with a classification problem, which justifies the term one-class classification. In the statistical framework the quest for such subset is referred as *density estimation* stressing that the target is to estimate how objects are drawn from an unknow, but existent, probability density function (pdf). The Support Vector Data Description (SVDD) is one of the available approach to density estimation inspired to SVM. It was developed by Tax and Duin (2004) and it is aimed at finding in the feature space the smallest hypersphere with center $\vec{c}$ and radius R that contains the given l objects $\vec{x_i}$. The formulation of the problem solved by SVDD is the following:

$$\begin{aligned} &\text{Minimize}_{R,\vec{c},\vec{\xi}} \quad R^2 + C\sum_{i=1}^{l}\xi_i \\ &\text{subject to :} \quad \left\|\vec{c} - \vec{x_i}\right\|^2 \leq R^2 + \xi_i \\ &\qquad\qquad\quad \xi_i \geq 0. \end{aligned} \tag{3}$$

As we can see, SVDD exploits slack variables allowing to keep some objects outside the positive region (with non zero associated ξ_i) in order to assure a smoother representation of the boundary. The parameter C controls the fraction of the objects that can be kept outside the hypersphere. To solve the problem we can switch to the following dual formulation which makes use of the Gaussian kernel:

$$\begin{aligned} &\text{Maximize}_{\vec{\alpha}} \sum_{i=1}^{l}\alpha_i - \frac{1}{2}\sum_{i,j=1}^{l}\alpha_i\alpha_j K(x_i, x_j) \\ &\text{subject to : } 0 \leq \alpha_i \leq C, \qquad i = 1, \ldots, l \\ &\qquad\qquad\quad \sum_{i=1}^{l}\alpha_i = 1. \end{aligned} \tag{4}$$

From this dual formulation it becomes clear why SVDD is considered as an extension of the standard SVM (see formula 1) for the case where labels y_i are not available. Indeed, this result is very important since it unleashes the exploitation of the key features of SV methods. As in the 2-class SVM, the region where the function is positive is expressed using only the objects at the boundary plus those outside the region that are together the Support Vectors (both with $\alpha_i > 0$). It is worth recalling that has been proved by Schölkopf et al (2001) that their standard extension of SVM and the SVDD approach are completely equivalent in the case of Gaussian kernel. One goal of this work is to show that 1-class SVM can be a valuable tool helping the search of groups in social network analysis. As previously presented, the main task of SVDD is to find a hypersurface in the input space that separates the data in two classes: object inside the hypersurface (positive) and object outside the hypersurface (negative). Notably, the method does not use any information regarding the data and no priors have to be specified. Indeed, if we know more about the data we can incorporate this knowledge in the C parameter even if it can be experimentally tested that SVDD is very robust in the respect of this value. As in other approaches based on the optimization theory (e.g. the blockmodeling), the value of α_i can be interpreted as a price to pay in order to include the sample $\vec{x_i}$ in the solution and the trade-off can be managed via the C parameter which, in addition, has a clear interpretation. In respect to other methods, SVDD has many key features:

- the SLT works very well with very sparse matrix (indeed it can be seen as a Bayesian theory for small samples);
- the shape of the boundary of the pdf can be very complex;
- the possibility of reject examples permits to find rationale (smooth) pdf avoiding overfitting;
- the optimization problem has no local maxima;
- we do not have to specify a priori the type of the model as for example in the blockmodeling approach;
- it can be applied directly to one-mode and two-mode data.

We note that the estimated hypersurface is able to produce only a binary labeling of the objects which is not so useful in the clustering sense. A first attempt to export the SVDD approach to clustering was due to Benhur et al (2001). In this case, graph analysis is performed after the density estimation to find the number of clusters and their members. We propose a different approach for improving the detection of classes especially in the context of social network analysis. In the following, the overall method will be explained using the well-known two-mode DGG dataset.

3 Illustration

In this section we illustrate a way to cluster the DGG dataset, making use of the SVDD tool provided by Tax (2005). We start by the original dataset directly arranged by the authors in a matrix of 18 women $\times$ 14 events. Following a standard setting in the machine learning community, we consider the 14 values of one row as vector of features representing the attendance pattern of each woman. We apply the SVDD with Gaussian kernel in order to find the objects which are inside the hypersurface. The reject fraction parameter is chosen such as a large fraction (say $> 60\%$) of objects can be put outside the region boundary while the σ parameter of the Gaussian kernel is set to a common value of 4.5. Obtained the hypersurface, we can remove all the objects with negative label which are not support vectors (indeed this procedure assures to keep clusters with few members). In order to find if there is a more fine separation inside each of the found subsets we can iterate this process only on the extracted subsets. In every case, the resulting set of objects represents the core of the distribution. We yet do not have clusters nor cluster membership since all we know is that residual objects belong to the core of the pdf. For instance, given an unknown pdf composed by n Gaussian distributions, we are not able to estimate the value of n. What we can extract are those objects which exhibits large variance relatively to their distribution, often called *outliers*. These outliers could affect the application of standard clustering methods, as the blockmodeling, since they are likely to become *singletons*. By removing singletons, we can make use of simpler models which have more chances to avoid overfitting. The idea is now to create a new set of data where singletons and peripheral objects are not present and then to apply standard clustering techniques to the aim of finding clusters and their membership. In order to show this approach, we apply a standard

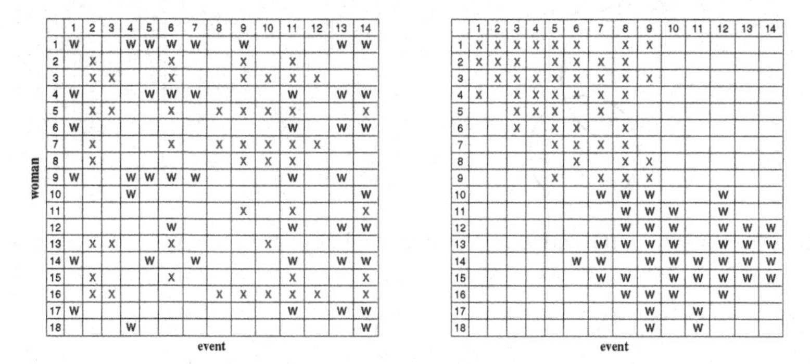

Fig. 1. The labeling of the scrambled DGG dataset (left) obtained by permutation of both rows and columns and the labeling obtained after the resuming of the original arrangement (right). The X and W represent the classes.

hierarchical clustering with adaptive cut-off parameter to this new dataset. The clustering algorithm, based on the Euclidean metric, founds two clusters and labels each objects according to the membership function. Finally, we reintroduce the removed objects and associate them to the cluster of the nearest labeled object (using Euclidean distance). Exploiting the interesting propriety of no local maxima of the QP problem we safety can permute the rows, the columns or both of the original dataset without changing the extracted subset. However, we must note that in the case of two or more identical samples (e.g. the case of women 17 and 18) the optimization problem can converge to different equivalent solutions with the same maximal value of the objective function. This is consistent with the fact that all these samples show the same attendance pattern and they are interchangeable both in an optimization and an interpretative sense. Figure 1 shows the labeling of a new dataset obtained by permutation of both rows and columns and the labeling obtained after the resuming of the original arrangement. As figure 2 shows, the resulting labeling is compatible with those found by several other approaches (Breiger (1974); Freeman (2003)).

4 Conclusions

This paper proposes a new approach for the analysis of relational data which relies on a novel machine learning technique based on the SVM, a powerful implementation of the Statistical Learning Theory. To achieve the novelty detection - a natural extension of the binary classification - we adopted an SVM-like algorithm known as SVDD. It finds the minimal hypersphere that

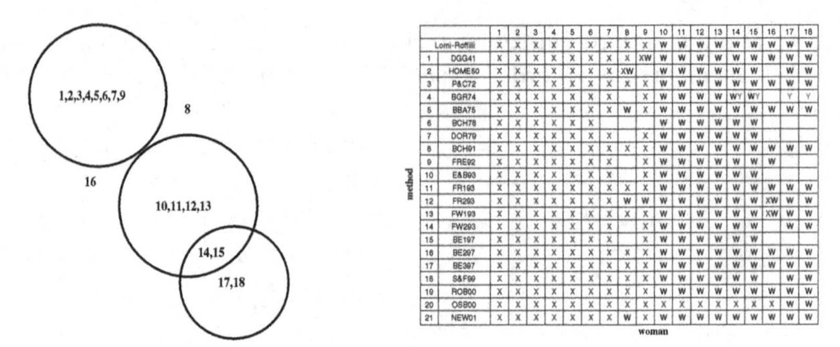

		1	2	3	4	5	6	7	8	9	10	11	12	13	14	15	16	17	18
Lomi-Roffilli		X	X	X	X	X	X	X	X	X	W	W	W	W	W	W	W	W	W
1	DGG41	X	X	X	X	X	X	X	X	XW	W	W	W	W	W	W	W	W	W
2	HOME50	X	X	X	X	X	X	X	XW		W	W	W	W	W	W		W	W
3	P&C72	X	X	X	X	X	X	X	X	X	W	W	W	W	W	W	W	W	W
4	BGR74	X	X	X	X	X	X	X		X	W	W	W	W	WY	WY		Y	Y
5	BBA75	X	X	X	X	X	X	X	W	X	W	W	W	W	W	W	W	W	W
6	BCH78	X	X	X	X	X	X				W	W	W	W	W	W			
7	DOR79	X	X	X	X	X	X	X		X	W	W	W	W	W	W			
8	BCH91	X	X	X	X	X	X	X	X	X	W	W	W	W	W	W	W	W	W
9	FRE92	X	X	X	X	X	X	X		X	W	W	W	W	W	W	W		
10	E&B93	X	X	X	X	X	X	X		X	W	W	W	W	W	W			
11	FR193	X	X	X	X	X	X	X	X	X	W	W	W	W	W	W	W	W	W
12	FR293	X	X	X	X	X	X	X	W	W	W	W	W	W	W	W	XW	W	W
13	FW193	X	X	X	X	X	X	X	X	X	W	W	W	W	W	W	XW	W	W
14	FW293	X	X	X	X	X	X	X		X	W	W	W	W	W	W		W	W
15	BE197	X	X	X	X	X	X	X		X	W	W	W	W	W	W			
16	BE297	X	X	X	X	X	X	X	X	X	W	W	W	W	W	W	W	W	W
17	BE397	X	X	X	X	X	X	X	X	X	W	W	W	W	W	W	W	W	W
18	S&F99	X	X	X	X	X	X	X	X	X	W	W	W	W	W	W		W	W
19	ROB00	X	X	X	X	X	X	X	X	X	W	W	W	W	W	W	W	W	W
20	OSB00	X	X	X	X	X	X	X	X	X	X	X	X	X	X	X	X	W	W
21	NEW01	X	X	X	X	X	X	X	W	X	W	W	W	W	W	W	W	W	W

Fig. 2. The Breiger (left) and the Freeman (right) partition of the DGG dataset.

encloses the data in a more representative hyperspace called feature space. When we back project the hypersphere from feature to input space we obtain a hypersurface that encloses some of the objects. In addition SVDD finds the examples, called Support Vectors, which are responsible for the shape of the pdf. As parameter, the SVDD requires to specify a priori the fraction of examples which can be rejected during the optimization. In this scenario, we proposed to use SVDD as a preprocessing step before applying a standard clustering technique. The idea is to find a subset of data that is easier to cluster, then to apply clustering to it, and finally to labels the data kept outside the subset according to the results. To do so, firstly we use SVDD to find such subset, then we remove objects outside the boundary. Reiterating this procedure we can control the sharpness of the boundary shape. In this way, we are likely to produce an easier clustering problem where all the original clusters are guarantied to be represented. Indeed, SVDD assures that clusters with few members are kept in the subset since they are considered as SVs. At this point we can use classical clustering techniques (for instance hierarchical clustering or blockmodeling) to find out the number of clusters and their membership. We now reintroduce the objects previously taken apart, and we associate them to the right cluster. To assess the performance of our approach, we tested the proposed method on the well-known two-mode Davis-Gardner-Gardner dataset. We found that the SVDD is able to efficiently extract a subset of original data which facilitate the clustering step. We showed that obtained cluster memberships are in line with the finds of other methods. However, we note that DGG is only a illustrative testbed because of its extremely simplicity. We are currently testing our approach on different one-mode and two-mode datasets with multiple clusters following the insight offered by recent work on generalized blockmodeling techniques applied to two-mode data (Doreian, Batagelj and Ferligoj, 2003). Preliminary experiments not reported in this paper, suggest that the ability of 1-class SVM of managing high dimensional spaces lowly populated is of primarily importance to reach successful results.

References

BENHUR A., HORH D., SIEGELMANN H.T. and VAPNIK V. (2001): Support Vector Clustering. *Journal of Machine Learning, 2, 125-137.*

BORGATTI S. P. and EVERETT M. G. (1997): Network analysis of 2-mode data. *Social Networks, 19/3, 243-269.*

BOORMAN S. and WHITE H. (1976): Social Structure from Multiple Networks II. Role Structures. *American Journal of Sociology 81: 1384-1446.*

BREIGER R. (1974): The Duality of Persons and Groups. *Social Forces, 53, 181-90.*

DAVIS A., GARDNER B. B. and GARDNER M. R. (1941): Deep South. *Chicago: The University of Chicago Press.*

DOREIAN P., BATAGELJ V. and FERLIGOJ A. (2004): Generalized blockmodeling of two-mode network data. *Social Networks 26, 29-53.*

DOREIAN P., BATAGELJ V. and FERLIGOJ A. (2005): *Generalized blockmodeling.* Cambridge University Press.

FREEMAN L. C. and WHITE D. R. (1994): Using Galois lattices to present network data. In P. Marsden (Ed.): *Sociological Methodology (pp. 127-146). Cambridge: Blackwell.*

FREEMAN L. (2003): Finding social groups: A meta-analysis of the southern women data. In R. Breiger, K. Carley and P. Pattison (Eds.): *Dynamic Social Network Modeling and Analysis. The National Academies Press.*

GREENACRE M. J. (1984): Theory and Applications of Correspondence Analysis. *London: Academic Press.*

PATTISON P. E. and BREIGER R. (2002): Lattices and Dimensional Representations: Matrix Decompositions and Ordering Structures. *Social Networks, 24, 423-444.*

SCHOLKOPF B., PLATT J.C., SHAWE-TAYLOR J., and SMOLA A.J. (2001): Estimating the support of a high-dimensional distribution. *Neural Computation, 13(7):14431471.*

TAX D.M.J. (2005): *DDtools, the Data Description Toolbox for Matlab.* version 1.4.1.

TAX D.M.J. and DUIN R.P.W. (2004): Support Vector Data Description. *Journal of Machine Learning, 54/1, 45-46.*

VAPNIK V. (1995): *The Nature of Statistical Learning Theory.* Springer Verlag.

WHITE H., BOORMAN S. and BREIGER R. (1976): Social Structure from Multiple Networks I. Blockmodels of Roles and Positions. *American Journal of Sociology 81: 730-779.*

Part IV

Analysis of Symbolic Data

Multidimensional Scaling of Histogram Dissimilarities

Patrick J.F. Groenen[1] and Suzanne Winsberg[2]

[1] Econometric Institute, Erasmus University Rotterdam,
P.O. Box 1738, 3000 DR Rotterdam, The Netherlands
email: groenen@few.eur.nl
[2] Predisoft, San Pedro, Costa Rica
email: SuzanneWinsberg@predisoft.com

Abstract. Multidimensional scaling aims at reconstructing dissimilarities between pairs of objects by distances in a low dimensional space. However, in some cases the dissimilarity itself is unknown, but the range, or a histogram of the dissimilarities is given. This type of data fall in the wider class of symbolic data (see Bock and Diday (2000)). We model a histogram of dissimilarities by a histogram of the distances defined as the minimum and maximum distance between two sets of embedded rectangles representing the objects. In this paper, we provide a new algorithm called Hist-Scal using iterative majorization, that is based on an algorithm, I-Scal developed for the case where the dissimilarities are given by a range of values ie an interval (see Groenen et al. (in press)). The advantage of iterative majorization is that each iteration is guaranteed to improve the solution until no improvement is possible. We present the results on an empirical data set on synthetic musical tones.

1 Introduction

Ordinary multidimensional scaling (MDS) represents the dissimilarities among a set of objects as distances between points in a low dimensional space. The aim of these MDS methods is to reveal relationships among the objects and to uncover the dimensions giving rise to the space. For example, the goal in many MDS studies in the fields of psychoacoustics and marketing is to visualize the objects and the distances among them and to discover the dimensions underlying the dissimilarity ratings.

Sometimes the proximity data are collected for n objects yielding a single dissimilarity matrix with the entry for the i-th row and the j-th column being the dissimilarity between the i-th and j-th object (with $i = 1, \ldots, n$ and $j = 1, \ldots, n$). Techniques for analyzing this form of data (two-way one-mode) have been developed (see, e.g., Kruskal (1964), Winsberg and Carroll (1989), or Borg and Groenen (2005)). Sometimes the proximity data are collected from K sources such as a panel of K judges or under K different conditions, yielding three way two mode data and an $n \times n \times K$ array. Techniques have been developed to deal with this form of data permitting the study of individual or group differences underlying the dissimilarity ratings (see, e.g., Carroll and Chang (1972), Winsberg and DeSoete (1993)).

These MDS techniques require that each entry of the dissimilarity matrix be a single numerical value. It may be that the objects in the set under consideration are of such a complex nature that the dissimilarity between each pair of them is better represented by a range, that is, an interval of values, or a histogram of values rather than a single value. For example, if the number of objects under study becomes very large, it may be unreasonable to collect pairwise dissimilarities from each judge and one may wish to aggregate the ratings from many judges where each judge has rated the dissimilarities from a subset of all the pairs. In such cases, rather than using an average value of dissimilarity for each object pair the researcher may wish to retain the information contained in the histogram of dissimilarities obtained for each pair of objects. Or it may be interesting to collect data reflecting the imprecision or fuzziness of the dissimilarity between each object pair.

Then, the ij-th entry in the $n \times n$ data matrix, that is, the dissimilarity between objects i and j, is an empirical distribution of values or, equivalently, a histogram. For example, we may have enough detailed information so that we can represent the empirical distribution of the dissimilarity as a histogram, for example by $.10[0,1], .30[1,2], .40[2,3], .20[3,4]$, where the first number indicates the relative frequency and values between the brackets define the bin. A special case would be when the resulting entry of the $n \times n$ dissimilarity matrix would be an interval of values $[a, b]$ corresponding to $[\delta_{ij}^{(L)}, \delta_{ij}^{(U)}]$. Of course, if a given entry of the matrix was single valued this data type is also a special case of the histogram data under consideration here.

The case where the dissimilarity between each object pair is represented by a range or interval of values has been treated. Denœux and Masson (2000) and Masson and Denœux (2002) have developed MDS techniques that treat dissimilarity matrices composed of interval data. These techniques model each object as alternatively a hyperbox (hypercube) or a hypersphere in a low dimensional space; they use a gradient descent algorithm. Groenen et al. (in press) have developed an MDS technique for interval data which yields a representation of the objects as hyperboxes in a low-dimensional Euclidean space rather than hyperspheres because the hyperbox representation is reflected as a conjunction of p properties where p is the dimensionality of the space. We follow this latter approach here.

This representation as a conjunction is appealing for two reasons. First linguistically, in everyday language, if we have objects consisting of repeated sound bursts differing with respect to loudness and the number of bursts per second, a given sound, might be referred to as having a loudness lying between 2 and 3 dbSPR and a repetition rate between 300 and 400 milliseconds between bursts, that is, as a conjunction of two properties. We would not refer to a sound as a hypersphere with a loudness and repetition rate centered at 2.5 dbSPR and 350 msec and a radius of 10 to be expressed in just what units. Perceptually a sound might not have a precise loudness or repetition rate to a listener. Second, since one of the principal aims of MDS is to reveal relationships among the objects in terms of the underlying dimensions, it is most useful for this type of data to express the location of each object in terms of a range, or histogram of each of these underlying attributes or dimensions.

We have extended the method developed by Groenen et al. (in press) to deal with the case in which the dissimilarity between object i and object j is an empirical distribution of values or, equivalently, a histogram. We can represent the results of our MDS analyses in two ways: a plot for each pair of dimensions displaying each object as a series of embedded rectangles, one for each bin of the histogram; and a graph for each underlying dimension displaying the location and histogram for each object on that dimension.

In the next section, we review briefly the I-SCAL algorithm developed by Groenen et al. (in press) for MDS of interval dissimilarities based on iterative majorization. Then, we present the extension to histogram data, and the HIST-SCAL algorithm. We have analyzed some empirical data sets dealing with dissimilarities of sounds. We end the paper with some conclusions and suggestions for continued research.

2 MDS of interval dissimilarities

To develop MDS for interval dissimilarities, the ranges of dissimilarities must be represented by ranges of distances. Here, we choose to represent the objects by rectangles and approximate the upper bound of the dissimilarity by the maximum distance between the rectangles and the lower bound by the minimum distance between the rectangles. Figure 1 shows an example of rectangle representation and how the minimum and maximum distance between two rectangles is defined.

Not only the distances are represented by ranges, the coordinates themselves are also ranges. Let the rows of the $n \times p$ matrix $\mathbf{X}$ contain the coordinates of the center of the rectangles, where n is the number of objects and p the dimensionality. The distance from the center of rectangle i along axis s, denoted the spread, is represented by r_{is}. Note that $r_{is} \geq 0$. The maximum Euclidean distance between rectangles i and j is given by

$$d_{ij}^{(U)}(\mathbf{X}, \mathbf{R}) = \left(\sum_{s=1}^{p} [|x_{is} - x_{js}| + (r_{is} + r_{js})]^2 \right)^{1/2} \tag{1}$$

and the minimum Euclidean distance by

$$d_{ij}^{(L)}(\mathbf{X}, \mathbf{R}) = \left(\sum_{s=1}^{p} \max[0, |x_{is} - x_{js}| - (r_{is} + r_{js})]^2 \right)^{1/2} . \tag{2}$$

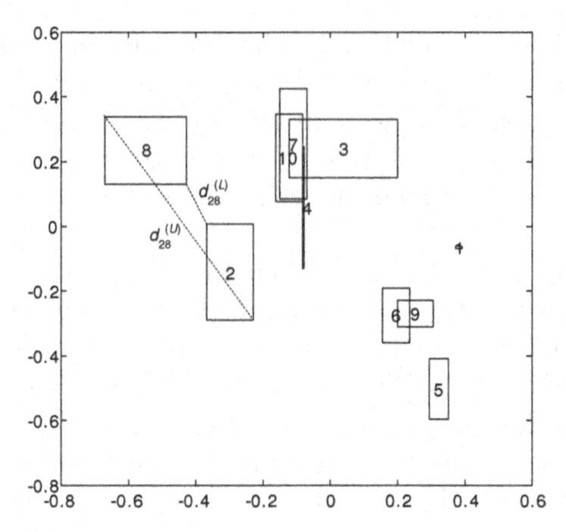

Fig. 1. Example of distances in MDS for interval dissimilarities where the objects are represented by rectangles.

Even though Euclidean distances are used between the hyperboxes, the lower and upper distances change when the solution is rotated. The reason is that the hyperboxes are defined with respect to the axes. For a dimensional interpretation, the property rotational uniqueness can be seen as an advantage of symbolic MDS. Of course, if $\mathbf{R} = \mathbf{0}$ all hyperboxes shrink to points and then symbolic MDS simplifies into ordinary MDS, which can be freely rotated.

The objective of symbolic MDS for interval dissimilarities is to represent the lower and upper bounds of the dissimilarities by minimum and maximum distances between rectangles as well as possible in least-squares sense. The I-Stress loss function that models this objective and needs to be minimized over $\mathbf{X}$ and $\mathbf{R}$ is given by

$$\sigma_{\mathrm{I}}^2(\mathbf{X}, \mathbf{R}) = \sum_{i<j}^{n} w_{ij} \left[\delta_{ij}^{(U)} - d_{ij}^{(U)}(\mathbf{X}, \mathbf{R})\right]^2 + \sum_{i<j}^{n} w_{ij} \left[\delta_{ij}^{(L)} - d_{ij}^{(L)}(\mathbf{X}, \mathbf{R})\right]^2,$$

where $\delta_{ij}^{(U)}$ is the upper bound of the dissimilarity of objects i and j, $\delta_{ij}^{(L)}$ is the lower bound , and w_{ij} is a given nonnegative weight.

One can also obtain diagnostics such as the variance for each dimension and fit per object. Groenen et al. (in press) have derived a majorization algorithm called I-SCAL to minimize I-Stress for two reasons. First, iterative majorization is guaranteed to reduce I-Stress in each iteration from any starting configuration until a stationary point is obtained. Although majorization cannot exclude the possibility that the stationary point is a saddle point, its monotone descent property ensures in practice that the stationary point is a local minimum indeed. Second, as in each iteration the algorithm operates on a quadratic function in $\mathbf{X}$ and $\mathbf{R}$ it is easy to impose constraints that have well known solutions for quadratic functions. This property can be useful for extensions of the algorithm that require constraints.

The majorization algorithm minimizes I-Stress over $\mathbf{X}$ and $\mathbf{R}$. The basic idea of iterative majorization is that the original loss function is replaced in each iteration by an auxiliary function that is easier to handle. The auxiliary function, the so called majorizing function, needs to satisfy two requirements: (i) the majorizing function is equal to the original function at the current estimate, and (ii) the majorizing function is always larger than or equal to the original function. Usually, the majorizing function is linear or quadratic so that the minimum of the majorizing function can be calculated easily. From the requirements it can be derived that (a) the loss of the majorizing function and the original loss function is equal at the current estimate, (b) at the update the majorizing function is smaller than at the current estimate, so that (c) the original loss function is smaller at the update since the original loss function is never larger than the majorizing function. This reasoning proves that if the conditions (i) and (ii) are satisfied, the iterative majorization algorithm yields a series of nonincreasing function values. In addition, constraints with an easy solution for quadratic loss functions can be easily handled. For more details on iterative majorization, see, for example, Borg and Groenen (2005).

Groenen et al. (in press) have derived the quadratic majorizing function for $\sigma_{\mathrm{I}}^2(\mathbf{X}, \mathbf{R})$ at the right hand side of

$$\begin{aligned}\sigma_{\mathrm{I}}^2(\mathbf{X}, \mathbf{R}) \leq & \sum_{s=1}^{p} (\mathbf{x}_s' \mathbf{A}_s^{(1)} \mathbf{x}_s - 2\mathbf{x}_s' \mathbf{B}_s^{(1)} \mathbf{y}_s) \\ & + \sum_{s=1}^{p} (\mathbf{r}_s' \mathbf{A}_s^{(2)} \mathbf{r}_s - 2\mathbf{r}_s' \mathbf{b}_s^{(2)}) + \sum_{s=1}^{p} \sum_{i<j} (\gamma_{ijs}^{(1)} + \gamma_{ijs}^{(2)}), \qquad (3)\end{aligned}$$

where $\mathbf{x}_s$ is column s of $\mathbf{X}$, $\mathbf{r}_s$ is column s of $\mathbf{R}$, $\mathbf{y}_s$ is column s of $\mathbf{Y}$ (the previous estimate of $\mathbf{X}$). The expression above contains several matrices ($\mathbf{A}_s^{(1)}, \mathbf{B}_s^{(1)}, \mathbf{A}_s^{(2)}$), vectors ($\mathbf{b}_s^{(2)}$), and scalars ($\gamma_{ijs}^{(1)}, \gamma_{ijs}^{(2)}$) that all depend dependent on previous estimates of $\mathbf{X}$ and $\mathbf{R}$, thus are known at the present iteration and their definition can be found in Groenen et al. (in press). The important thing is to realize that the majorizing function (3) is quadratic in $\mathbf{X}$ and $\mathbf{R}$, so that an update can be readily derived by setting the derivatives equal to zero. The I-SCAL algorithm consists of iteratively minimizing the quadratic majorizing function in (3), so that a converging sequence of decreasing I-Stress values is obtained.

The I-SCAL algorithm has been validated by analyzing several artificial data sets, and by investigating the local minimum problem (see Groenen et al. (in press)). The algorithm permits considering both the rational start described in Groenen et al. (in press) and many random starts and then chooses the best global solution. It was shown that majorization combined with this multistart strategy, ie the I-SCAL algorithm, performs better than the gradient descent approach used by Denœux and Masson (2002).

Table 1. Example of upper and lower bounds of the distribution of δ_{ij} used in symbolic MDS for histograms for $\alpha' = [.20, .30, .40]$.

		Lower bound		Upper bound	
ℓ	α_ℓ	Percentile	$\delta_{ij\ell}^{(L)}$	Percentile	$\delta_{ij\ell}^{(U)}$
1	.20	20	$\delta_{ij1}^{(L)}$	80	$\delta_{ij1}^{(U)}$
2	.30	30	$\delta_{ij2}^{(L)}$	70	$\delta_{ij2}^{(U)}$
3	.40	40	$\delta_{ij3}^{(L)}$	60	$\delta_{ij3}^{(U)}$

3 Symbolic MDS for histogram data

Consider the case that instead of having a single interval available for each δ_{ij}, we have its empirical distribution so that a histogram can be made for each dissimilarity. Then, the objective of symbolic MDS is to find a distribution of the MDS coordinates. In particular, we compute several percentiles α_ℓ for the distribution of each δ_{ij}. For example, choose $\alpha' = [.20, .30, .40]$ yields $\delta_{ij1}^{(L)}$ to be the $\alpha_1 \times 100$ (= 20-th) percentile of the distribution of δ_{ij} and $\delta_{ij1}^{(U)}$ the $(1-\alpha_1)\times 100$ (= 80-th) percentile. Table 1 shows the bounds for this little example. Then, symbolic MDS for histogram data models the lower and upper bounds for each percentile by concentric rectangles, that is, each object has the same center coordinate x_{is} but has different width $r_{is\ell}$ for each percentile. In addition, we want the rectangle of the next percentile to be larger than (or equal to) the previous rectangle. These objectives are minimized by the Hist-Stress loss function

$$
\begin{aligned}
\sigma^2_{\text{Hist}}(\mathbf{X}, \mathbf{R}_1, \ldots, \mathbf{R}_L) = & \sum_\ell \sum_{i<j}^n w_{ij}\left[\delta_{ij\ell}^{(U)} - d_{ij}^{(U)}(\mathbf{X}, \mathbf{R}_\ell)\right]^2 \\
& + \sum_\ell \sum_{i<j}^n w_{ij}\left[\delta_{ij\ell}^{(L)} - d_{ij}^{(L)}(\mathbf{X}, \mathbf{R}_\ell)\right]^2, \\
& \text{subject to } 0 \le r_{is1} \le r_{is2} \le \ldots \le r_{isL}.
\end{aligned}
$$

where the subscript ℓ is added to indicate dependence on the percentile α_ℓ and L is the number of percentiles. Note that σ^2_{Hist} has been proposed earlier by Masson and Denœux (2002) in the somewhat different context of α-cuts for fuzzy dissimilarities but without the inequality restrictions on the radii.

Hist-Stress can be seen as several separate I-Stress functions that have the restriction that there is a single $\mathbf{X}$. The I-SCAL algorithm can be easily adapted for Hist-Stress. The reason is that the I-SCAL algorithm is based on iterative majorization with quadratic majorizing functions for $\mathbf{X}$ and $\mathbf{R}_\ell$. Now, for each percentile α_ℓ, we can compute a majorizing function as in the I-SCAL case; and just as Hist-Stress is a sum of I-Stress functions, so the majorizing function for Hist-Stress is a sum of the majorizing functions for

I-Stress, that is,

$$\sigma^2_{\text{Hist}}(\mathbf{X}, \mathbf{R}_{\text{All}}) \leq$$
$$\sum_{s=1}^{p} \left(\mathbf{x}'_s \left[\sum_{\ell} \mathbf{A}^{(1)}_{s\ell} \right] \mathbf{x}_s - 2\mathbf{x}'_s \left[\sum_{\ell} \mathbf{B}^{(1)}_{s\ell} \right] \mathbf{y}_s \right)$$
$$+ \sum_{\ell} \sum_{s=1}^{p} (\mathbf{r}'_{s\ell} \mathbf{A}^{(2)}_{s\ell} \mathbf{r}_{s\ell} - 2\mathbf{r}'_{s\ell} \mathbf{b}^{(2)}_{s\ell}) + \sum_{\ell} \sum_{s=1}^{p} \sum_{i<j} (\gamma^{(1)}_{ijs\ell} + \gamma^{(2)}_{ijs\ell}), \tag{4}$$

where the subscript ℓ refers to percentile α_ℓ and $\mathbf{R}_{\text{All}} = [\mathbf{R}_1 \mid \ldots \mid \mathbf{R}_L]$. As (4) is again a quadratic function in $\mathbf{X}$ and $\mathbf{R}_\ell$, the updates can be obtained by setting the derivatives to zero.

For updating $\mathbf{X}$, the linear system that should be solved for $\mathbf{x}^+_s$ is

$$(\sum_{\ell} \mathbf{A}^{(1)}_{s\ell})\mathbf{x}^+_s = (\sum_{\ell} \mathbf{B}^{(1)}_{s\ell})\mathbf{y}_s.$$

It turns out that by construction $\mathbf{A}^{(1)}_{s\ell}$ is not of full rank and has a zero eigenvalue corresponding to the eigenvector $n^{-1/2}\mathbf{1}$ (which also holds for $\mathbf{B}^{(1)}_{s\ell}$ so that $(\sum_\ell \mathbf{B}^{(1)}_{s\ell})\mathbf{y}_s$ is column centered). Therefore, the update for $\mathbf{X}$ can be computed by adding the null space $\mathbf{11}'$ to $\mathbf{A}^{(1)}_{s\ell}$ and solving

$$(\mathbf{11}' + \sum_{\ell} \mathbf{A}^{(1)}_{s\ell})\mathbf{x}^+_s = (\sum_{\ell} \mathbf{B}^{(1)}_{s\ell})\mathbf{y}_s. \tag{5}$$

The unconstrained update of $\mathbf{R}$ is given by $\bar{r}_{is\ell} = b^{(2)}_{is\ell}/a^{(2)}_{is\ell}$ and those for object i in dimension s are gathered in the $L \times 1$ vector $\bar{\mathbf{r}}_{is}$. Let $\mathbf{A}_{is}$ be a diagonal matrix with the weights. Then, the part of the majorizing function that is a function of $\mathbf{R}_{\text{All}}$ can be minimized by

$$\mathbf{r}^+_{is} = \text{argmin} \|\mathbf{r}_{is} - \bar{\mathbf{r}}_{is}\|^2_{\mathbf{A}_{is}} \text{ subject to } 0 \leq r_{is1} \leq r_{is2} \leq \ldots \leq r_{isL}, \tag{6}$$

that is, by applying for each combination of i and s Kruskal's (1964) weighted monotone regression with $\bar{\mathbf{r}}_{is}$ and weights the diagonal values of $\mathbf{A}_{is}$. Note that the order in which $\mathbf{X}$ and $\mathbf{R}$ are updated is unimportant because the majorizing function in (4) has separate quadratic terms for $\mathbf{X}$ and $\mathbf{R}$.

The HIST-SCAL algorithm for minimizing $\sigma^2_{\text{Hist}}(\mathbf{X}, \mathbf{R}_{\text{All}})$ using iterative majorization is shown in Figure 2.

4 Synthesized musical instruments

We have considered an empirical data set where the entries in the dissimilarity matrix are a histogram of values. The objects in the study (see Table 2) correspond to sounds from 18 synthesized musical instruments some of which

```
1 Initialize X^(0) and R_All^(0).
  Set k := 0, X^(-1) := X^(0), R_All^(-1) := R_All^(0) for all ℓ.
  Set ε to a small positive value.
2 While k = 0 or σ²_Hist(X^(k-1), R_All^(k-1)) - σ²_Hist(X^(k), R_All^(k)) ≤ ε
3     k := k + 1
5     For s = 1 to p
6         Compute the update of x_s by (5).
7         For ℓ = 1 to L
8             Compute the update of r_sℓ by (6).
9         End for
10    End for
11    Set X_k := X and R_All^(k) := R_All.
12 End
```

Fig. 2. The HIST-SCAL algorithm.

are hybrids of natural instruments (e.g., the guitarnet is a combination of specific features of the guitar and the clarinet). The pitch, duration, and loudness of the sounds were equalized. We applied the Hist-Scal algorithm to 50 random starts so that bad local minima are avoided. The first two dimensions of the best 3D solution having $\sigma^2_{\text{Hist}} = 0.01407128$ are given in Figure 3. The first three recovered dimensions are respectively, log rise time (attack time), spectral centroid, and spectral irregularity, (spectral flux). Higher values on dimension 1 correspond to shorter rise times, higher values on dimension 2 to higher spectral centroids, and higher values on dimension 3 (not shown) correspond to lesser spectral flux. The solution has the following two properties: the rise times of sounds with longer rise times are more difficult to localize, than shorter rise times, sounds with long rise times and low spectral centroids are easier to localize on the spectral centroid dimension. In addition, the 3d solution indicates that the spectral flux dimension is more defined or localized for sounds of greater spectral flux. These three dimensions are those that have been recovered in many previous studies of musical timbre.

The results for rise time are consistent with those observed for a single judge, an expert, reporting an interval of values for the dissimilarity of each pair of sounds which were synthesized so as to have the two dimensions rise time and spectral centroid corresponding to the general range found in musical instruments (see Groenen et al. (in press)).

5 Discussion and conclusions

We have presented an MDS technique for symbolic data that deals with fuzzy dissimilarities consisting of a histogram of values observed for each pair of objects. In this technique, each object is represented as a series of embedded

Table 2. Eighteen synthesized musical instruments and their abbreviation.

Instrument		Instrument	
Bassoon	bsn	Obochord (oboe/harpsichord)	obc
Bowed string	stg	Oboleste(oboe/celeste)	ols
Clarinet	cnt	Piano	pno
English horn	ehn	Striano(bowed string/piano)	sno
French horn	hrn	Trombone	tbn
Guitar	gtr	Trumpar(trumpet/guitar)	tpr
Guitarnet (guitar/clarinet)	gnt	Trumpet	tpt
Harp	hrp	Vibraphone	vbs
Harpsichord	hcd	Vibrone (vibraphone/trombone)	vbn

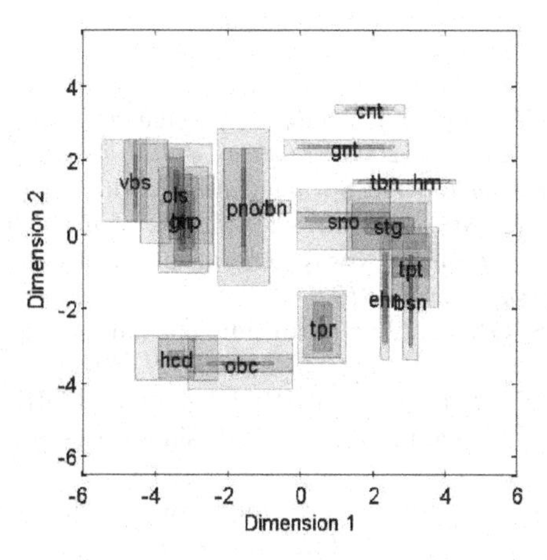

Fig. 3. First two dimensions of a 3D Hist-Scal solution of the synthesized musical instruments data having $\sigma^2_{\text{Hist}} = 0.01407128$.

hyperboxes in a p dimensional space. By representing the objects as embedded hypercubes, we are able to convey information contained when the dissimilarity between the objects or for any object pair needs to be expressed as a histogram of values not a single value. It may be so, moreover, that the precision inherent in the dissimilarities is such that the precision in one recovered dimension is worse than that for the other dimensions. Our technique is able to tease out and highlight this kind of information.

The HIST-SCAL algorithm for MDS of histogram dissimilarities is based on iterative majorization, and the I-SCAL algorithm created to deal with the case when dissimilarities are given by a range or interval of values. The advantage is that each iteration yields better Hist-Stress until no improvement is possible. Simulation studies have shown that I-SCAL and HIST-SCAL combined with multiple random start and a rational start yields good quality solutions.

Denœux and Masson (2000) discuss an extension for interval data that allows the upper and lower bounds to be transformed. Although it is technically feasible to do so in our case, we do not believe that transformations are useful for symbolic MDS with interval or histogram data. The reason is that by having the available information of a given interval for each dissimilarity, it seems unnatural to destroy this information. Therefore, we recommend applying symbolic MDS without any transformation.

As the HIST-SCAL algorithm is based on iterative majorization, each majorizing function is quadratic in the parameters. Therefore, restrictions as the extension of symbolic MDS to three-way data (by, for example, the weighted Euclidean model) can be easily derived combined with this algorithm. We intend to pursue these extensions in future publications.

References

BOCK, H.H. and DIDAY, E. (2000): *Analysis of Symbolic Data* Springer, Berlin.

BORG, I. and GROENEN, P.J.F. (2005): *Modern Multidimensional Scaling: Theory and Applicatons, Second Edition* Springer, New York.

CARROLL, J.D. and CHANG, J.J. (1972): Analysis of individual differences in multidimensinal scaling via an N-way generalization of Eckart-Young decomposition. *Psyhometika, 35, 283-319.*

DENŒUX, T. and MASSON, M. (2000): Multidimensinal scaling of interval-valued dissimilarity data. *Pattern Recognition Letters, 21, 83-92.*

GROENEN, P.J.F., WINSBERG, S., RODRIGUEZ, O. and DIDAY, E. (in press): I-Scal: Multidimensionl scaling of interval dissimilarities. *Computational Statistics and Data Analysis.*

KRUSKAL, J.B. (1964): Multidimensional scaling by optimizing goodnes of fit to a nonmetric hypothesis. *Psychometrika, 29, 1-27.*

MASSON, M. and DENŒUX, T. (2002): Multidimensional scaling of fuzzy dissimilarity data; *Fuzzy Sets and Systems, 128, 339-352.*

WINSBERG, S. and CARROLL, J.D. (1989): A quasi-nonmetric method for multidimenional scaling via an extended Euclidean model. *Psychomtrika, 54, 217-229.*

WINSBERG, S. and DESOETE, G. (1993): A latent class approach to fitting the weighted Euclidean model, CLASCAL. *Psychometika, 58, 31-330.*

Dependence and Interdependence Analysis for Interval-Valued Variables

Carlo Lauro[1] and Federica Gioia[2]

Dipartimento di Matematica e Statistica
Università di Napoli Federico II
Complesso Universitario di Monte S. Angelo, Via Cinthia
I-80126 Napoli, Italy

Abstract. Data analysis is often affected by different types of errors as: measurement errors, computation errors, imprecision related to the method adopted for estimating the data. The methods which have been proposed for treating errors in the data, may also be applied to different kinds of data that in real life are of interval type. The uncertainty in the data, which is strictly connected to the above errors, may be treated by considering, rather than a single value for each data, the interval of values in which it may fall: *the interval data*. The purpose of the present paper is to introduce methods for analyzing the *interdependence* and *dependence* among *interval-valued* variables. Statistical units described by interval-valued variables can be assumed as a special case of Symbolic Object (SO). In Symbolic Data Analysis (SDA), these data are represented as boxes. Accordingly, the purpose of the present work is the extension of the *Principal Component Analysis* to obtain a visualization of such boxes, on a lower dimensional space. Furthermore, a new method for fitting an *interval simple linear regression* equation is developed. With difference to other approaches proposed in the literature that work on scalar recoding of the intervals using classical tools of analysis, we make extensively use of the interval algebra tools combined with some optimization techniques.
keywords: interval-valued variable, interval algebra, visualization.

1 Introduction

The statistical modelling of many problems must account in the majority of cases of "errors" both in the data and in the solution. These errors may be for example: *measurement errors, computation errors, errors due to uncertainty in estimating parameters.* Interval algebra provides a powerful tool for determining the effects of uncertainties or errors and for accounting them in the final solution. Interval mathematics deals with numbers which are not single values but sets of numbers ranging between a maximum and a minimum value. Those sets of numbers are the sets all possible determinations of the errors. Modern developments of such an algebra were started by R.E. Moore (Moore (1966)). Main results may be found in Alefeld-Herzerberger (1983). The methods which have been proposed for treating errors in the data, may be as well applied to different kind of data that in real life are of interval type. For example: *financial data* opening value and closing value in a session), *customer satisfaction data* (expected or perceived characteristic of the

quality of a product), *tolerance limits in quality control, confidence intervals* of estimates from sample surveys, *query on a database.* Statistical indexes for interval-valued variables have been defined in Canal-Pereira (1998) as scalar statistical summaries. These scalar indexes, may cause loss of information inherent in the interval data. For preserving the information contained in the interval data many researchers and in particular Diday and his school of Symbolic Data Analysis (SDA) have developed some methodologies for interval data which provide interval index solutions that sometimes appear oversized as they include unsuitable elements. An approach, which is typical for handling imprecise data, is proposed by Marino-Palumbo (2003). *Interdependence analysis* among interval-valued variables aims to extend classical methods for analyzing the relations existing among more than two interval-valued variables. Methods for Factorial Analysis and in particular for Principal Component Analysis (PCA) on interval data, has been proposed by Cazes et al. (1997), Lauro-Palumbo (2000), Gioia (2001), Palumbo-Lauro (2003), Gioia-Lauro (2005b). Statistical units described by interval data can be assumed as a special case of Symbolic Object (SO). In Symbolic Data Analysis (SDA), these data are represented as boxes. The purpose of the present work is the extension of Principal Component Analysis (PCA) to obtain a visualisation of such boxes, in a lower dimensional space pointing out the relationships among the variables, the units, and between both of them. The introduced methodology, named *Interval Principal Component Analysis* (IPCA) will embrace classical PCA as special case.
Dependence analysis between two interval-valued variables has been treated in Billard-Diday (2000) and Billard-Diday (2002): methods for studying the linear relationship between two interval-valued variables are introduced. The authors derive the results as a combination of two different regression equations for single-valued variables. An alternative methodology, is proposed by Marino-Palumbo (2003) with an approach which takes into account the centre and the radius of each considered interval and the relations between these two quantities. In the present work, a new method for fitting a simple linear regression equation for interval-valued variables (IRA) is developed. Following our previous paper Gioia-Lauro (2005a), both IPCA and RIA are based on interval algebra and optimization tools whereas existing methods are based on scalar recoding of the intervals and on classical tools of analysis. The proposed method has been tested on real data sets and the numerical results have been compared with the one already in the literature.

2 Interval algebra

An interval $[a, b]$ with $a \leq b$, is defined as the set of real numbers between a and b: $[a, b] = \{x : a \leq x \leq b\}$. Degenerate intervals of the form $[a, a]$, also named *thin* intervals, are equivalent to real numbers. Let $\Im$ be the set of intervals. Thus if $I \in \Im$ then $I = [a, b]$ for some $a \leq b$. Let us introduce an arithmetic on the elements of $\Im$. The arithmetic will be an extension of real arithmetic. If $\bullet$ is one of the symbols $+, -, \cdot, /$, we define arithmetic operations on intervals by:

$$[a,b] \bullet [c,d] = \{x \bullet y: \; a \leq x \leq b, \; c \leq y \leq d\} \tag{1}$$

except that we do not define $[a,b] \setminus [c,d]$ if $0 \in [c,d]$.
The sum, the difference, the product, and the ratio (when defined) between two intervals is the set of the sums, the differences, the products, and the ratios between any two numbers from the first and the second interval respectively (Moore (1966)).

Definition 2.1
An $n \times p$ interval matrix is the following set:

$$X^I = [\overline{X}, \underline{X}] = \{X : \underline{X} \leq X \leq \overline{X}\}$$

where $\underline{X}$ and $\overline{X}$ are $n \times p$ matrices satisfying: $\underline{X} \leq \overline{X}$; the inequalities are understood to be component wise. Operations between interval matrices are defined in Alefeld-Herzerberger (1983).
Given an $n \times p$ interval data matrix X^I, a lot of research has been done in characterizing solutions of the following interval eigenvalues problem:

$$X^I \mathbf{u}^I = \lambda \mathbf{u}^I \tag{2}$$

which has interesting properties (Deif (1991a)), and it serves a wide range of applications in physics and engineering.
The interval eigenvalue problem (2) is solved by determining two sets λ_α^I and $\mathbf{u}_\alpha^I$ given by:

$$\lambda_\alpha^I = [\lambda_\alpha(X) : \; X \in X^I] \text{ and } \mathbf{u}_\alpha^I = [\mathbf{u}_\alpha(X) : \; X \in X^I]$$

where $(\lambda_\alpha^I, \mathbf{u}_\alpha^I)$ is an eigenpair of $X \in X^I$. The couple $(\lambda_\alpha^I, \mathbf{u}_\alpha^I)$ will be the α-th eigenpair of X^I and it represents the set of all α-th eigenvalues and the set of the corresponding eigenvectors of all matrices belonging to the interval matrix X^I. Results for computing interval eigenvalues, interval singular values and interval eigenvectors of an interval matrix may be found in Deif (1991a), Deif (1991b), Seif et al. (1992).

3 Principal component analysis on interval data

The task of the present section is to adapt the mathematical models, on the basis of the classical PCA, to the case in which an $n \times p$ interval data matrix matrix X^I is given.

Let us suppose that the interval-valued variables have been previously standardized (Gioia-Lauro (2005b)).

It is known that the classical PCA on a real matrix X, in the space spanned by the variables, solves the problem of determining $m \leq p$ axes $\mathbf{u}_\alpha$, $\alpha = 1, \ldots, m$ such that the sum of the squared projections of the point-units on $\mathbf{u}_\alpha$ is maximum:

$$\mathbf{u}_\alpha^\mathsf{T} X^\mathsf{T} X \mathbf{u}_\alpha = Max, \quad 1 \leq \alpha \leq m \tag{3}$$

under the constraints:

$$\begin{cases} \mathbf{u}_\alpha^\mathsf{T} \mathbf{u}_\beta = 0 & \alpha \neq \beta \\ \mathbf{u}_\alpha^\mathsf{T} \mathbf{u}_\beta = 1 & \alpha = \beta \end{cases}$$

The above optimization problem may be reduced to the eigenvalue problem:

$$X^\mathsf{T} X \mathbf{u}_\alpha = \lambda \mathbf{u}_\alpha \quad 1 \leq \alpha \leq m \tag{4}$$

When the data are of interval type, X^I may be substituted in (4) and the interval algebra may be used for the products; equation (4) becomes an *interval eigenvalue problem* of the form:

$$(X^I)^\mathsf{T} X^I \mathbf{u}_\alpha = \lambda^I \mathbf{u}_\alpha \tag{5}$$

which has the following interval solutions:

$$\left[\lambda_\alpha(Z) : Z \in (X^I)^\mathsf{T} X^I\right], \left[\mathbf{u}_\alpha(Z) : Z \in (X^I)^\mathsf{T} X^I\right], \quad \alpha = 1, \ldots, p \tag{6}$$

i.e., the set of α-th eigenvalues of any matrix Z contained in the interval product $(X^I)^\mathsf{T} X^I$, and the set of the corresponding eigenvectors respectively. The intervals in (6) may be computed as in Deif (1991a).

Using the interval algebra for solving problem (5), the *interval solutions* will be computed but, refer to worse, those intervals are *oversized* with respect to the *intervals of solutions* that we are searching for as it will be discussed below. For the sake of simplicity, let us consider the case $p = 2$, thus two interval-valued variables:

$$X_1^I = (X_{i1} = [\underline{x}_{i1}, \overline{x}_{i1}]), i = 1, \ldots, n, X_2^I = (X_{i2} = [\underline{x}_{i2}, \overline{x}_{i2}]), i = 1, \ldots, n$$

have been observed on the n considered units. X_1^I and X_2^I assume an *interval of values* on each statistical unit: we do not know the exact value of the components x_{i1} or x_{i2} for $i = 1, \ldots, n$, but only the *range* in which this value falls. In the proposed approach the task is to contemplate *all possible values* of the components x_{i1}, x_{i2} each of which in its own interval of values $X_{i1} = [\underline{x}_{i1}, \overline{x}_{i1}])$, $X_{i2} = [\underline{x}_{i2}, \overline{x}_{i2}])$ for $i = 1, \ldots, n$. Furthermore for each different set of values $x_{11}, x_{21}, \ldots, x_{n1}$ and $x_{12}, x_{22}, \ldots, x_{n2}$, where $x_{ij} \in \left[\underline{x}_{ij}, \overline{x}_{ij}\right]$, $i = 1, \ldots, n, j = 1, 2$ a different cloud of points in the plane is uniquely determined and the PCA on that set of points must be computed. Thus, with interval PCA (IPCA) we must determine the set of solutions of the classical PCA on each set of point-units, set which is uniquely determined for any different choice of the point-units each of which is in its own rectangle of variation.

Therefore, the *interval of solutions* which we are looking for are the set of the α-th axes, each of which maximize the sum of square projections of a set of points in the plane, and the set of the *variances* of those sets of points respectively. This is equivalent to solve the optimization problem (3), and so the eigenvalue problem (4) for each matrix $X \in X^I$. In the light of the above considerations, the background in approaching directly the interval eigenvalue problem (5), comes out by observing that the following inclusion holds:

$$(X^I)^\mathsf{T} X^I = \{XY \ : \ X \in (X^{\mathsf{I}})^T, Y \in X^I\} \supset \{X^\mathsf{T} X \ : \ X \in X^I\} \qquad (7)$$

this means that in the interval matrix $(X^I)^\mathsf{T} X^I$ are contained also matrices which *are not* of the form $X^\mathsf{T} X$. Thus the *interval eigenvalues* and the *interval eigenvectors* of (7) will be *oversized* and in particular will *include* the set of all eigenvalues and the set of the corresponding eigenvectors of any matrix of the form $X^\mathsf{T} X$ contained in $(X^I)^\mathsf{T} X^I$. This drawback may be solved by computing an interval eigenvalue problem considering instead of the product:

$$(X^I)^\mathsf{T} X^I = \{XY \ : \ X \in (X^{\mathsf{I}})^T, Y \in X^I\}$$

the following set of matrices:

$$\Theta^I = \{(X^I)^\mathsf{T} X^I \ : \ X \in (X^{\mathsf{I}})^T\}$$

i.e., the set of all matrices given by the product of a matrix multiplied by its transpose. For computing the α-th eigenvalue and the corresponding eigenvector of set Θ, that will still be denoted by λ_α^I and $\mathbf{u}_\alpha^I$, the singular values of X^I may be computed as in Deif (1991b). The α-th interval axis or interval factor will be the α-th interval eigenvector associated with the α-th interval eigenvalue in decreasing order [1]. The *orthonormality* between couples of interval axes must be interpreted according to:

$$\forall \mathbf{u}_\alpha \in \mathbf{u}_\alpha^I \ \textit{such that} \ \mathbf{u}_\alpha^\mathsf{T} \mathbf{u}_\alpha = 1, \exists \mathbf{u}_\beta \in \mathbf{u}_\beta^I \ \textit{with} \ \alpha \neq \beta \ \textit{such that} \ \mathbf{u}_\beta^\mathsf{T} \mathbf{u}_\beta = 1 \ : \ \mathbf{u}_\alpha^\mathsf{T} \mathbf{u}_\beta = 0$$

Thus two interval axes are orthonormal to one another if, taking a unitary vector in the first interval axis there exists a unitary vector in the second one so that their scalar product is zero. In the classical case the importance

[1] Considering that the α-th eigenvalue of Θ^I is computed by perturbing the α-th eigenvalue of $(X^c)^\mathsf{T} X^c$, the ordering on the interval eigenvalues is given by natural ordering of the corresponding scalar eigenvalues of $(X^c)^\mathsf{T} X^c$

explained by the α-th factor is computed by: $\frac{\lambda_\alpha}{\sum_{\beta=1}^{p}\lambda_\beta}$. In the interval case the importance of each interval factor is the interval:

$$\left[\frac{\underline{\lambda}_\alpha}{\underline{\lambda}_\alpha+\sum_{\beta=1,\beta\neq\alpha}^{p}\overline{\lambda}_\beta}, \frac{\overline{\lambda}_\alpha}{\overline{\lambda}_\alpha+\sum_{\beta=1,\beta\neq\alpha}^{p}\underline{\lambda}_\beta}\right] \tag{8}$$

i.e., the set of all ratios of variance explained by each real factor $\mathbf{u}_\alpha$ belonging to the interval factor $\mathbf{u}_\alpha^I$.

Analogously to what already seen in the space $\Re^p$, in the space spanned by the units ($\Re^n$), the eigenvalues and the eigenvectors of the set

$$(\Theta^\mathsf{T})^I = \{XX^\mathsf{T} \ : \ X \in X^I\}$$

must be computed as in Deif (1991b); the α-th interval axis will be the α-th interval eigenvector associated with the α-th interval eigenvalue in decreasing order. It is known that a real matrix and its transpose have the same eigenvalues and the corresponding eigenvectors connected by a particular relationship. Let us indicate again with $\lambda_1^I, \lambda_2^I, \ldots, \lambda_p^I$ the interval eigenvalues of $(\Theta^\mathsf{T})^I$ and with $\mathbf{v}_1^I, \mathbf{v}_2^I, \ldots, \mathbf{v}_p^I$ the corresponding eigenvectors, and let us see how the above relationship applies also for the "interval" case. Let us consider for example the α-th interval eigenvalue λ_α^I and let $\mathbf{u}_\alpha^I$, $\mathbf{v}_\alpha^I$ be the corresponding eigenvectors of Θ^I and $(\Theta^\mathsf{T})^I$ associated with λ_α^I respectively. Taking an eigenvector of some $X^\mathsf{T}X : \mathbf{v}_\alpha \in \mathbf{v}_\alpha^I$, then:

$$\exists \mathbf{u} \in \mathbf{u}_\alpha^I \ : \ \mathbf{u}_\alpha = k_\alpha X^\mathsf{T}\mathbf{u}_\alpha$$

where the constant k_α is introduced for the condition of unitary norm of the vector $X^\mathsf{T}\mathbf{v}_\alpha$.

Representation and interpretation

From classical theory, given an $n \times p$ real matrix X we know that the α-th principal component $\mathbf{c}_\alpha$ is the vector of the coordinates of the n units on the α-th axis. Two different approaches may be used to compute $\mathbf{c}_\alpha$:

1. $\mathbf{c}_\alpha$ may be computed by multiplying the standardized matrix X by the $\alpha - th$ computed axis $\mathbf{u}_\alpha : X\mathbf{u}_\alpha$
2. from the relationship (3) among the eigenvectors of $X^\mathsf{T}X$ and XX^T, $\mathbf{c}_\alpha$ may be computed by the product $\sqrt{\lambda_\alpha}\cdot\mathbf{v}_\alpha$ of the α-th eigenvalue of XX^T with the corresponding eigenvector.

When an $n \times p$ interval-valued matrix X^I is given, the *interval coordinate* of the i-th interval unit on the α-th interval axis, is a representation of an interval which comes out from a linear combination of the original intervals of the i-th unit by p interval weights; the weights are the interval components of the α-th interval eigenvector. A box in a bi-dimensional space of representation, is a rectangle having for dimensions the *interval coordinates* of the corresponding unit on the couple of computed interval axis. For computing the α-th interval principal component $\mathbf{c}_\alpha^I = (c_{1\alpha}^I, c_{2\alpha}^I, \ldots, c_{n\alpha}^I)$ two different approaches may be used:

1. compute by the interval row-column product: $\mathbf{c}_\alpha^I = X^I \mathbf{u}_\alpha^I$,
2. compute the product between a constant interval and an interval vector: $\mathbf{c}_\alpha^I = \sqrt{\lambda_\alpha^I} \mathbf{v}_\alpha^I$

In both cases, the interval algebra product is used thus, the i-th component $c_{i\alpha}^I$ of $\mathbf{c}_\alpha^I$ will *include* the interval coordinate, as it has been defined above, of the i-th interval unit on the α-th interval axis. We refer to the first approach, for computing the principal components, when the theorem for solving the eigenvalue problems (for computing $\mathbf{v}_\alpha^I$) cannot be applied if its hypotheses are not verified. Classical PCA gives a representation of the results by means of graphs, which permit us to represent the units on projection planes spanned by couples of factors. The methodology (IPCA), that we have introduced, permits to visualize on planes how the coordinates of the units vary when each component, of the considered interval-valued variable, ranges in its own interval of values, or equivalently when each point-unit describes the boxes to which it belongs. Indicating with U^I the interval matrix whose j-th column is the interval eigenvector $\mathbf{u}_\alpha^I$ $(a = 1, \ldots, p)$, the coordinates of all the interval-units on the computed interval axis are represented by the interval product $X^I U^I$. In the classical case, the coordinate of the i-th variable on the α-th axis is the correlation coefficient between the considered variable and the α-th principal component. Thus variables with greater coordinates (in absolute value) are those which best characterize the factor under consideration. Furthermore, the standardization of each variable makes the variables, represented in the factorial plane, fall inside the correlation circle. In the interval case the interval coordinate of the i-th interval-valued variable on the α-th interval axis is the interval correlation coefficient (Gioia-Lauro (2005)) between the variable and the α-th interval principal component. The interval variables in the factorial plane however, are represented, not in the circle but in the rectangle of correlations. In fact, computing all possible couple of elements, each of which in its own interval correlation, may happens that couples with the coordinates that are not in relation one another would be also represented; i.e. couples of elements which are correlations of different *realizations* of the two single-valued variables for which the correlation would be considered. The interval coordinate of the i-th interval-valued variable on the first two interval axes $\mathbf{u}_\alpha^I \mathbf{u}_\beta^I$, namely, the interval correlation between the considered variable and the first and second interval principal component respectively, will be computed according to the procedure in Gioia-Lauro (2005) and indicated as follow:

$$\begin{aligned} corr((X\mathbf{u}_\alpha)^I, X_i^I) &= [\overline{corr}(\mathbf{u}_\alpha, i), \underline{corr}(\mathbf{u}_\alpha, i)] \\ corr((X\mathbf{u}_\beta)^I, X_i^I) &= [\overline{corr}(\mathbf{u}_\beta, i), \underline{corr}(\mathbf{u}_\beta, i)] \end{aligned} \quad (9)$$

Naturally the rectangle of correlations will be restricted, in the representation plane, to its intersection with the circle with center in the origin and unitary radius.

Numerical results

This section shows an example of the proposed methodology on a real data set: the Oil data set (Ichino 1988) (**Table 1**). The data set presents eight different classes of oils described by four quantitative interval-valued variables: "Specific gravity", "Freezing point", "Iodine value", "Saponification".

Table 1. The Oil data set

	Spec. gravity	*Freezing point*	*Iodine value*	*Saponification*
Linseed	0.93 0.94	-27 -18	170 204	118 196
Perilla	0.93 0.94	-5 -4	192 208	188 197
Cotton	0.92 0.92	-6 -1	99 113	189 198
Sesame	0.92 0.93	-6 -4	104 116	187 193
Camellia	0.92 0.92	-21 -15	80 82	189 193
Olive	0.91 0.92	0 6	79 90	187 196
Beef	0.86 0.87	30 38	40 48	190 199
Hog	0.86 0.86	22 32	53 77	190 202

The first step of the IPCA consists in calculating the following interval correlation matrix:

Table 2. The interval correlation matrix

	Spec. gravity	*Freezing point*	*Iodine value*	*Saponification*
Spec. gravity	[1.00, 1.00]			
Freezing point	[-0.97, -0.80]	[1.00, 1.00]		
Iodine value	[0.62, 0.88]	[-0.77, -0.52]	[1.00, 1.00]	
Saponification	[-0.64, -0.16]	[0.30, 0.75]	[-0.77, -0.34]	[1.00, 1.00]

The interpretation of the interval correlations must take into account both the location and the span of the intervals. Intervals containing the zero are not of interest because they indicate that "everything may happen". An interval with a radius smaller than that of another one is more interpretable. In fact as the radius of the interval correlations decreases, the stability of the correlations improves and a better interpretation of the results is possible. In the considered example, the interval correlations are well interpretable because all intervals do not contain the zero, thus each couple of interval-valued

variables are positively correlated or negatively correlated. For example we observe a strong positive correlation between Iodine and Specific gravity and a strong negative correlation between Freezing point and Specific gravity. At equal lower bounds, the interval correlation between Iodine value and Freezing point is more stable than that between Iodine value and Saponification.
Eigenvalues and explained variance:
$\lambda_1 = [2.45, 3.40]$, *Explained Variance on the* 1*st axes*: $[61\%, 86\%]$
$\lambda_2 = [0.68, 1.11]$, *Explained Variance on the* 2*nd axes*: $[15\%, 32\%]$
$\lambda_3 = [0.22, 0.33]$, *Explained Variance on the* 1*st axes*: $[4\%, 9\%]$
$\lambda_4 = [0.00, 0.08]$, *Explained Variance on the* 1*st axes*: $[0\%, 2\%]$.
The choice of the eigenvalues and so of the interval principal components may be done using the interval eigenvalue-one criterion $[1, 1]$. In the numerical example, only the first principal component is of interest because the lower bound of the corresponding eigenvalue is greater than 1. The second eigenvalue respects the condition of the interval eigenvalue-one partially and, moreover, it is not symmetric with respect to 1. Thus the representation on the second axis is not of great interest even though the two first eigenvalues reconstruct most part of the initial variance. Thus, the second axis is not well interpretable. *Interval variables representation*: The principal components representation is made analyzing the correlations among the interval-valued variables and the axes, as illustrated below:

Table 3. Correlations variables/axes

	Spec. gravity	*Freezing point*	*Iodine value*	*Saponification*
Correlations Vars/1st axes	[-0.99, -0.34]	[0.37, 0.99]	[-0.99, -0.20]	[-0.25, 0.99]
Correlations Vars/2st axes	[-0.99, 0.99]	[-0.99, 0.99]	[-0.99, 0.99]	[-0.99, 0.99]

The first axis is well explained by the contraposition of the variable *Freezing point*, on the positive quadrant, with respect to the variables *Specific gravity* and *Iodine value* on the negative quadrant. The second axis is less interpretable because all the correlations vary from -0.99 and 0.99. Here below, the graphical results achieved by IPCA on the input data table are shown. In **Figure 1** the graphical representation of the units is presented; in **Figure 2** *Specific gravity* and *Freezing point* are represented.

The objects (**Figure 1**) have a position on the first axis which is strictly connected to the "influence" that the considered variables have on that axis. It can be noticed that *Beef* and *Hog* are strongly influenced by *Saponification* and *Freezing point*; on the contrary *Linseed* and *Perilla* are strongly influenced by *Specific gravity* and *Iodine value*. The other oils Camilla and Olive, are positioned in the central zone so they are not particularly characterized by the interval-valued variables.

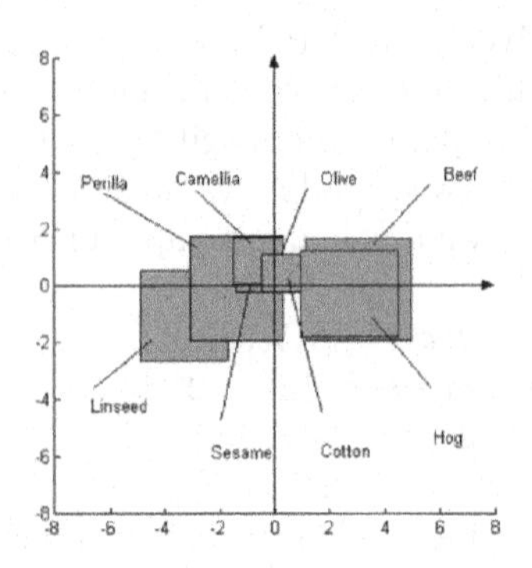

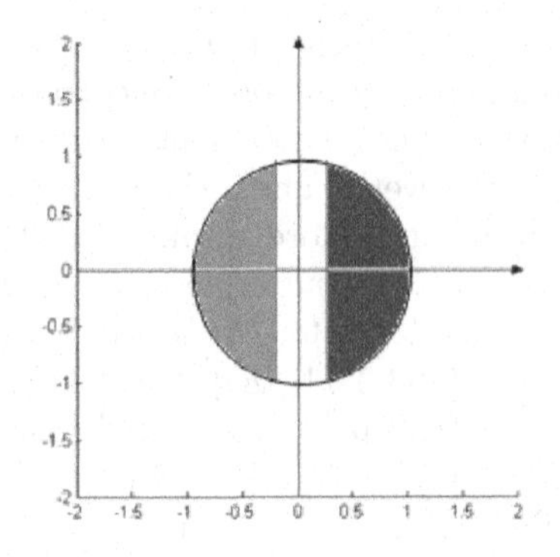

Fig. 1. Left figure: Representation of the units on the 1st factorial plane; **Right figure:** Representation of the variables on the first factorial plane.

It is important to remark that the different oils are characterized not only by the positions of the boxes but also by their *size* and *shape*. A bigger size of a box with respect to the first axis, remarks a greater variability of the characteristics of the oil represented by the first axis. However also the shape and the position of the box can give information on the variability of the characteristics of the oil, with respect to the first and second axis.

4 Interval simple regression

In this section we propose an extension of simple linear regression to the case of interval data. Let us indicate with X^I and Y^I the independent and the dependent interval-valued variables respectively, which assume the following interval values on the n statistical units chosen for our experiment:

$$X^I = (X_i = [\underline{x}_i, \bar{x}_i]), \quad i = 1, \cdots, n$$

$$Y^I = \left(Y_i = [\underline{y}_i, \bar{y}_i]\right), \quad i = 1, \cdots, n$$

The aim is to take into account *all possible values* of the components x_i y_i each of which is in its interval of values $[\underline{x}_i, \bar{x}_i]$, $[\underline{y}_i, \bar{y}_i]$ for $i = 1, \cdots n$. Thus making regression between two interval-valued variables means to compute the *set of regression lines* each of which realizes the best fit, in the Minimum Least Square sense, of a set of points in the plane. This set of points is univocally determined each time the components $x_1, x_2, \cdots, x_n$, $y_1, y_2, \cdots, y_n$ take a particular value in their own interval of variation. Mathematically computing the interval regression line between two interval-valued variables X^I and Y^I is equivalent to compute the following two sets:

$$\hat{\beta}^I = \left\{ \hat{\beta}(X,Y) = \frac{\sum_{i=1}^{n}(x_i - \overline{x})(y_i - \overline{y})}{\sum_{i=1}^{n}(x_i - \overline{x})^2}, \quad X \in X^I, Y \in Y^I \right\} \tag{10}$$

$$\hat{\alpha}^I = \left\{ \hat{\alpha}(X,Y) = \bar{y} - \hat{\beta}\bar{x}, \quad X \in X^I, Y \in Y^I \right\} \tag{11}$$

where and $\bar{x}$ and $\bar{y}$, regarded as functions of $x_1, x_2, \cdots, x_n$, $y_1, y_2, \cdots, y_n$, are given by:

$$\bar{x} = \frac{1}{n}\sum_{i=1}^{n} x_i \quad ; \quad \bar{y} = \frac{1}{n}\sum_{i=1}^{n} y_i$$

Sets (10) and (11) are respectively the set described by the slopes and the set described by the intercepts of all regression lines

$$y = \hat{\alpha} + \hat{\beta}x$$

varying x_i, y_i in their own interval of values $X_i = [\underline{x}_i, \bar{x}_i]$, $Y_i = \left[\underline{y}_i, \bar{y}_i\right]$ for $i = 1, \cdots, n$. These sets may be computed numerically by solving some optimization problems; i.e., searching for the minimum and for the maximum of functions $\hat{\alpha}(X,Y)$ and $\hat{\beta}(X,Y)$ in (10) and (11). These functions are both continuous[2] on a connected and compact set and this assures that sets (10) and (11) are the following closed intervals:

$$\hat{\beta}^I = \left[\min_{\substack{X \in X^I \\ Y \in Y^I}} \beta(\hat{X,Y}), \max_{\substack{X \in X^I \\ Y \in Y^I}} \beta(\hat{X,Y}) \right] \tag{12}$$

$$\hat{\alpha}^I = \left[\min_{\substack{X \in X^I \\ Y \in Y^I}} \alpha(\hat{X,Y}), \max_{\substack{X \in X^I \\ Y \in Y^I}} \alpha(\hat{X,Y}) \right] \tag{13}$$

The *interval regression line* may be written as:

$$[\underline{y}, \overline{y}] = \hat{\alpha}^I + \hat{\beta}^I [\underline{x}, \overline{x}] \tag{14}$$

[2] The quantity $\sum_{i=1}^{n}(x_i - \overline{x})^2$ could be nil only in the case in which: $x_1 = x_2 = \cdots = x_n$. This is in contradiction with the classic hypothesis that at least two different observations must be available in the experiment.

and may be interpreted as follow: *chosen* an intercept $\hat{\alpha}$ in the interval $\hat{\alpha}^I$ it exists a slope $\hat{\beta}$ in the interval $\hat{\beta}^I$ so that the regression line:

$$y = \hat{\alpha} + \hat{\beta}x \tag{15}$$

is the unique line that realizes the *best fit*, in the of Minimum Least Square sense, of a given set of points $(x_1, y_1), (x_2, y_2), \ldots, (x_n, y_n)$ in the plane $(x_i \in [\underline{x}_i, \bar{x}_i],\ y_i \in [\underline{y}_i, \bar{y}_i],\ i = 1, \ldots, n)$.
Given the interval $[\underline{x}_i, \overline{x}_i]$ of the independent variable X^I, the prevision $[\underline{y}_i, \bar{y}_i]$ of the interval value assumed by the dependent variable Y^I is: $[\underline{y}_i, \bar{y}_i] = \hat{\alpha} + \hat{\beta}[\underline{x}_i, \bar{x}_i]$.

Numerical results

In the following example (Marino-Palumbo (2003)) we will take into account the relationship between the weaving of ground and the water retention. We will compute a simple regression involving the water retention as dependent variable and the bulk density (for the weaving of ground) as independent one. Let us consider the (49 × 2) interval data matrix of **Table 4**, we find the following interval intercepts and interval slope: $\beta_0 = [0.79, 1.05], \beta_1 = [-0.46, -0.25]$ which are narrower than those calculated in Marino-Palumbo (2003).

Table 4. The data set

water retention min	*water retention max*	*bulk density min*	*bulk density max*
0.5885	0.6127	0.9609	0.9632
0.6261	0.6261	0.9350	0.9350
.	.	.	.
.	.	.	.
0.4482	0.4656	1.2156	1.2418
0.4931	0.4931	1.1345	1.1345

References

ALEFELD, G. and HERZERBERGER, J. (1983): *Introduction to Interval computation.* Academic Press, New York.

BILLARD, L. and DIDAY, E. (2000): Regression Analysis for Interval-Valued Data. In: H.-H. Bock and E. Diday (Eds.): *Data Analysis, Classification and Related Methods.* Springer, Berlin, 123-124.

BILLARD, L. and DIDAY, E. (2002): Symbolic regression Analysis. In: Proceedings IFCS. Krzysztof Jajuga et al (Eds.): *Data Analysis, Classification and Clustering Methods Heidelberg.* Springer-Verlag.

CANAL, L. and PEREIRA, M. (1998): In: Proceedings of the NTTS'98 Seminar: *Towards statistical indices for numeroid data.* Sorrento, Italy.

CAZES, P. CHOUAKRIA, A. DIDAY, E. and SCHEKTMAN, Y. (1997): Extension de l'analyse en composantes principales des donnes de type intervalle. *Revue de Statistique Applique, XIV, 3, 5-24.*

DEIF, A.S. (1991a): The Interval Eigenvalue Problem. *ZAMM, 71, 1.61-64.* Akademic-Verlag Berlin.

DEIF, A.S. (1991b): Singular Values of an Interval Matrix. *Linear Algebra and its Applications, 151, 125-133.*

GIOIA, F. (2001): Statistical Methods for Interval Variables, Ph.D. thesis. Dep. of Mathematics and Statistics-University Federico II Naples, in Italian.

GIOIA, F. and LAURO, C.N. (2005a): Basic Statistical Methods for Interval Data. *Statistica Applicata, In press.*

GIOIA, F. and LAURO, C.N. (2005b): Principal Component Analysis on Interval Data. *Computational Statistics, special issue on: "Statistical Analysis of Interval Data". In press.*

LAURO, C.N. and PALUMBO, F. (2000): Principal component analysis of interval data: A symbolic data analysis approach. *Computational Statistics, 15, 1, 73-87.*

MARINO, M. and PALUMBO, F. (2002): Interval arithmetic for the evaluation of imprecise data effects in least squares linear regression. *Statistica Applicata, 14(3), 277-291.*

MOORE, R.E. (1966): *Interval Analysis.* Prentice Hall, Englewood Cliffs, NJ.

PALUMBO, F. and LAURO, C.N. (2003): In: New developments in Psychometrics, Yanai H. et al. (Eds.): *A PCA for interval valued data based on midpoints and radii.* Psychometric Society, Springer-Verlag, Tokyo.

SEIF, N.P. HASHEM, S. and DEIF, A.S. (1992): Bounding the Eigenvectors for Symmetric Interval Matrices. *ZAMM, 72, 233-236.*

A New Wasserstein Based Distance for the Hierarchical Clustering of Histogram Symbolic Data

Antonio Irpino and Rosanna Verde

Facoltá di Studi Politici e per l'Alta Formazione Europea e Mediterranea
"Jean Monnet", Seconda Universitá degli Studi di Napoli,
Caserta, I-81020, Italy

Abstract. Symbolic Data Analysis (SDA) aims to to describe and analyze complex and structured data extracted, for example, from large databases. Such data, which can be expressed as concepts, are modeled by symbolic objects described by multivalued variables. In the present paper we present a new distance, based on the Wasserstein metric, in order to cluster a set of data described by distributions with finite continue support, or, as called in SDA, by "histograms". The proposed distance permits us to define a measure of inertia of data with respect to a barycenter that satisfies the Huygens theorem of decomposition of inertia. We propose to use this measure for an agglomerative hierarchical clustering of histogram data based on the Ward criterion. An application to real data validates the procedure.

1 Introduction

Symbolic Data Analysis (Bock and Diday (2000), Billard and Diday (2003)) is a new approach in statistical analysis that aims to supply techniques and methodologies for the analysis of complex data. The complexity of data is related to their description. While classical data are described by a single value for each variable, symbolic data are described by multiple values for each descriptor. Further, in symbolic data description it is possible to take into account extra information expressed in terms of relationships between descriptors (hierarchical or dependence rules) or within descriptors (taxonomies). Such way to describe data may have two main sources. The most common (from a statistical point of view) is a compositional or a classification source: for example, when a group of records is extracted from a database, it needs a suitable representation by means of intervals of values, frequency distributions, and so on. For example, the weather conditions across a day can be described by the temperature, by the atmospheric conditions (cloudy or sunny) and by the directions and the forces of winds during the 24 hours. These three characteristics cannot be suitably synthesized by a single value. For example, temperature can be described by the interval of the observed values, or by the distribution of values observed at given times.
In the present paper we consider symbolic data described by a "histogram" of values, that is a particular kind of symbolic description (Bock and Diday

(2000)). Gibbs and Su (2002) present a good review about metrics between probability measures (histograms can be considered as the representation an empirical frequency distribution). In a different context of analysis, Chavent et al.(2003) propose two measure for the comparison of histograms: the L^2 norm and a *two component* dissimilarity. L^2 norm is simply computed considering the weights of the elementary intervals but not their width. While the *two component* is a dissimilarity which does not satisfy the usual properties of distance measures. In section 2, after recalling the definition of histogram data we present an extension of the Wassertein distance in order to compare two histograms descriptions. We also prove that is possible to define an inertia measure among data that satisfies the Huygens theorem. The last result allows to use the Ward criterion to cluster data according to a hierarchical agglomerative procedure. In section 3, we present some results for a real dataset. Section 4 reports some concluding remarks.

2 Wasserstein metric for histogram data

According to Billard and Diday (2003) histogram data can be considered as a special case of compositional data. Compositional data (Aitchison, 1986) are vectors of nonnegative real components having a constant sum; probability measures or histogram data can be considered as a special case of compositional data having sum equal to one. They can arise, for example, when it is necessary to synthesize information about a group of individuals. These types of data can be written as symbolic data by taking into account the variation inside a class of units assuming this class as the new statistical unit. In SDA, an histogram variable is a special kind of symbolic descriptor. It is a multi-valued descriptor with a frequency, probability, or weight associated with each of the values observed on the individuals. I.e., given a set of units $i \in \Omega$ a weighted variable Y is a mapping

$$Y(i) = \{S(i), \pi_i\}$$

for $i \in \Omega$, where π_i is a nonnegative measure or a distribution on the domain of Y and $S(i)$ is the support of π_i. In the case of histogram description it is possible to assume that $S(i) = [\underline{z}_i; \overline{z}_i]$, where $z_i \in \Re$. Considering a set of uniformly dense intervals $I_{hi} = [\underline{z}_{hi}, \overline{z}_{hi})$ such that

$$\begin{array}{ll} i. & I_{li} \cap I_{mi} = \emptyset; \;\; l \neq m \;; \\ ii. & \bigcup\limits_{s=1,\dots,n_i} I_{si} = [\underline{z}_i; \overline{z}_i] \end{array}$$

the support can be written also as $S(i) = \{I_{1i}, ..., I_{ui}, ..., I_{n_i i}\}$. In the present paper we denote by $\psi_i(z)$ the density function associated with the description of i and by $\Psi_i(z)$ its distribution function. It is possible to define the modal description of i as:

$$Y(i) = \{(I_{ui}, \pi_{ui}) \mid \forall I_{ui} \in S(i);\ \pi_{ui} = \int_{I_{ui}} \psi_i(z)dz \geq 0\} \text{ where } \int_{S(i)} \psi_i(z)dz = 1.$$

An example of histogram data of its graphical representation (Fig. 1) of the mean monthly temperatures recorded in Alabama from 1895 to 2004[1] is the following:

$$\begin{aligned} Y_{Jan}(Alab.) = \{ & ([32.5; 37.5], \mathbf{0.03}); ([37.5; 40], \mathbf{0.08}); ([40; 42.5], \mathbf{0.11}); \\ & ([42.5; 45], \mathbf{0.25}); ([45; 47.5], \mathbf{0.18}); ([47.5; 50], \mathbf{0.20}); \\ & ([50; 55], \mathbf{0.13}); ([55; 62.5], \mathbf{0.03})\} \end{aligned}$$

In the present paper, our main aim is to obtain a distance measure for

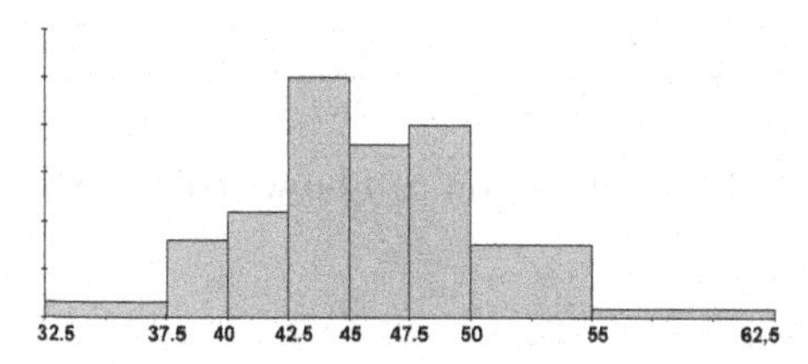

Fig. 1. Temperature in Fahrenheit degrees observed in Alabama in January from 1895 to 2004

comparing two histogram descriptions in order to perform a clustering of histogram data.

Among several measures presented in the literature (Gibbs and Su (2002)), we adopt the Wasserstein (or Kantorivich) metric as the most suitable for our aims, having some interesting properties. If F and G are the distribution functions of μ and ν respectively, the Kantorovich metric is defined by

$$d_W(\mu, \nu) := \int_{-\infty}^{+\infty} |F(x) - G(x)|\, dx = \int_0^1 \left|F^{-1}(t) - G^{-1}(t)\right| dt.$$

We are interested in computing the following distance, which derives from the previous one on considering a Euclidean norm:

$$d_W(Y(i), Y(j)) := \sqrt{\int_0^1 \left(\Psi_i^{-1}(w) - \Psi_j^{-1}(w)\right)^2 dw}. \tag{1}$$

[1] Data are freely available at http://www1.ncdc.noaa.gov/pub/data/cirs/

It is the well known Mallow's (1972) distance in L^2, derived from the Kantorovich metric, and it can be considered as the expected value of the squared Euclidean distance between homologous points of the supports of the two distributions.
Given an histogram description of i by means of n_i weighted intervals as follows:

$$Y(i) = \{(I_{1i}, \pi_{1i}), ..., (I_{ui}, \pi_{ui}), ..., (I_{n_i i}, \pi_{n_i i})\}$$

we may define the following function in order to represent the cumulative weights associated with the elementary intervals of $Y(i)$:

$$w_{li} = \begin{cases} 0 & if\ l = 0 \\ \sum_{h=1}^{l} \pi_{hi} & if\ l = 1, ..., n_i \end{cases} \tag{2}$$

Using (2), and assuming uniformity within the intervals, we may describe the distribution function as:

$$\Psi_i(z) = w_i + (z - \underline{z}_{li}) \frac{w_{li} - w_{l-1i}}{\underline{z}_{li} - \overline{z}_{li}} \quad iff\ \underline{z}_{li} \leq z \leq \overline{z}_{li}$$

Then the inverse distribution function can be written as the following piecewise function:

$$\Psi_i^{-1}(t) = \begin{cases} \underline{z}_{1i} + \frac{t}{w_{1i}} (\overline{z}_{1i} - \underline{z}_{1i}) & 0 \leq t < w_{1i} \\ \vdots & \vdots \\ \underline{z}_{li} + \frac{t - w_{l-1i}}{w_{li} - w_{l-1i}} (\overline{z}_{li} - \underline{z}_{li}) & w_{l-1i} \leq t < w_{li} \\ \vdots & \vdots \\ \underline{z}_{n_i i} + \frac{t - w_{n_i - 1i}}{1 - w_{n_i - 1i}} (\overline{z}_{n_i i} - \underline{z}_{n_i i}) & w_{n_i - 1i} \leq t < 1 \end{cases}$$

In order to compute distance between two histogram descriptions $Y(i)$ and $Y(j)$ we need to identify a set of uniformly dense intervals to compare. Let w be the set of the cumulative weights of the two distributions:

$$w = \left\{w_{0i}, ..., w_{ui}, ..., w_{n_i i}, w_{0j}, ..., w_{vi}, ..., w_{n_j j}\right\}.$$

In order to compute the distance, we need to sort w without the repetitions. The sorted values can be represented by a vector

$$\mathbf{w} = [w_0, ..., w_l, ..., w_m]$$

where: $w_0 = 0 \quad w_m = 1 \quad \max(n_i, n_j) \leq m \leq (n_i + n_j - 1)$.
The last vector permits us to prove that the computation of the square of (1) can be decomposed as the following sum:

$$d_W^2(Y(i), Y(j)) := \sum_{l=1}^{m} \int_{w_{l-1}}^{w_l} \left(\Psi_i^{-1}(t) - \Psi_j^{-1}(t)\right)^2 dt. \tag{3}$$

Each couple (w_{l-1}, w_l) permits to identify two uniformly dense intervals, one for i and one for j, having respectively the following bounds:

$$I_{li} = [\Psi_i^{-1}(w_{l-1}); \Psi_i^{-1}(w_l)] \quad \text{and} \quad I_{lj} = [\Psi_j^{-1}(w_{l-1}); \Psi_j^{-1}(w_l)].$$

As intervals are uniformly distributed, in order to show that the distance is decomposable into quadratic terms, we consider each interval according to a transformation into a function of t and of its center and radius as follows:

$$I = [a, b] \Leftrightarrow I(t) = c + r(2t-1) \ \ for \ 0 \leq t \leq 1 \ where \ c = \frac{a+b}{2} \ and \ r = \frac{b-a}{2}.$$

Using $\mathbf{w}$, we compute a vector of m weights $\pi = [\pi_1, \ldots, \pi_l, \ldots, \pi_m]$ where $\pi_l = w_l - w_{l-1}$. After some algebra, it is possible to rewrite the equation (3) as:

$$d_W^2(Y(i), Y(j)) := \sum_{l=1}^{m} \pi_l \int_0^1 \left[(c_{li} + r_{li}(2t-1)) - (c_{lj} + r_{lj}(2t-1))\right]^2 dt$$

$$d_W^2(Y(i), Y(j)) := \sum_{l=1}^{m} \pi_l \left[(c_{li} - c_{lj})^2 + \frac{1}{3}(r_{li} - r_{lj})^2\right]. \tag{4}$$

Given p histogram variables for the description of i and j, under the hypothesis that the variables are independent (as is usually assumed in SDA) it is possible to express a multivariate version of $d_W^2(Y(i), Y(j))$ as follows:

$$d_W^2(Y(i), Y(j)) := \sum_{k=1}^{p} \sum_{l=1}^{m_k} \pi_l^{(k)} \left[\left(c_{li}^{(k)} - c_{lj}^{(k)}\right)^2 + \frac{1}{3}\left(r_{li}^{(k)} - r_{lj}^{(k)}\right)^2\right]. \tag{5}$$

In the classification context, d_W has useful properties for the classification of histograms, especially when we need to define a measure of inertia of a set of histogram data. First of all, given a cluster of histograms data, it is possible to obtain the "barycenter" expressed again as an histogram.

Once we fix m (and hence also π_m) equal to the cardinality of the elementary intervals of the union of the supports of the $Y(i)$'s, the support of $Y(b)$ can be expressed as a vector of m couples of (c_{jb}, r_{jb}). The barycentric histogram can be computed as the solution of the minimization of the following function:

$$f(Y(b)|Y(1), \ldots, Y(i), \ldots, Y(n)) = f(c_{1b}, r_{1b}, \ldots, c_{mb}, r_{mb}) = \tag{6}$$

$$= \sum_{i=1}^{n} d^2(Y(i), Y(b)) = \sum_{i=1}^{n} \sum_{j=1}^{m} \pi_j \left[(c_{ji} - c_{jb})^2 + \frac{1}{3}(r_{ji} - r_{jb})^2\right]$$

that is minimized when the usual first order conditions are satisfied:

$$\begin{cases} \frac{\partial f}{\partial c_{jb}} = -2\pi_j \sum_{i=1}^{n} \left[(c_{ji} - c_{jb})\right] = 0 \\ \frac{\partial f}{\partial r_{jb}} = -\frac{2}{3}\pi_j \sum_{i=1}^{n} \left[(r_{ji} - r_{jb})\right] = 0 \end{cases}$$

for each $j = 1, \ldots, m$. Then, the function (6) has a minimum when:

$$c_{jb} = n^{-1} \sum_{i=1}^{n} c_{ji} \quad ; \quad r_{jb} = n^{-1} \sum_{i=1}^{n} r_{ji}.$$

We may write the barycentric histogram description of n histogram data as:

$$\begin{aligned} Y(b) = \{([c_{1b} - r_{1b}; c_{1b} + r_{1b}], \pi_1); \ldots; ([c_{jb} - r_{jb}; c_{jb} + r_{jb}], \pi_j); \ldots; \\ ; \ldots; ([c_{mb} - r_{mb}; c_{mb} + r_{mb}], \pi_m)\}. \end{aligned}$$

The identification of the barycenter permits us to show a second property of the proposed distance. It is possible to express a measure of inertia of data using d_W^2. The total inertia(TI), with respect a barycentric description $Y(b)$ of a set of n histogram data, is given by:

$$TI = \sum_{i=1}^{n} d_W^2(Y(i), Y(b)).$$

We here show that TI can be decomposed into the within (WI) and between (BI) clusters inertia, according to the Huygens theorem. Let us consider a partition of Ω into k clusters. For each cluster C_h, $h = 1, .., k$, a barycenter denoted as $Y(b_h)$ is computed by a local optimization of (6). Minimizing the following function:

$$\begin{aligned} & f(Y(b)|Y(b_1), \ldots, Y(b_h), \ldots, Y(b_k)) = f(c_{1b_1}, r_{1b_1}, \ldots, c_{mb_k}, r_{mb_k}) = \\ & = \sum_{h=1}^{k} \tfrac{|C_h|}{n} d^2(Y(b_h), Y(b)) = \sum_{h=1}^{k} \tfrac{|C_h|}{n} \sum_{j=1}^{m} \pi_j \left[(c_{jb_h} - c_{jb})^2 + \tfrac{1}{3} (r_{jb_h} - r_{jb})^2 \right] \end{aligned} \quad (7)$$

where $|C_h|$ is the cardinality of cluster C_h, it is possible to prove that the problem (6) and (7) have the same solution for $Y(b)$. The last result permits us to obtain a decomposition of the total inertia[2] as follows:

$$\begin{aligned} TI = WI + BI = \\ = \sum_{h=1}^{k} \sum_{i \in C_h} d_W^2(Y(i), Y(b_h)) + \sum_{h=1}^{k} |C_h| d_W^2(Y(b_h), Y(b)). \end{aligned} \quad (8)$$

In SDA, this result assumes a great interest because few distances proposed in the literature for this kind of data respects this property.

In order to exploit the properties of the proposed distance, we propose a hierarchical agglomerative clustering algorithm of histogram data using the Ward criterion (Ward (1963)). Indeed, given two disjoint clusters C_s and C_t the inertia of their union can be computed as follows:

$$TI(C_s \cup C_t) = TI(C_s) + TI(C_t) + \frac{|C_s||C_t|}{|C_s| + |C_t|} d_W^2(Y(b_s), Y(b_t))$$

[2] For the sake of brevity, we do not report the whole proof.

from which we find that the Ward criterion to join two clusters can be computed as:

$$d_{Ward}(C_s, C_t) = \frac{|C_s||C_t|}{|C_s| + |C_t|} d_W^2(Y(b_s), Y(b_t)).$$

In a hierarchical clustering agglomerative procedure, in order to pass from n to $n-1$ clusters, the two clusters corresponding to the minimum d_{Ward} are joined.

3 Application

In the present section we present some results of an analysis performed on a climatic dataset. The original dataset `drd964x.tmpst.txt`[3] contains the

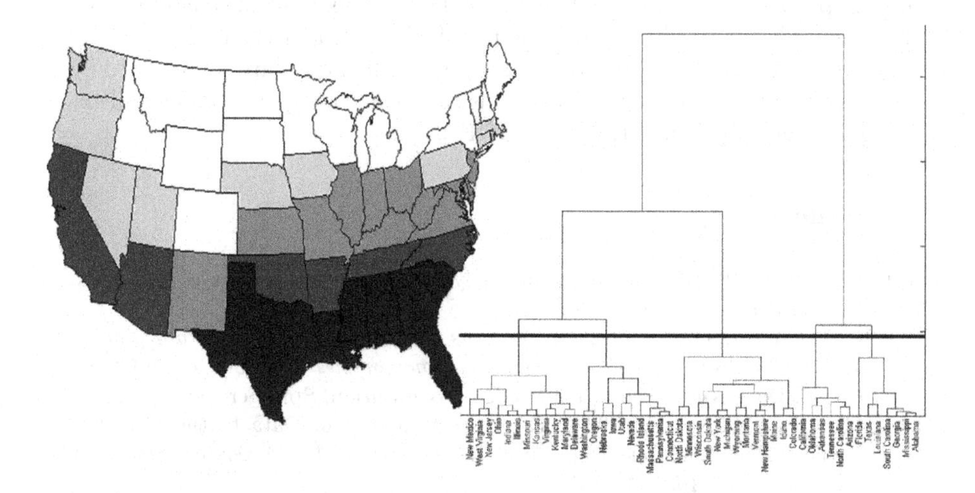

Fig. 2. Ward clustering tree of USA temperature dataset. On the left, the five clusters are colored in different levels of gray

sequential "Time Biased Corrected" state climatic division monthly Average Temperatures recorded in the 48 states of US from 1895 to 2004 (Hawaii and Alaska are not present in the dataset).
The analysis consists of the following three steps:

1. we have represented the distributions of temperatures of each of the 48 states for each month by means of histograms;
2. we have computed the distance matrix using d_W;

[3] freely available at the National Climatic Data Center website of US `http://www1.ncdc.noaa.gov/pub/data/cirs/`

3. we have performed a hierarchical clustering procedure based on the Ward criterion .

The main results are shown in figure 2 and seem to be consistent with the geographic characteristics of the clustered states.

4 Conclusions

In this paper we have presented a new distance for clustering symbolic data represented by histograms. As histograms are bounded and locally uniform distributed, we have shown how to compute the distance. The proposed distance can be considered also as an inertia measure satisfying the Huygens theorem of decomposition of inertia. We have shown also a coherent way to identify the barycenter of a cluster of histograms as an histogram itself. Further, as proposed by Irpino and Verde (2005), the same distance can be used also when data are intervals and it is possible to consider them as uniformly dense. The distance is theoretically useful also when data are represented by continuous density functions, but, in this case, some computational problems may arise due to the invertibility of CDF's.

References

AITCHISON, J. (1986): The Statistical Analysis of Compositional Data, New York: Chapman Hall.

BOCK, H.H. and DIDAY, E. (2000): *Analysis of Symbolic Data, Exploratory methods for extracting statistical information from complex data*, Studies in Classification, Data Analysis and Knowledge Organisation, Springer-Verlag.

BILLARD, L., DIDAY, E. (2003): From the Statistics of Data to the Statistics of Knowledge: Symbolic Data Analysis *Journal of the American Statistical Association, 98, 462, 470-487.*

CHAVENT, M., DE CARVALHO, F.A.T., LECHEVALLIER, Y., and VERDE, R. (2003): Trois nouvelles méthodes de classification automatique des données symbolique de type intervalle, *Revue de Statistique Appliquée, LI, 4, 5–29.*

GIBBS, A.L. and SU, F.E. (2002): On choosing and bounding probability metrics, *International Statistical Review, 70, 419.*

IRPINO, A. and VERDE, R.(2005): A New Distance for Symbolic Data Clustering, *CLADAG 2005, Book of short papers, MUP, 393-396.*

MALLOWS, C. L. (1972): A note on asymptotic joint normality. *Annals of Mathematical Statistics, 43(2),508-515.*

WARD, J.H. (1963): Hierarchical Grouping to Optimize an Objective Function,*Journal of the American Statistical Association, vol. 58, 238-244.*

Symbolic Clustering of Large Datasets

Yves Lechevallier[1], Rosanna Verde[2], and Francisco de A.T. de Carvalho[3]

[1] Domaine de Voluceau, Rocquencourt B.P. 105,
78153 Le Chesnay Cedex,France (yves.lechevallier@inria.fr)
[2] Dip. di Strategie Aziendali e Metod. Quantitative
Seconda Universitá di Napoli - Piazza Umberto I, 81043 Capua (CE), Italy
(rosanna.verde@unina2.it)
[3] Centro de Informatica - CIn/UFPE, Av. Prof. Luiz Freire, s/n, Cidade
Universitaria, CEP 50740-540, Recife-PE, Brasil (fatc@cin.ufpe.br)

Abstract. We present an approach to cluster large datasets that integrates the Kohonen Self Organizing Maps (SOM) with a dynamic clustering algorithm of symbolic data (SCLUST). A preliminary data reduction using SOM algorithm is performed. As a result, the individual measurements are replaced by micro-clusters. These micro-clusters are then grouped in a few clusters which are modeled by symbolic objects. By computing the extension of these symbolic objects, symbolic clustering algorithm allows discovering the natural classes. An application on a real data set shows the usefulness of this methodology.

1 Introduction

Cluster analysis aims to organize a set of items (usually represented as a vector of quantitative values) into clusters, such that items within a given cluster have a high degree of similarity, whereas items belonging to different clusters have a high degree of dissimilarity. The most usual structures furnished by the clustering algorithms are hierarchies and partitions.

Dynamic Clustering Algorithms (DCA) are based on an iterative two-step relocation procedure. At each iteration, DCA's look for clusters and a suitable representation of them by prototype (means, axes, probability laws, groups of elements, etc.). The criterion optimized is an adequacy measure between the clusters and their corresponding representation (Diday and Simon (1976)). DCA's perform an allocation step in order to assign the objects to the classes according to their proximity to the prototypes. This is followed by a representation step where the prototypes are updated according to the new objects assigned to classes at the allocation step, until achieving the convergence of the algorithm, when the adequacy criterion reaches a stationary value.

Objects to be clustered are usually represented as a vector of quantitative measurements. Nevertheless this model is too restrictive to represent complex data which, in their descriptions, take into account the variability and/or the uncertainty inherent to the data. Moreover, it can be interesting to create typologies of objects homogeneous from a conceptual point of view. That is typical of classifications in the natural world as well as in the definition of types of behaviors in economics, political and social analysis. So, according to Symbolic Data Analysis (SDA), we assume that the objects to be clustered are described by a set of categories or intervals, possibly even with associated a system of weights.

Nowadays, many clustering methods for complex data (i.e. symbolic data) have been proposed (Michalski et al. (1981), De Carvalho et al. (1999); Verde et al. (2001); Chavent et al. (2002, 2003); Lechevallier et al. (2004)) which differ in the type of the data descriptors, in their cluster structures and/or in the clustering criteria. In particular, this paper deals with a method for clustering a set of objects which represent not simple individuals but concepts. Such concepts are modeled by symbolic objects (SO's) and described by multi-valued variables: intervals, multi-categorical, modal (Bock and Diday (2000)). The number of clusters in which the set have to be partitioned is predefined. The classes are suitably interpreted and represented by class prototypes. The proposed partitioning algorithm is based on a generalization of the classical "Dynamic Clustering Method". The general optimized criterion is a measure of the best fitting between the partition and the representation of the classes. The prototype is a model of a class, and its representation can be an element of the same space of representation of the concepts to be clustered which generalizes the characteristics of the elements belonging to the class. Therefore, the prototype is a concept itself and it is modeled as a symbolic object too.

The allocation function for the assignment of the objects to the classes depends on the nature of the variables which describe the SO's. The choice of the allocation function must be related to the particular type of prototype taken as a representation model of the class.

The peculiarity of this symbolic clustering method is in the interpretation of the classes as concepts. Modeling the concepts by prototypes, defined as SO's, it makes possible to give a symbolic meaning at the elements of the partition at each step of the algorithm and not only at the end of the procedure.

In spite of the usefulness of clustering algorithms in many application areas, the rapid collection of large data sets of increasing complexity poses a number of new problems that traditional algorithms are not equipped to address. One important feature of modern data collection is the ever increasing size of a typical data base: it is not so unusual to work with databases containing few millions of individuals and hundreds or thousands of variables. Now, most classical clustering algorithms are severely limited as to the number of individuals they can handle (from a few hundred to a few thousands).

This paper presents an approach to cluster large datasets that integrates the Kohonen Self Organizing Maps (SOM) with the dynamic clustering algorithm of symbolic data (SCLUST). A preliminary data reduction using SOM algorithm is performed. As a result, the individual measurements are replaced by micro-clusters. These micro-clusters are then grouped in a few clusters which are modeled by symbolic objects. The calculation of the extension of these symbolic objects allows the symbolic clustering algorithm to discover the natural classes. An application on a real database shows the usefulness of this methodology.

2 The dynamic clustering approach

The proposed Dynamic Clustering Method generalizes the standard clustering method in order to partition a set of individuals E, modeled by symbolic descriptions or a set of concepts C, modeled by SO's in k classes.

The algorithm is based on: a) the choice of the prototypes for representing the classes; b) the choice of a proximity function to allocate the concepts to the classes at each step. The clustering criterion to be optimized is a measure of best fitting between the partition of the set of concepts and the prototype descriptions.

2.1 The input data

We recall that a SO s_c is defined (Bock and Diday (2000)) by the triple (a_c, R, d_c), where $d_c = (d_c^1, \ldots, d_c^p) \in D$ is its description, R is a binary or fuzzy relation between descriptions, and a_c is the mapping from a set of individuals Ω to the interval [0,1] or to the set $\{0,1\}$.

The clustering method runs on a *Symbolic Data Table*, denoted X (Bock and Diday (2000)). The columns of the input data table represent a set of p *variables* $y_1, \ldots, y_p$ and the rows contain the descriptions d_c of the concepts c of C, which are modeled by SO's.

The prototype g_i of a class $P_i \in P$ is modeled as a symbolic object by the triple (a_{g_i}, R, G_i). We denote Λ as the space of the prototypes $g \in \Lambda$ and Γ the space of their descriptions. If the space D of description of the elements of C is the same space Γ of the descriptions of the prototypes g_i then we have: $\psi(c, g_i) = R(y(c), y(g_i)) = [d_c R G_i]$.

2.2 The symbolic dynamic clustering algorithm (SCLUST)

Symbolic Dynamic Clustering method is a generalization of the standard Dynamic Clustering method to cluster a set of concepts $c \in C$, modeled by SO's $s_c \in E$ into k homogeneous classes.

In particular, the algorithm is here performed on symbolic data described by symbolic variables of two types: multi-categorical and interval ones. This method allows us to construct a partition of symbolic objects using dissimilarity or distance measures defined for both types of variables.

The criterion $\Delta(P, L)$ optimizes by the algorithm is defined as the sum of the dissimilarity values $\psi(c, g_i)$ computed between each concept c belonging to a class $P_i \in P$ and the prototype $g_i \in \Lambda$ of P_i :

$$\Delta(P, L) = \sum_{i=1}^{k} \sum_{c \in P_i} \psi(c, g_i),$$

The criterion $\Delta(P, L)$ is also additive with respect to the data descriptors. Therefore, the convergence of the algorithm to a optimum of the function Δ is guaranteed by the consistence between the representation of the classes by prototypes and the allocation function $\psi(.)$.

The general scheme of symbolic dynamic clustering algorithm

- *Initialization*
  Let $P^{(o)} = \{P_1^{(o)}, \ldots, P_k^{(o)}\}$ be the initial random partition of C in k classes.
- *Representation step t*
  For $i = 1, \ldots, k$, compute a prototype $g_i^{(t)}$ as the SO representing the class $P_i \in P^{(t)}$
- *Allocation step t*
  Any concept $c \in C$ is assigned to the class P_i, iff $\psi(c, g_i)$ is a minimum:
  $P_i^{(t+1)} = \{c \in \mathrm{C} \mid i = argmin\{\psi(c, g_l)/l = 1, \ldots, k\}$
- *Stopping rule or stability*
  If $P^{(t+1)} = P^{(t)}$ then STOP else GO TO *Representation step*

To define the prototype which represents the classes of the partition P, we can distinguish between two cases: the prototype, expressed by a single element of the class (e.g., the element at the minimum average distance from all the elements of the class as well as by the element which minimizes the criterion function); a prototype, chosen as a summarizing function of the elements of the class. In this last case, the prototype can be suitably modeled by a *modal SO* (Bock and Diday (2000)). The description of a modal SO is given by frequency (or probability) distributions associated with the categories of the p multi-categorical descriptors.

In the *representation step*, according to the nature of the descriptors of the set C of SO's we distinguish different cases: i) all the SO descriptors are intervals; ii) all the SO descriptors are multi-categorical variables.

In the first case, Chavent et al. (2002, 2003) demonstrated that is possible to represent a class by an interval chosen as the one at minimum average distance from all the others intervals belonging to the class. The optimal solution was found analytically, assuming as distance a suitable measure defined on intervals: the *Hausdorff distance.*

The second case can be considered according to two different approaches: a) the prototype is expressed by the most representative element of the class (or by a virtual element v_b) according to the *allocation* function; b) the prototype is a high order SO's, described by the distribution function associated with the multi-nominal variables.

In particular, in the first approach the prototype is selected as the neighbour of all the elements of the class. Given a suitable *allocation* function $\psi(c_h, g_i)$ the prototype g_i of the class P_i is chosen as the symbolic object associated with the concept c_h where $h = argmin\{\sum_{h' \in P_i} \psi(c_{h''}, c_{h'}) : c_{h''} \in P_i\}$. Similarly, a prototype g of P_i can be constructed considering all the descriptions of the SO's $c_h \in P_i$ and associating with them a set of descriptions corresponding to the most representatives among the elements of P_i, such that $\sum_{h \in P_i} \psi(c_h, g) = min\left\{\sum_{h \in P_i} \psi(c_h, g') : g' \in \Gamma\right\}$. A similar criterion has been followed by Chavent et al. (2002, 2003) in the choice of the description of the prototypes as interval data, according to the Hausdorff distance.

Nevertheless, we can point out that if the virtual prototype g is not a SO associated with a concept of C, its description could be inconsistent with a conceptual meaning own of a symbolic object. So that, instead of taking g to represent the class P_i it is more appropriate to choose the nearest SO of $c_h \in P_i$, according to the allocation function value. This choice is a generalization of the nearest neighbours algorithm criterion in the dynamical clustering. However, it respects the numerical criterion of the minimum dissimilarity measure and guarantees coherence with the allocation function.

In the second approach, we can also assume to associate with multi-categorical variables a uniform distribution in order to SO descriptors as *modal* ones. The prototype g_i of cluster P_i is described by the minimum generalization of the descriptions of the elements belonging to the class P_i, for all the categorical multi-valued variables (De Carvalho et al. (1999)).

In the *allocation step* the coherence between the prototype and the allocation function guarantees the convergence of the partitioning algorithm. Thus, we distinguish two different situations: *(1).* the SO and prototype description space is the same; *(2).* the prototypes are modal SO's.

In the first case, both prototypes and SO's are modeled by vectors of intervals for interval descriptors as well as, by sets of categories for categorical multi-valued variables. Finally, they are also in the same space of representation whenever both are described by *modal* variables. The second case corresponds to situation where the prototypes are modeled by modal variables, whereas the SO's are described by interval and/or categorical multi-valued variables.

Partitioning for modal symbolic data. When the set of elements to be clustered are modal objects and $D = \Gamma$, suitable dissimilarity measures can be proposed as an allocation function. When both concepts and prototypes are modeled by distributions a classical ϕ^2 distance can be proposed as an allocation function. As noted above, modal data can be derived by imposing a system of weights (pseudo-frequencies, probabilities, beliefs) on the domain of categorical multi-valued or interval descriptors. These transformations of the SO's description space are requested wherever prototypes have been chosen as modal SO's. Simple Euclidean distances between profiles can be suitably used too.

Partitioning for mixed symbolic data. All the proposed distance functions for p variables are determined by sums of dissimilarities corresponding to the univariate symbolic component descriptors y_j. In practice, however, symbolic data to be clustered are typically described by different types of variables. In order to compute an overall dissimilarity measure two approaches are proposed here:

1) *Weighted linear combination of dissimilarity measures:* if the symbolic object associated with c is described by different type of variables the overall dissimilarity between c and g_i is obtained by a linear combination of the proximity measures computed with respect to the different (classical or symbolic) variables.
2) *Categorization (discretization, segmentation, ...)*: in this case, all the variables y_j are transformed to the same type.

3 An application of the integrated methods: SOM and SCLUST

Our approach to clustering large datasets is to integrate the Symbolic Dynamic Clustering (SCLUST) algorithm and the Kohonen Self Organizing Map (SOM) algorithm. The strategy starts with a pre-processing step where the dataset is partitioned in a few hundreds micro-classes by the SOM algorithm (Murtagh (1995), Ambroise et al. (2000)), then to describe each micro-class by symbolic interval variables, to cluster these micro-class symbolic descriptions in a few number of clusters by using the Hausdorff distance.

These clusters are then modeled by symbolic objects whose symbolic descriptions generalize the characteristics of the micro-classes belonging to the respective classes. The extension of the clusters, that is the elements which are assigned to each cluster, is obtained using the Hausdorff distance as matching function.

To show the usefulness of the proposed approach to cluster large datasets, we have applied the proposed strategy to the well known waveform dataset (Breiman et al (1984), pp $49-55$). For each class a set of individuals is generated using a model for data generation. Each model uses two basic waveforms (each one being a shifted triangular distribution) and Gaussian noise. The number of the individuals of each class is 20000 and the number of variables describing the individuals is 21. All variables are quantitative continuous. The Bayes misclassification rate for this three-class classification problem is approximately 0.14.

Our aim is to compare the partition quality of the partitioning of the waveform dataset in six classes, measured by the error rate of classification, given by a direct application of the Kohonen Self Organizing Map (SOM) algorithm with the presented approach which integrates the Symbolic Dynamic Clustering (SCLUST) algorithm and SOM algorithm. SCLUST clustering algorithm is implemented in the SODAS software (www.info.fundp.ac.be/asso/), a computer program developed in the framework of the ASSO European project (17 European and 1 Brazilian teams) whose aim was to develop new data analysis methods and algorithms to manage symbolic data.

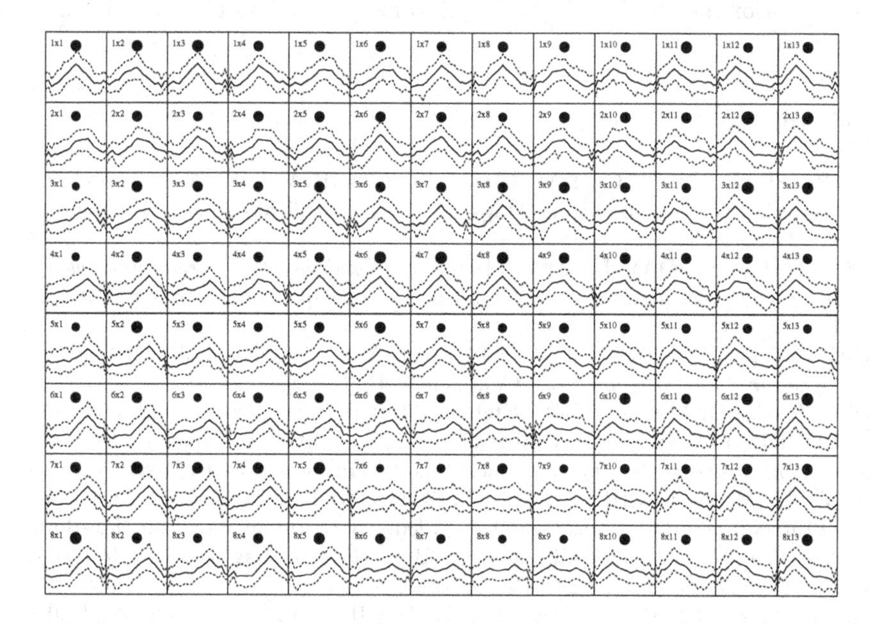

Fig. 1. Kohonen Self Organizing Maps for 104 micro-classes

The error rate of classification was 0.39, 0.305, 0.292 and 0.286 respectively for the partitioning of the waveform dataset in six classes by the direct application of the SOM algorithm and the partitioning of this dataset in 104 (Fig.1), 150 and 200 micro-classes in the pre-processing step using the SOM algorithm. These results showed that the proposed approach, which integrates the SOM algorithm and the SCLUST symbolic clustering algorithm clearly outperforms the direct application of the SOM algorithm concerning the partitioning of large datasets. The classes are represented by intervals, given by the minimum and maximum curves (plotted in Fig. 1 by dash lines), rather than by the average curve (plotted in Fig. 1 by continuous line) as in the classical SOM approach.

4 Final remarks and conclusions

This paper presents an approach to partitioning large datasets in a few number of homogeneous classes through the integration of Kohonen Self Organizing Map (SOM) algorithm and the Symbolic Dynamic Cluster (SCLUST)

algorithm. To show the usefulness of the proposed approach it considered the recognition of a three classes synthetic waveform dataset by the SOM algorithm alone versus the integration of the SOM and SCLUST algorithms. The error rate of classification was calculated and the clustering results showed that the integration of SOM and SCLUST algorithms outperforms the direct application of the SOM algorithm concerning the partitioning of this large dataset.

References

AMBROISE, C., SEŻE, G., BADRAN, F. and THIRIA, S. (2000): Hierarchical clustering of Self-Organizing Maps for cloud classification. *Neurocomputing, 30, 47–52.*

BOCK, H.H. and DIDAY, E. (2000): *Analysis of Symbolic Data, Exploratory methods for extracting statistical information from complex data.* Springer, Heidelberg.

BREIMAN, L., FRIEDMAN, J.H., OSLHEN, R.A. and STONE, C.J. (1984): *Classification and regression trees.* Chapman & Hall/CRC.

CELEUX, G. , DIDAY, E. , GOVAERT, G. , LECHEVALLIER, Y. and RALAMBONDRAINY, H. (1988): *Classification Automatique des Données : Environnement Statistique et Informatique.* Dunod, Gauthier-Villards, Paris.

CHAVENT, M. and LECHEVALLIER, Y. (2002). Dynamical Clustering Algorithm of Interval Data: Optimization of an Adequacy Criterion Based on Hausdorff Distance. In: A. Sokolowski and H.-H. Bock (Eds.): *Classification, Clustering and Data Analysis.* Springer, Heidelberg, 53-59.

CHAVENT, M., DE CARVALHO, F.A.T., LECHEVALLIER, Y. and VERDE, R. (2003). Trois nouvelles mthodes de classification automatique de donnes symboliques de type intervalle. *Revue de Statistique Applique, v. LI, n. 4, p. 5-29.*

DE CARVALHO, F.A.T., VERDE, R. and LECHEVALLIER, Y. (1999). A dynamical clustering of symbolic objcts based on a context dependent proximity measure. In : *Proceedings of the IX International Symposium on Applied Stochastic Models and Data analysis. Lisboa, p. 237-242.*

DIDAY, E. and SIMON, J.J. (1976): Clustering Analysis. In: Fu, K. S. (Eds): *Digital Pattern Recognition.* Springer-Verlag, Heidelberg, 47-94.

DIDAY, E. (2001). An Introduction to Symbolic Data Analysis and SODAS software. *Tutorial on Symbolic Data Analysis. GfKl 2001, Munich.*

GORDON, A.D. (1999): *Classification.* Chapman and Hall/CRC, Florida.

ICHINO, M. and YAGUCHI, H. (1994). Generalized Minkowski Metrics for Mixed Feature Type Data Analysis. *IEEE Trans. Systems Man and Cybernetics, 1, 494–497.*

LECHEVALLIER, Y. and CIAMPI A. (2004): Clustering large and Multi-levels Data Sets. In: *International Conference on Statistics in Heath Sciences 2004*, Nantes.

MICHALSKI, R.S., DIDAY, E. and STEPP, R.E.(1981). A recent advance in data analysis: Clustering Objects into classes characterized by conjunctive concepts. In: Kanal L. N., Rosenfeld A. (Eds.): *Progress in pattern recognition.* North-Holland, 33–56.

MURTAGH, F. (1995): Interpreting the Kohonen self-organizing feature map using contiguity-constrained clustering. *Patterns Recognition Letters, 16, 399–408.*

VERDE, R., LECHEVALLIER, Y. and DE CARVALHO, F.A.T. (2001): A dynamical clustering algorithm for symbolic data. *Tutorial Symbolic Data Analysis, GfKl*, Munich.

A Dynamic Clustering Method for Mixed Feature-Type Symbolic Data

Renata M.C.R. de Souza[1], Francisco de A.T. de Carvalho[1], and Daniel Ferrari Pizzato[1]

Centro de Informatica - CIn/UFPE, Av. Prof. Luiz Freire, s/n, Cidade Universitaria, CEP 50740-540, Recife-PE, Brasil, {rmcrs,fatc,dfp}@cin.ufpe.br

Abstract. A dynamic clustering method for mixed feature-type symbolic data is presented. The proposed method needs a previous pre-processing step to transform Boolean symbolic data into modal symbolic data. The presented dynamic clustering method has then as input a set of vectors of modal symbolic data and furnishes a partition and a prototype to each class by optimizing an adequacy criterion based on a suitable squared Euclidean distance. To show the usefulness of this method, examples with symbolic data sets are considered.

1 Introduction

Clustering aims to summarize data sets in homogeneous clusters that may be organized according to different structures (Gordon (1999), Everitt (2001)): hierarchical methods yield complete hierarchy, i.e., a nested sequence of partitions of the input data, whereas partitioning methods seek to obtain a single partition of the input data into a fixed number of clusters by, usually, optimizing a criterion function.

In classical data analysis, the items to be grouped are usually represented as a vector of quantitative or qualitative measurements where each column represents a variable. In particular, each individual takes just one single value for each variable. In practice, however, this model is too restrictive to represent complex data since to take into account variability and/or uncertainty inherent to the data, variables must assume sets of categories or intervals, possibly even with frequencies or weights.

The aim of *Symbolic Data Analysis* (SDA) is to extend classical data analysis techniques (clustering, factorial techniques, decision trees, etc.) to these kinds of data (sets of categories, intervals, or weight (probability) distributions) called symbolic data (Bock and Diday (2000)). SDA is a domain in the area of knowledge discovery and data management related to multivariate analysis, pattern recognition and artificial intelligence.

This paper addresses the partitioning of mixed feature-type symbolic data using the dynamic clustering methodology. Dynamical clustering is an iterative two steps relocation partitioning algorithm involving, at each iteration, the construction of clusters and the identification of a suitable representation or prototype (means, axes, probability laws, groups of elements, etc.) for each cluster by locally optimizing an adequacy criterion between the clusters and their corresponding representatives.

SDA has provided partitioning methods in which different types of symbolic data are considered. Ralambondrany (1995) extended the classical k-means clustering method in order to deal with data characterized by numerical and categorical variables. El-Sonbaty and Ismail (1998) have presented a fuzzy k-means algorithm to cluster data on the basis of different types of symbolic variables. Bock (2002) has proposed several clustering algorithms for symbolic data described by interval variables, based on a clustering criterion and thereby generalized similar approaches in classical data analysis. Chavent and Lechevallier (2002) proposed a dynamic clustering algorithm for interval data where the class representatives are defined based on a modified Hausdorff distance. Souza and De Carvalho (2004) have proposed partitioning clustering methods for interval data based on city-block distances.

In this paper, we introduce a partitioning dynamic clustering method for mixed feature-type symbolic data. To be able to manage ordered and non-ordered mixed feature-type symbolic data, this method assumes a previous pre-processing step the aim of which is to obtain a suitable homogenization of mixed symbolic data into modal symbolic data. Section 2 presents the data homogenization pre-processing step. Section 3 presents the dynamic clustering algorithm for mixed feature-type symbolic data. To evaluate this method, section 4 shows experiments with real symbolic data sets, the clustering quality being measured by an external cluster validity index. In Section 5, the concluding remarks are given.

2 Data homogenization pre-processing step

Usual data allow exactly one value for each variable. However, this data is not able to describe complex information, which must take into account variability and/or uncertainty. It is why symbolic variables have been introduced: multi-valued variables, interval variables and modal variables (Bock and Diday (2000)).

Let $\Omega = \{1, \dots, n\}$ be a set of n items indexed by i described by p symbolic variables $X_1, \dots, X_p$. A symbolic variable X_j is categorical multivalued if, given an item i, $X_j(i) = x_i^j \subseteq \mathcal{A}_j$ where $\mathcal{A}_j = \{t_1^j, \dots, t_{H_j}^j\}$ is a set of categories. A symbolic variable X_j is an interval variable when, given un item i, $X_j(i) = x_i^j = [a_i^j, b_i^j] \subseteq \mathcal{A}_j$ where $\mathcal{A}_j = [a, b]$ is an interval. A symbolic variable X_j is a modal variable if, given un item i, $X_j(i) = (S(i), \mathbf{q}(i))$ where $\mathbf{q}(i)$ is a vector of weights defined in $S(i)$ such that a weight $w(m)$ corresponds to each category $m \in S(i)$. $S(i)$ is the support of the measure $\mathbf{q}(i)$.

Each object i $(i = 1, \dots, n)$ is represented as a vector of mixed feature-type symbolic data $\mathbf{x}_i = (x_i^1, \dots, x_i^p)$. This means that $x_i^j = X_j(i)$ can be a (ordered or non ordered) set of categories, an interval or a weight distribution according to the type of the corresponding symbolic variable.

Concerning the methods described in Chavent et al. (2003), there is one of them which is a dynamic clustering algorithm based on a suitable squared Euclidean distance to cluster interval data. This method assumes a pre-processing step which transform interval data into modal data. However, the approach considered to accomplish this data transformation is not able to take into consideration the ordered nature inherent to interval data.

In this paper we consider a new data transformation pre-processing approach, the aim of which is to obtain a suitable homogenization of mixed symbolic data into modal symbolic data, which is able to manage ordered and non-ordered mixed feature-type symbolic data in the framework of a dynamic clustering algorithm. In this way, the presented dynamic cluster algorithm has as input data only vectors of weight distributions.

The data homogenization is accomplished according to type of symbolic variable: categorical non-ordered or ordered multivalued variables, interval variables.

Categorical multivalued variables. If X_j is a categorical non-ordered multivalued variable, its transformation into a modal symbolic variable $\widetilde{X}_j$ is accomplished in the following way: $\widetilde{X}_j(i) = \widetilde{x}_i^j = (\mathcal{A}_j, \mathbf{q}^j(i))$, where $\mathbf{q}^j(i) = (q_1^j(i), \ldots, q_{H_j}^j(i))$ is a vector of weights $q_h^j(i)\,(h = 1, \ldots, H_j)$, a weight being defined as (De Carvalho (1995)):

$$q_h^j(i) = \frac{c(\{t_h^j\} \cap x_i^j)}{c(x_i^j)} \tag{1}$$

$c(A)$ being the cardinality of a finity set A.

If X_j is a categorical ordered multivalued variable, its transformation into a modal symbolic variable $\widetilde{X}_j$ is accomplished in the following way: $\widetilde{X}_j(i) = \widetilde{x}_i^j = (\mathcal{A}_j, \mathbf{Q}^j(i))$, where $\mathbf{Q}^j(i) = (Q_1^j(i), \ldots, Q_{H_j}^j(i))$ is a vector of cumulative weights $Q_h^j(i)\,(h = 1, \ldots, H_j)$, a cumulative weight being defined as:

$$Q_h^j(i) = \sum_{r=1}^{h} q_r^j(i), \quad \text{where } q_r^j(i) = \frac{c(\{t_r^j\} \cap x_i^j)}{c(x_i^j)} \tag{2}$$

It can be shown (De Carvalho (1995)) that $0 \leq q_h^j(i) \leq 1 \quad (h = 1, \ldots, H_j)$ and $\sum_{h=1}^{H_j} q_h^j(i) = 1$. Moreover, $q_1^j(i) = Q_1^j(i)$ and $q_h^j(i) = Q_h^j(i) - Q_{h-1}^j(i)\ (h = 2, \ldots, H_j)$.

Interval variables. In this case, the variable X_j is transformed into a modal symbolic variable $\widetilde{X}_j$ in the following way (De Carvalho (1995), De Carvalho et al. (1999), Chavent et al. (2003)): $\widetilde{X}_j(i) = \widetilde{x}_i^j = (\widetilde{\mathcal{A}}_j, \mathbf{Q}^j(i))$, where $\widetilde{\mathcal{A}}_j =$

$\{I_1^j, \ldots, I_{H_j}^j\}$ is a set of elementary intervals, $\mathbf{Q}^j(i) = (Q_1^j(i), \ldots, Q_{H_j}^j(i))$ and $Q_h^j(i)$ $(h = 1, \ldots, H_j)$ is defined as:

$$Q_h^j(i) = \sum_{r=1}^{h} q_r^j(i), \text{ where } q_r^j(i) = \frac{l(I_r^j \cap x_i^j)}{l(x_i^j)} \tag{3}$$

$l(I)$ being the length of a closed interval I.

The bounds of these elementary intervals I_h^j $(h = 1, \ldots, H_j)$ are obtained from the ordered bounds of the $n+1$ intervals $\{x_1^j, \ldots, x_n^j, [a,b]\}$. They have the following properties:

1. $\bigcup_{h=1}^{H_j} I_h^j = [a,b]$
2. $I_h^j \bigcap I_{h'}^j = \emptyset$ if $h \neq h'$
3. $\forall h \; \exists i \in \Omega$ such that $I_h^j \bigcap x_i^j \neq \emptyset$
4. $\forall i \; \exists S_i^j \subset \{1, \ldots, H_j\} : \bigcup_{h \in S_i^j} I_h^j = x_i^j$

It can be shown (De Carvalho (1995)) that also in this case $0 \leq q_h^j(i) \leq 1$ $(h = 1, \ldots, H_j)$ and $\sum_{h=1}^{H_j} q_h^j(i) = 1$. Moreover, again $q_1^j(i) = Q_1^j(i)$ and $q_h^j(i) = Q_h^j(i) - Q_{h-1}^j(i)$ $(h = 2, \ldots, H_j)$.

3 A dynamic clustering algorithm for mixed feature-type symbolic data

This section presents a dynamic clustering method which allows to cluster mixed feature-type symbolic data. The aim of this method is to determine a partition $P = \{C_1, \ldots, C_K\}$ of Ω into K classes such that the resulting partition P is (locally) optimum with respect to a given clustering criteria.

Let $\Omega = \{1, \ldots, n\}$ be a set of n items. After the pre-processing step, each object i $(i = 1, \ldots, n)$ is represented by a vector of modal symbolic data $\widetilde{\mathbf{x}}_i = (\widetilde{x}_i^1, \ldots, \widetilde{x}_i^p)$, $\widetilde{x}_i^j = (\mathcal{D}_j, \mathbf{u}^j(i))$, where $\mathcal{D}_j$ is a (ordered or non-ordered) set of categories if $\widetilde{X}_j$ is a modal variable, $\mathcal{D}_j$ is a non-ordered set of categories if $\widetilde{X}_j$ is a categorical non-ordered multivalued variable, $\mathcal{D}_j$ is an ordered set of categories if $\widetilde{X}_j$ is a categorical ordered multivalued variable and $\mathcal{D}_j$ is a set of elementary intervals if $\widetilde{X}_j$ is an interval variable. Moreover, $\mathbf{u}^j(i) = (u_1^j(i), \ldots, u_{H_j}^j(i))$ is a vector of weights if $\mathcal{D}_j$ is a non-ordered set of categories and $\mathbf{u}^j(i)$ is a vector of cumulative weights if $\mathcal{D}_j$ is an ordered set of categories or a set of elementary intervals.

As in the standard dynamical clustering algorithm (Diday and Simon (1976)), this clustering method for symbolic data aims to provide a partition of Ω in a fixed number K of clusters $P = \{C_1, \ldots, C_K\}$ and a corresponding set of prototypes $L = \{L_1, \ldots, L_K\}$ by locally minimizing a criterion W that evaluates the fit between the clusters and their representatives.

Here, each prototype L_k of C_k $(k = 1, \ldots, K)$ is also represented as a vector of modal symbolic data $\mathbf{g}_k = (g_k^1, \ldots, g_k^p)$, $g_k^j = (\mathcal{D}_j, \mathbf{v}^j(k))$ $(j = 1, \ldots, p)$, where $\mathbf{v}^j(k) = (v_1^j(k), \ldots, v_{H_j}^j(k))$ is a vector of weights if $\mathcal{D}_j$ is a non-ordered set of categories and $\mathbf{v}^j(k)$ is a vector of cumulative weights if $\mathcal{D}_j$ is an ordered set of categories or a set of elementary intervals. Notice that for each variable the modal symbolic data presents the same support $\mathcal{D}_j$ for all individuals and prototypes. The criterion W is then defined as:

$$W(P, L) = \sum_{k=1}^{K} \sum_{i \in C_k} \phi(\widetilde{\mathbf{x}}_i, \mathbf{g}_k) \tag{4}$$

where

$$\phi(\widetilde{\mathbf{x}}_i, \mathbf{g}_k) = \sum_{j=1}^{p} d^2(\mathbf{u}^j(i), \mathbf{v}^j(k)) \tag{5}$$

The comparison between the two vectors of (non-cumulative or cumulative) weights $\mathbf{u}^j(i)$ and $\mathbf{v}^j(k)$ for the variable j is accomplished by a suitable squared Euclidean distance:

$$d^2(\mathbf{u}^j(i), \mathbf{v}^j(k)) = \sum_{h=1}^{H_j} (u_h^j(i) - v_h^j(k))^2 \tag{6}$$

The cumulative weights obtained in the pre-processing step will allow the dynamic clustering algorithm to take into account the order inherent to the categorical multivalued or interval symbolic data.

As in the standard dynamical clustering algorithm (Diday and Simon (1976)), this algorithm starts from an initial partition and alternates a *representation step* and an *allocation step* until convergence when the criterion W reaches a stationary value representing a local minimum.

3.1 Representation step

In the representation step, each cluster C_k is fixed and the algorithm looks for the prototype $\mathbf{g}_k = (g_k^1, \ldots, g_k^p)$ of class C_k $(k = 1, \ldots, K)$ which minimizes the clustering criterion W in equation (4).

As the criterion W is additive, the optimization problem becomes to find for $k = 1, \ldots, K, j = 1, \ldots, p$ and $h = 1, \ldots, H_j$, the weight $v_h^j(k)$ minimizing

$$W(C_k, L_k) = \sum_{i \in C_k} (u_h^j(i) - v_h^j(k))^2 \tag{7}$$

The solution for $v_h^j(k)$ is :

$$\hat{v}_h^j(k) = \frac{1}{n_k} \sum_{i \in C_k} u_h^j(i) \tag{8}$$

where n_k is the cardinality of the class C_k. The prototype of class C_k is then $\hat{\mathbf{g}}_k = (\hat{g}_k^1, \ldots, \hat{g}_k^p)$, where $\hat{g}_k^j = (\mathcal{D}_j, \hat{v}_h^j(k))$.

3.2 Allocation step

In this step, the vector of prototypes $L = (L_1, \ldots, L_K)$ is fixed. The algorithm finds for each $k \in \{1, \ldots, K\}$ the class

$$C_k = \{i \in \Omega : \phi(\widetilde{\mathbf{x}}_i, \mathbf{g}_k) \leq \phi(\widetilde{\mathbf{x}}_i, \mathbf{g}_m), \forall m \in \{1, \ldots, K\}\} \tag{9}$$

3.3 The algorithm

The dynamic cluster algorithm for mixed feature-type symbolic data has the following steps:
SCHEMA OF DYNAMIC CLUSTERING ALGORITHM FOR WEIGHT DISTRIBUTIONS OF SYMBOLIC DATA

1. **Initialization.**
   Randomly choose a partition $\{C_1 \ldots, C_K\}$ of Ω or randomly choose K distinct objects $L_1, \ldots, L_K$ belonging to Ω and assign each objects i to the closest prototype L_{k*}, where $k* =$ arg $\min_{k=1,\ldots,K} \phi(\widetilde{\mathbf{x}}_i, \mathbf{g}_k)$.
2. **Representation step.**
   (the partition P is fixed)
   For $k = 1, \ldots, K$, compute the vector of modal symbolic data prototype $\mathbf{g}_k = (g_k^1, \ldots, g_k^p)$, $g_k^j = (\mathcal{D}_j, \mathbf{v}^j(k))$ $(j = 1, \ldots, p)$, where $\mathbf{v}^j(k) = (v_1^j(k), \ldots, v_{H_j}^j(k))$ and $v_h^j(k)$ $(h = 1, \ldots, H_j)$ is given by equation (8).
3. **Allocation step.**
   (the set of prototypes L is fixed)
   $test \leftarrow 0$
   for $i = 1$ to n do
   define the cluster C_{k*} such that
   $k* = arg\ \min_{k=1,\ldots,K} \phi(\widetilde{\mathbf{x}}_i, \mathbf{g}_k)$
   if $i \in C_k$ and $k* \neq k$
   $test \leftarrow 1$
   $C_{k*} \leftarrow C_{k*} \cup \{i\}$
   $C_k \leftarrow C_k \setminus \{i\}$
4. **Stopping criterion.**
   If $test = 0$ then STOP, else go to (2).

4 Experimental evaluation

In order show the usefulness of the proposed dynamic clustering method, this section presents the clustering results furnished by it on a real symbolic data set. To evaluate the clustering results furnished by this dynamic clustering method an external index, the adjusted Rand index (CR), will be considered (Hubert and Arabie (1985)). The CR index measures the similarity between an a priori partition and a partition furnished by the clustering algorithm. CR takes its values on the interval [-1,1], where the value 1 indicates perfect agreement between partitions, whereas values near 0 (or negatives) correspond to cluster agreement found by chance.

We apply the proposed dynamic clustering algorithm on a real symbolic interval data set. Our aim is to compare the approach presented in (Chavent et al. (2003)), which transforms interval symbolic data on modal symbolic data represented by non-cumulative weight distributions, with the approach presented in this paper, which transforms interval symbolic data on modal symbolic data represented by cumulative weight distributions.

The car data set consists of a set of 33 car models described by 8 interval, 3 nominal variables (see Table 1). In this application, the 8 interval variables - *Price, Engine Capacity, Top Speed, Acceleration, Step, Length, Width* and *Height* and 2 categorical non-ordered multi-valued variables - *Alimentation* and *Traction* - have been considered for clustering purposes, the nominal variable *Car Category* has been used as a *a priori* classification.

Table 1. Car data set with 8 interval and one nominal variables

	Price	Engine Capacity	...	Height	Category
Alfa 145	[27806, 33596]	[1370, 1910]	...	[143, 143]	Utilitarian
Alfa 156	[41593, 62291]	[1598, 2492]	...	[142, 142]	Berlina
...	...	...	...	...	...
Porsche 25	[147704, 246412]	[3387, 3600]	...	[130, 131]	Sporting
Rover 25	[21492, 33042]	[1119, 1994]	...	[142, 142]	Utilitarian
Passat	[39676, 63455]	[1595, 2496]	...	[146, 146]	Luxury

Concerning this data set, the CR indices taken with respect to Car category were 0.11 and 0.55, repectively, for the approach presented in (Chavent et al. (2003)) and for the approach presented in this paper. This indicates that, for this data set, the accumulated version outperforms the non-accumulated.

5 Concluding remarks

A partitioning clustering method for mixed feature-type symbolic data using a dynamic cluster algorithm based on the squared Euclidean distance was presented in this paper. To be able to manage ordered and non-ordered mixed feature-type symbolic data, it was introduced a previous pre-processing step the aim of which is to obtain a suitable homogenization of mixed symbolic data into modal symbolic data. An application considering a real interval symbolic data set allowed to compare the results of this clustering algorithm using non-cumulative and cumulative weight vectors to represent this interval data set. The accuracy of the results furnished by the clustering method introduced in this paper was assessed by the adjusted Rand index. These results clearly show that the accuracy of the clustering method using cumulative weight vectors of interval data is superior to that which uses non-cumulative weight vectors.

Acknowledgments: The authors would like to thank CNPq and FACEPE (Brazilian Agencies) for their financial support.

References

BOCK, H.H. (2002): Clustering algorithms and kohonen maps for symbolic data. Proc. ICNCB, Osaka, 203-215. J. Jpn. Soc. Comp. Statistic, **15**, 1–13.

BOCK, H.H. and DIDAY, E. (2000): *Analysis of Symbolic Data, Exploratory methods for extracting statistical information from complex data.* Springer, Heidelberg.

CHAVENT, M. and LECHEVALLIER, Y. (2002). Dynamical Clustering Algorithm of Interval Data: Optimization of an Adequacy Criterion Based on Hausdorff Distance. In: A. Sokolowski and H.-H. Bock (Eds.): *Classification, Clustering and Data Analysis.* Springer, Heidelberg, 53–59.

CHAVENT, M., DE CARVALHO, F.A.T., LECHEVALLIER, Y. and VERDE, R. (2003). Trois nouvelles mthodes de classification automatique de données symboliques de type intervalle. *Revue de Statistique Appliquée, v. LI, n. 4, p. 5–29.*

DE CARVALHO, F.A.T. (1995). Histograms in Symbolic Data Analysis. *Annals of Operations Research, v. 55, p. 229–322.*

DE CARVALHO, F.A.T., VERDE, R. and LECHEVALLIER, Y. (1999). A dynamical clustering of symbolic objcts based on a context dependent proximity measure. In : *Proceedings of the IX International Symposium on Applied Stochastic Models and Data analysis. Lisboa : Universidade de Lisboa, p. 237–242.*

DIDAY, E. and SIMON, J.J. (1976): Clustering Analysis. In: Fu, K. S. (Eds): *Digital Pattern Recognition.* Springer-Verlag, Heidelberg, 47-94.

EL-SONBATY, Y. and ISMAIL, M.A. (1998): Fuzzy Clustering for Symbolic Data. *IEEE Transactions on Fuzzy Systems **6**, 195-204.*

EVERITT, B. (2001): *Cluster Analysis.* Halsted, New York.

GORDON, A.D. (1999): *Classification.* Chapman and Hall/CRC, Boca Raton, Florida.

HUBERT, L. and ARABIE. P. (1985): Comparing Partitions. *Journal of Classification, **2**, 193-218.*

RALAMBONDRAINY, H. (1995): A conceptual version of the k-means algorithm. *Pattern Recognition Letters **16**, 1147-1157.*

SOUZA, R.M.C.R. and DE CARVALHO, F.A.T. (2004): Clustering of interval data based on city-block distances. *Pattern Recognition Letters, **25** (3), 353-365.*

Part V

General Data Analysis Methods

Iterated Boosting for Outlier Detection

Nathalie Cheze[1,2] and Jean-Michel Poggi[1,3]

[1] Laboratoire de Mathématique – U.M.R. C 8628, "Probabilités, Statistique et Modélisation", Université Paris-Sud, Bât. 425, 91405 Orsay cedex, France
[2] Université Paris 10-Nanterre, Modal'X, France
[3] Université Paris 5, France

Abstract. A procedure for detecting outliers in regression problems based on information provided by boosting trees is proposed. Boosting is meant for dealing with observations that are hard to predict, by giving them extra weights. In the present paper, such observations are considered to be possible outliers, and a procedure is proposed that uses the boosting results to diagnose which observations could be outliers. The key idea is to select the most frequently resampled observation along the boosting iterations and reiterate boosting after removing it. A lot of well-known bench data sets are considered and a comparative study against two classical competitors allows to show the value of the method.

1 Introduction

The book by Rousseeuw and Leroy (1987) contains an overview of outlier detection problems in the regression context. The underlying model, the estimation method and the number of outliers with respect to the number of observations lead to define various kinds of outliers. For example, one can consider different ways of contamination: outliers in the response space, outliers in the covariate space or outliers in both spaces. Many methods have been developed to cope with such situations. They are essentially supported by robustness ideas and are based on linear modeling (for a short software-oriented review, see Verboven and Hubert (2005)).
Of course, these approaches suffer from the restriction of the outlier definition related to deviations with respect to the linear model. More generally, the outlier definition depends on a given parametric regression design method. Here we consider the generalized regression model $Y = f(X) + \xi$.
The aim of this paper is to propose a procedure based on boosting and such that the regression design method is nonparametric and able to explore different features of the data by adaptive resampling; the detection is entirely automatic and the associated parameters are data-driven; it is possible to detect outliers in the response direction as well as in the covariate space.

2 Boosting regression trees

A classical remark about the boosting procedure AdaBoost (introduced for classification problems by Freund and Schapire (1997) and then for regression ones by Drucker (1997)) and its variants, is its sensitivity to outliers.

This property is in general identified as a drawback, but it can be used (see Gey and Poggi (2006)) to improve the model estimated by a given estimation method to be better adapted to particularly hard observations. The goal is here to use it to detect outliers. Our procedure is based on the information provided by the adaptive resampling process generated when boosting CART regression trees (see Breiman et al. (1984)) is used. This adaptive process tells us a lot about the data set and this is one of the most attractive features of the boosting from the data analytic point of view.

Table 1. *Boosting algorithm* $[M, i_0] = boost(L, K)$.

Input: L the sample of size n and K the number of iterations

Initialization: Set $p_1 = D$ the uniform distribution on $\{1, \ldots, n\}$

Loop: for $k = 1$ to K do

step 1 - randomly draw from L with replacement, according to p_k, a sample L_k of size n,

step 2 - using CART, construct an estimator $\hat{f}_k$ of f from L_k,

step 3 - set from the original sample L: $i = 1, \ldots, n$

$$l_k(i) = \; Y_i - \hat{f}_k(X_i) \;^2 \quad \text{and} \quad \epsilon_{p_k} = {}^{n}_{i=1} p_k(i) l_k(i),$$

$$\beta_k = \frac{\epsilon_{p_k}}{\max_{1 \leq i \leq n} l_k(i) - \epsilon_{p_k}} \quad \text{and} \quad d_k(i) = \frac{l_k(i)}{\max_{1 \leq i \leq n} l_k(i)},$$

$$p_{k+1}(i) = \beta_k^{1-d_k(i)} p_k(i),$$

normalize p_{k+1} to be of sum 1

step 4 - compute $I_{i,k}$ the number of times when the observation i appears in L_k

Output: compute $S_i = \frac{1}{K} {}^{K}_{k=1} I_{i,k}$ and

$$M = max_{i \in L} \, S_i, \quad i_0 = argmax_{i \in L} \, S_i$$

The boosting algorithm generates a sequence of regression function estimates whose elements are fitted from a bootstrap sample obtained from the original training sample by adaptive resampling, highlighting the observations poorly predicted by its predecessor in the sequence. It turns out that such a resampling leads to focus on hard observations with respect to the chosen estimation method, that is to focus on more often badly predicted observations. Of course an outlier is such an observation.

The boosting algorithm used in this paper, proposed by Drucker (1997) and studied in Gey and Poggi (2006)) can be found in Table 1.

3 Outlier detection procedure

So the adopted strategy given in Table 2, is two stages: the first highlights the hard observations and the second selects among them the outliers.

Table 2. *Outlier detection algorithm.*

Input: J the number of applications of boosting,
L the initial sample,
α the indicative significance level of confidence interval and
K the number of iterations of each boosting.
Initialization: Set $L^1 = L$
Stage 1: for $j = 1$ to J do
$[M_j, i(j)] = boost(L^j, K)$;
$L^{j+1} = L^j \setminus i(j)$;
$H = L \setminus L^J$
Stage 2: Outliers are defined as the observations of index
$i(j) \in H$ such that $(M_j > C_\alpha)$

The key idea of the first stage is to retain the most frequently resampled observation along the boosting iterations and reiterate after removing it. So the final set H of Table 2 contains the J observations whose indices are $i(j)$ and which have appeared in average M_j times in the bootstrap samples. The second stage defines a data-driven confidence region to select outliers in H, the rationale for the selection rule is the following.
For each $j \in (1, ..., J)$, let us assimilate the outlier detection problem to the individual test of the null hypothesis: H_0 : *the observation* $i(j)$ *is not an outlier*, against the alternative hypothesis: H_1 : *the observation* $i(j)$ *is an outlier*. Since if $i(j)$ is associated to an outlier then M_j is large, it is consistent to choose the rejection region $W = (M_j > C_\alpha)$ for a given level of significance α. Using Tchebychev's inequality, we obtain $C_\alpha = \hat{m}_{rob} + \sqrt{\frac{\hat{\sigma^2}_{rob}}{\alpha}}$ where $\hat{m}_{rob}$ and $\hat{\sigma^2}_{rob}$ are robust estimators of m and σ^2 the expectation and variance of M_j under the hypothesis H_0. The gap between M_j and $m = E_{H_0}(M_j)$ under H_1 allows to circumvent the usual Tchebychev's inequality conservativeness. Indeed, even if $P_{H_0}\left(|M_j - m|/ > \sigma\alpha^{-1/2}\right) << \alpha$, leads to shrink the rejection region, the hypotheses to be tested are sufficiently separated to correctly select the outliers. So in the sequel, we use $\alpha = 5\%$ for all the computations. We emphasize that it is noise distribution free.
Let us make two additional remarks.
Why to reiterate boosting? As a matter of fact, the j_0 most frequently resampled observations along the iterations of a single boosting are different from

the first j_0 observations selected stepwise using J boosting reiterations. The reason is that for a given boosting application, the most frequently resampled observation would mask other hard observations.
How to choose the number of boosting iterations? Boosting is reiterated until all the outliers have been removed and in addition a sufficient number of observations non contaminated by outliers are available to estimate the mean and variance under H_0 to plug in the Tchebichev's inequality. When n is not too large, a convenient choice for J is to take the integer part of $0.75n$.

4 Outliers in well-known small size real data sets

We have examined various well-known bench data sets allowing to study the behavior of the proposed method for various kinds of outliers depending on the way of contamination, for various sample sizes including small ones (which could be critical for nonparametric estimation method) as well as larger ones, see Cheze and Poggi (2005) in which we focus on a lot of interesting and intensively studied real examples of small sample size from the book of Rousseeuw and Leroy (1987), examined during twenty years by many authors. We apply our method to all the examples and take the results given by the Least Trimmed Squares (LTS) method as a reference since it has been considered as a convenient one for such small data sets. The main conclusion is that in many cases, we obtain results very close to those obtained using Minimum Covariance Determinant (MCDCOV) and LTS methods in spite of the small sample size (around twenty for most of these data sets) and the parametric model. More precisely, we obtain unsuccessful results for only three examples among eighteen. For the others, we obtain always satisfactory detection with partial or total selection.
Let us illustrate the three typical situations. Each of them corresponds to a figure containing four plots: at the top left, the relevant data are displayed (a legend specifies the concerned useful data); at the top right, the plot represents the value of M_j for $1 \leq j \leq J$ (defined in Table 2) obtained by our method (using $\alpha = 5\%$); at the bottom, two plots give the results obtained by the two alternative methods based on the standardized LTS residuals and the robust distances respectively. Let us denote by n_{out}^{LTS} the number of outliers detected by the LTS method. The estimates $\hat{m}_{rob}$ and $\hat{\sigma}_{rob}$ needed to compute the rejection bound are obtained from the MCD estimators applied to $(M_j)_{1 \leq j \leq J}$. These estimates and the results obtained by these alternative methods have been carried out using the library LIBRA (see Verboven and Hubert (2005)) developed using MATLAB®. We use for each method, the default values for the corresponding parameters.
For our method and the MCDCOV one, outliers are indices associated with points located above the solid horizontal line while for the LTS method, outliers are located outside the interval delimited by two horizontal lines. In addition, we indicate, for simulated data sets, the indices of outliers and for

real data sets, those of some observations chosen to facilitate the interpretation of the plots. Let us remark that in the plot corresponding to our method the J points corresponding to the J boosting reiterations are drawn while for the two other methods, n points are drawn.

4.1 Why the method can fail?

First of all, let us focus on the three examples for which the method fails. A careful examination of the decision trees leads to easily explain this drawback: when CART creates a single node containing all the outliers, the method cannot highlight them.

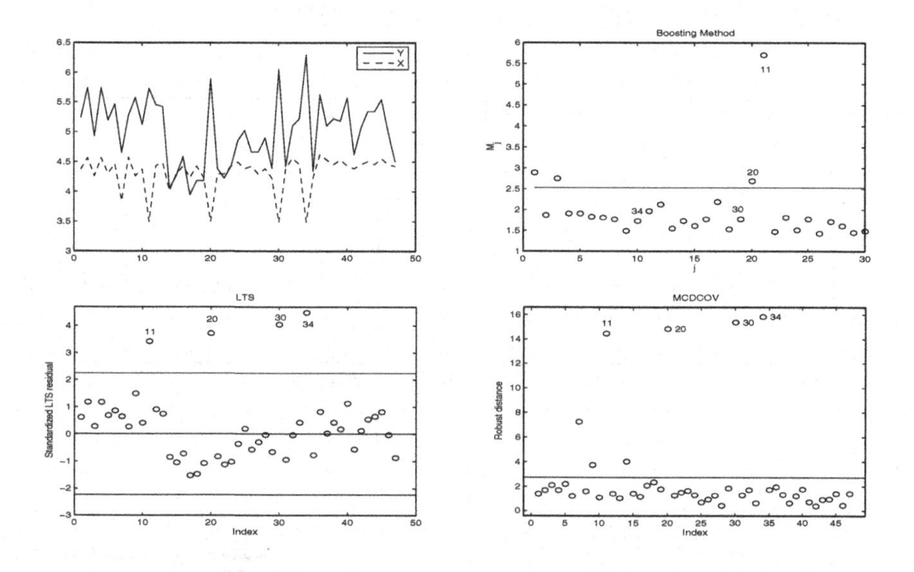

Fig. 1. Data set page 27 in Rousseeuw and Leroy (1987), $n = 47$, $p = 1$, $n_{out}^{LTS} = 4$.

Figure 1 illustrates such a situation: as it can be seen at the top left plot, the four outliers (identified using LTS) of indices 11, 20, 30 and 34, are atypical both in response and covariate directions. Let us mention that these data come from the Hertzsprung-Russell Diagram of the Star Cluster CYG OB1, giving the logarithm of the effective temperature at the surface of the star (X) and the logarithm of the light intensity of the star (Y). The four identified stars are the huge ones. We detect only two of them, LTS captures the four and MCDCOV identifies the four same observations plus three others. The explanation is that CART is sufficiently flexible to create a node containing the four outliers which are atypical in a similar way: their X-values are close to each other and far from the other observations, and their Y-values are the four first maxima. Let us observe that, along the iterations of the detection algorithm (see the top right plot), as soon as the observations 34 and 30 are suppressed from the learning sample, outliers of index 20 and 11 are then easily detected.

4.2 Examples of correct detection

Second, when the percentage of outliers is less than 10%, our method performs correctly except, of course, when the above mentioned drawback occurs. In particular, when the two other methods do not detect any outlier, our method also performs correctly.
The example of Figure 2 exhibits interesting behaviour and highlights an important difference with MCDCOV and LTS methods.

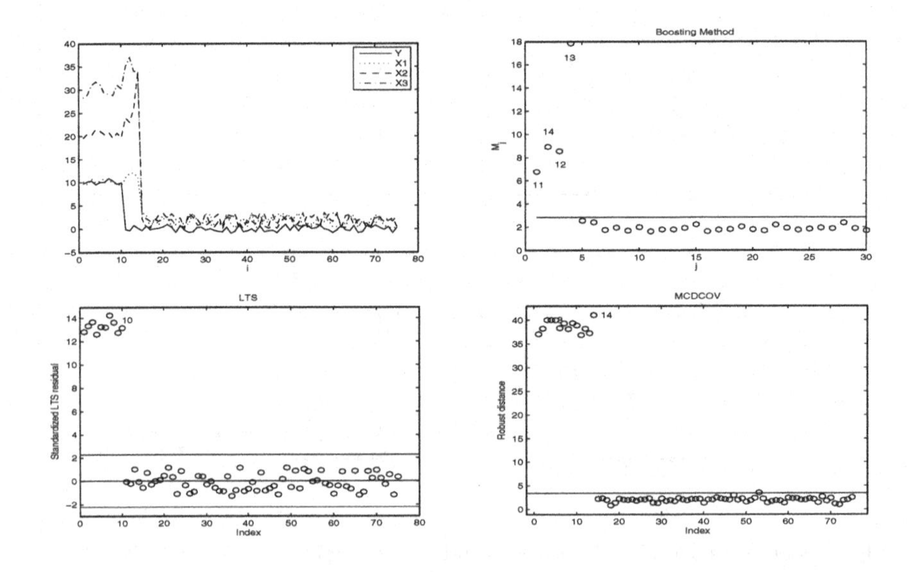

Fig. 2. Data set page 94 in Rousseeuw and Leroy (1987), $n = 75$, $p = 3$, $n_{out} = 4$.

Since it is the only one simulated in Rousseeuw and Leroy (1987), so the number of "true" outliers is known and equal to 4. The top left plot shows that the sample can be divided in three parts, two different populations and the outliers: the observations of index from 1 to 10, those of index greater than 15 and the four outliers from 11 to 14. Our method detects correctly the four outliers without any false detection while the two other methods assimilate the first population to outliers. MCDCOV detects the outliers and the first observations since it tracks outliers separately in each direction. LTS fails to detect outliers since it fits a single linear model for all the data which delivers a robust prediction close to zero.

4.3 Example of good selection but poor detection

Third, when the percentage of outliers is greater than 10%, the outliers are brought at the top of the set H but the threshold is too large to automatically select all the outliers.

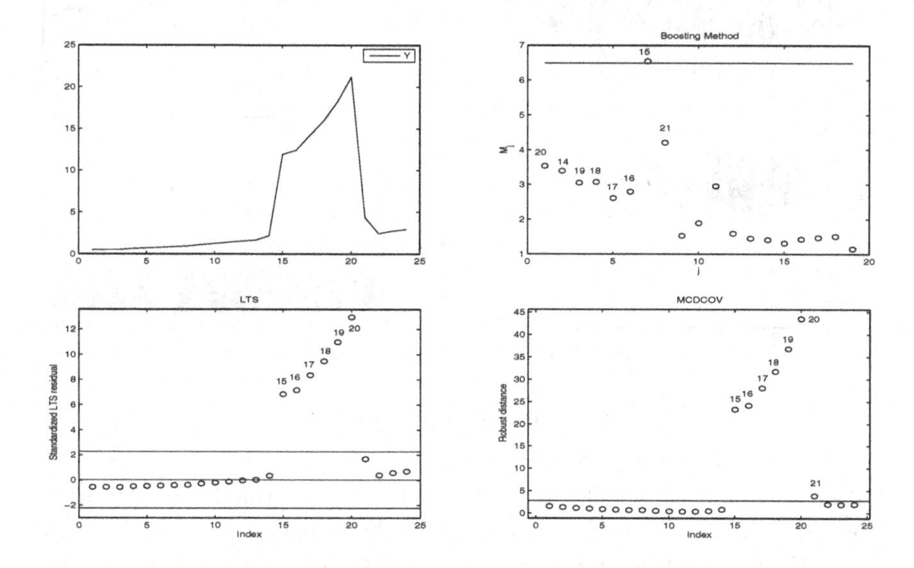

Fig. 3. Data set page 26 in Rousseeuw and Leroy (1987), $n = 24$, $p = 1$, $n_{out}^{LTS} = 6$.

Figure 3 shows a perfect detection for both MCDCOV and LTS methods, while our method fails to correctly detect the seven outliers which are the observations of index from 15 to 20, as it can be seen in the top left plot showing the sample values of the explained variable Y. Nevertheless, the boosting method selects correctly the outliers: the top eight values of the set H do contain all the outliers but the bound is too large. This comes from the following fact: $n = 24$ and $J - j_0 = 19 - 6$ are too small to have a sufficient number of observations to conveniently estimate the unknown parameters involved in the detection region definition.

5 A real-world example of large size data set

The Paris Pollution data are used to deal with the analysis and prediction of ozone concentration in Paris area (see Cheze et al. (2003)). Highly polluted days are often hard to predict and the underlying model becomes highly nonlinear for such observations. In Figure 4, it can be seen as expected, that the LTS and MCDCOV methods lead to very large numbers of false detections while the boosting one highlights only one day.

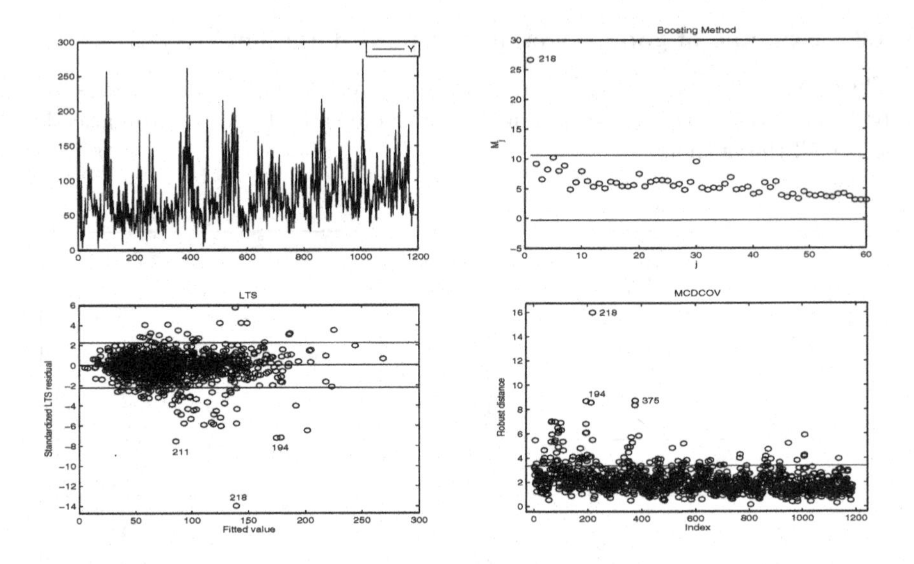

Fig. 4. Pollution real data set.

A deeper examination of this day selected by the boosting-based algorithm, shows that it corresponds to a day where the temperature is high (about 28°C), the day before is polluted (about 126 $\mu g/m^3$) and there is no wind, so the ozone concentration should be about 120 $\mu g/m^3$ but only 15 $\mu g/m^3$ is observed, which is particularly hard to predict and atypical with respect to the small set of explanatory variables considered in this model.

References

BREIMAN, L., FRIEDMAN, J. H., OLSHEN, R. A. and STONE, C. J. (1984): *Classification And Regression Trees.* Chapman & Hall.

CHEZE, N. and POGGI, J-M. (2005): Outlier Detection by Boosting Regression Trees. *Preprint 2005-17, Orsay* . www.math.u-psud.fr/ biblio/ppo/2005/

CHEZE, N., POGGI, J-M. and PORTIER, B. (2003): Partial and Recombined Estimators for Nonlinear Additive Models. *Stat. Inf. Stoch. Proc., 6, 155-197.*

DRUCKER, H. (1997): Improving Regressors using Boosting Techniques. In: Proc. of the 14th Int. Conf. on Machine Learning. Morgan Kaufmann, 107–115.

FREUND, Y. and SCHAPIRE, R. E. (1997): A Decision-Theoretic Generalization of On-line Learning and an Application to Boosting. *Journal of Computer and System Sciences, 55, 1, 119-139.*

GEY, S. and POGGI, J-M. (2006): Boosting and Instability for Regression Trees. *Computational Statistics & Data Analysis, 50, 2, 533-550.*

ROUSSEEUW, P.J. and LEROY, A. (1987): *Robust regression and outlier detection.* Wiley.

VERBOVEN, S. and HUBERT, M. (2005): LIBRA: a MATLAB library for robust analysis. *Chemometrics and Intelligent Laboratory Systems*, 75, 127-136.

Sub-species of *Homopus Areolatus*? Biplots and Small Class Inference with Analysis of Distance

Sugnet Gardner and Niël J. le Roux

Department of Statistics and Actuarial Science, Stellenbosch University, Private Bag X1, Matieland, 7602, South Africa

Abstract. A canonical variance analysis (CVA) biplot can visually portray a one-way MANOVA. Both techniques are subject to the assumption of equal class covariance matrices. In the application considered, very small sample sizes resulted in some singular class covariance matrix estimates and furthermore it seemed unlikely that the assumption of homogeneity of covariance matrices would hold. Analysis of distance (AOD) is employed as nonparametric inference tool. In particular, AOD biplots are introduced for a visual display of samples and variables, analogous to the CVA biplot.

1 Introduction

The biplot as introduced by Gabriel (1971) proved to be a valuable tool for exploring data visually. Moreover, the philosophy of Gower and Hand (1996) to view the biplot as the multivariate analogue of a scatterplot made biplot interpretation easily accessible to non-statistical audiences.

MANOVA is a popular inferential method for data with a class structure. This can be complemented by a canonical variate analysis (CVA) biplot for visual appraisal of class separation and overlap. Analysis of distance (AOD) (Gower and Krzanowski (1999)) is an alternative inferential method when class covariance matrices differ.

The question addressed in this paper is whether morphometric differences among tortoises might point to different subspecies. The data set consisting of carapace measurements of tortoise shells from different geographical regions contained very small subsample sizes for some regions. It seemed unlikely for the assumption of homogeneity of class covariance matrices to hold. Testing this MANOVA assumption was problematic due to some singular class covariance matrix estimates. However, AOD inference on classes with fewer observations than variables could be performed. The permutation tests suggested by Gower and Krzanowski (1999) provided an inferential procedure for testing class differences. These authors give pictorial representations of the class means and discuss the inclusion of individual samples.

In this paper AOD biplots are constructed. In contrast to the plots of Gower and Krzanowski (1999) only displaying information on the samples, the biplot contains information on both samples and variables providing a multivariate scatterplot for exploring class separation and overlap.

2 Analysis of distance

The basic variance formula can be written as the sum of squared Euclidean distances among all sample points, *viz.*

$$\frac{1}{n-1}\sum_{i=1}^{n}(\mathbf{x}_i-\overline{\mathbf{x}})^2=\frac{1}{2n(n-1)}\sum_{i=1}^{n}\sum_{j=1}^{n}(\mathbf{x}_i-\mathbf{x}_j)^2 .$$

Gower and Krzanowski (1999) introduced AOD analogous to analysis of variance by decomposing the total intersample squared distances into a within and a between component.

Consider a data matrix $\mathbf{X}$: $n \times p$ containing $n_1 + n_2 + \ldots + n_J = n$ observations on p variables where n_j observations come from the j-th class ($j = 1, 2, \ldots, J$) and an indicator matrix $\mathbf{G}$: $n \times J$ with ij-th element equal to 1 if observation i is in class j and zero otherwise. The matrix of class means is given by $\overline{\mathbf{X}} = \mathbf{N}^{-1}\mathbf{G}^T\mathbf{X}$ where $\mathbf{N} = diag(n_1, n_2, \ldots, n_J)$.

Let $\mathbf{D}$: $n \times n$ be the matrix with ih-th element $d_{ih} = \frac{1}{2}\delta^2_{ih}$ where $\delta_{ih} = \sqrt{\sum_{k=1}^{p}(x_{ik} - x_{hk})^2}$ is the Euclidean distance between the i-th and h-th observations. The sum of all the squared distances between samples $\mathbf{x}_i$ and $\mathbf{x}_h$ is twice $\mathbf{T} = \frac{1}{n}\mathbf{1}_n^T\mathbf{D}\mathbf{1}_n$ where $\mathbf{1}_n$ denotes a $n \times 1$ vector of ones. Parallel to the variance decomposition in MANOVA, Gower and Krzanowski (1999) show that $\mathbf{T} = \mathbf{W} + \mathbf{B}$ where $\mathbf{W}$ is the within-class component and $\mathbf{B}$ the between class component.

Partitioning $\mathbf{D}$ into J^2 submatrices where $\mathbf{D}_{rs}$ contains the squared distances (divided by two) between samples from class r and class s the within-class sum of squared distances $\mathbf{W} = \sum_{j=1}^{J}\frac{1}{n_j}\mathbf{1}_{n_j}^T\mathbf{D}_{jj}\mathbf{1}_{n_j}$. Similar to $\mathbf{D}$: $n \times n$, define the matrix $\overline{\mathbf{\Delta}}$: $J \times J$ with rs-th element the squared distance (divided by two) between the class means of classes r and s then the between class sum of squared distances $\mathbf{B} = \frac{1}{n}\mathbf{n}^T\overline{\mathbf{\Delta}}\mathbf{n}$ where $\mathbf{n}$: $J \times 1$ is a vector containing the class sizes $n_1, n_2, \ldots, n_J$.

Nonparametric inference based on permutation tests can be performed without making any distributional or homogeneity of covariance matrices assumptions. The only assumption is that the distance can be calculated between any two samples. Dividing the n observations randomly into J classes of sizes $n_1, n_2, \ldots, n_J$ respectively should reduce the between class sum of squared distances $\mathbf{B}$ if significant differences among the observed classes do exist.

3 AOD biplots

The nonlinear biplot described by Gower and Harding (1988), Gower and Hand (1996) and Gardner (2001) is a versatile general formulation for constructing a multidimensional scatterplot. The nonlinear biplot uses principal co-ordinate analysis (PCO), also known as classical multi-dimensional scaling, to produce a visual display of any distance matrix $\mathbf{D}$ containing Euclidean embeddable distances.

Two special cases of the nonlinear biplot are the principal component analysis (PCA) biplot where $\mathbf{D}$ is calculated from ordinary Euclidean distances and the CVA biplot where $\mathbf{D}$ is calculated from Mahalanobis distances using the pooled within class covariance matrix of the data.

It is well known that PCA aims to optimally represent the variability in a data set in a few, say r, dimensions. For a PCA biplot usually $r = 2$ representing as much variability as possible in two dimensions. In the situation where the data set consists of different classes, PCA does not take this into account. If the directions of maximum variation happen to coincide with the directions separating the classes, different classes can be distinguished in the PCA biplot. Since the PCA technique does not take cognisance of the class structure the directions of maximum variation might disguise the class structure and classes cannot be distinguished in the lower dimensional plot.

On the other hand, a CVA biplot operates on the Mahalanobis distances among the class means. CVA is closely linked to linear discriminant analysis and aims to find the linear combinations maximising the between to within class variance ratio. Under the assumption of equal class covariance matrices, permitting the pooling of class covariance matrices into a within class covariance matrix, a CVA biplot optimally separates the classes in a few (usually two) dimensions. Since the between class and pooled within class covariance matrices used for a CVA biplot are also employed in performing a MANOVA, the CVA biplot can be viewed as a (possibly reduced space) visual display of a MANOVA.

When the assumption of equal class covariance matrices does not hold, or when some classes' covariance matrices yield singular covariance matrix estimates, CVA is not applicable. Gower and Krzanowski (1999) suggest a PCO based on the matrix $\overline{\mathbf{\Delta}}$ defined in section 2 above to pictorially represent differences among classes.

An AOD biplot can be constructed using the relationship between PCO for nonlinear biplots and PCA biplots discussed above. It follows that an AOD biplot is a special case of the nonlinear biplot based on ordinary Euclidean distances between the class means without making any distributional or covariance matrix assumptions. Since the PCA biplot is a special case of a PCO where ordinary Euclidean distances are used, an AOD biplot is simply a PCA biplot of the matrix of class means $\overline{\mathbf{X}}$: $J \times p$.

The formulation of the AOD display in this way enables the construction of a well-defined AOD biplot. The representation of the original variables of measurement as calibrated biplot axes for PCA biplots is discussed in *e.g.* Gower and Hand (1996). The resulting biplot will display the class means and biplot axes of the original variables of measurement. To display the individual samples on the biplot, these are added as supplementary points. The PCO procedure is discussed by Gower and Krzanowski (1999) but the reformulation in terms of a PCA biplot facilitates this interpolation process by a simple matrix multiplication.

4 Principal component analysis in morphometrics

It is customary to analyse the natural logarithm of measurements in morphometric data. The data matrix $\mathbf{X}$: $n \times p$ is therefore assumed to contain the logarithm of each observation.

When analysing morphometric data with PCA, the first principal component usually plays the role of a size vector. This is due to the fact that all measurements are positively correlated, leading to a first principal component with all coefficients of similar sign.

Let $\widetilde{\mathbf{X}}$: $n \times p$ be the centred data matrix calculated from $\mathbf{X}$: $n \times p$, the principal components are calculated from the singular value decomposition $\widetilde{\mathbf{X}}^T\widetilde{\mathbf{X}} = \mathbf{V\Lambda V}^T$ where the s-th column of $\mathbf{V}$: $p \times p$ contains the coefficients of the s-th principal component when the eigenvalues in the diagonal matrix $\mathbf{\Lambda}$ are ordered in decreasing order. Since the eigenvectors are only unique up to multiplication by -1, it can be assumed that the coefficients of the first principal component are all positive.

Following Flury (1997) the one dimensional principal component approximation of the data matrix $\mathbf{X}$ is given by $\mathbf{Y}_{(1)} = \mathbf{1}\overline{\mathbf{x}}^T + (\mathbf{X} - \mathbf{1}\overline{\mathbf{x}}^T)\mathbf{v}_1\mathbf{v}_1^T$ where $\mathbf{v}_1$ is the first column of $\mathbf{V}$. The p-dimensional principal component approximation has a similar form and can be decomposed as follows:

$$
\begin{aligned}
\mathbf{Y}_{(p)} &= \mathbf{1}\overline{\mathbf{x}}^T + (\mathbf{X} - \mathbf{1}\overline{\mathbf{x}}^T)\mathbf{V}\mathbf{V}^T \\
&= \mathbf{X} \\
&= \mathbf{1}\overline{\mathbf{x}}^T + (\mathbf{X} - \mathbf{1}\overline{\mathbf{x}}^T)\begin{bmatrix} \mathbf{v}_1 & \mathbf{V}^* \end{bmatrix}\begin{bmatrix} \mathbf{v}_1^T \\ \mathbf{V}^{*T} \end{bmatrix} \\
&= \mathbf{1}\overline{\mathbf{x}}^T + (\mathbf{X} - \mathbf{1}\overline{\mathbf{x}}^T)\mathbf{v}_1\mathbf{v}_1^T + (\mathbf{X} - \mathbf{1}\overline{\mathbf{x}}^T)\mathbf{V}^*\mathbf{V}^{*T}.
\end{aligned}
$$

The first two terms in the row above, can be viewed as a size component with the remaining third term containing the shape component as well as random error. When investigating differences in shape among classes of observations, the size component is removed and the analysis of

$$\mathbf{X}_{(-size)} = (\mathbf{X} - \mathbf{1}\overline{\mathbf{x}}^T)\mathbf{V}^*\mathbf{V}^{*T}$$

focuses on differences in shape only. Each of the variables represented in $\mathbf{X}_{(-size)}$ can be interpreted as the deviation in that variable from the value of an 'average shaped tortoise' of that particular size.

5 Application: *Homopus Areolatus*

The common padloper, *Homopus Areolatus*, is a small terrestrial tortoise species endemic to South Africa. Although taxonomically *H. areolatus* is recognised as a single species with little geographical variation (Branch (1998); Boycott and Bourquin (2000); Varhol (1998)) analysed the molecular systematics of these tortoises and found higher than expected sequence variation within the genus, suggesting distinct population structuring and possibly cryptic species.

Subtle morphometric differences between tortoise shells were noticed in *H. areolatus* specimens from different geographical regions. In an effort to compare the possible morphometric differences to the conclusion of Varhol (1998) that inland *H. areolatus* warrants taxonomic recognition, the sample of 109 shells originating from six geographical regions were analysed statistically. The number of tortoise shells available are summarised in Table 1.

Table 1. Number of tortoise shells available for analysis from the six geographical regions Fynbos Western Cape (FWC), Fynbos Central Mountains (FCM), Fynbos Southern Cape (FSC), Eastern Cape (EC), Great Karoo (GK) and Little Karoo (LK). The Karoo regions are inland, with all other regions along the coast.

Region	Males	Females	Juveniles	Total
FWC	17	12	4	33
FCM	8	10	0	18
FSC	1	5	5	11
EC	20	12	1	33
GK	4	3	1	8
LK	3	2	1	6

It is clear from Table 1 that in some cases very few observations are available, specifically for the regions of greatest interest, namely the inland Karoo regions. For each tortoise shell 17 variables were measured in accordance to the measurements used in the study of Germano (1993). A list of the variables is given in Table 2.

Because *H. areolatus* is sexually dimorphic with the females being larger than the male tortoises, possible shape differences could occur between the sexes and influence tests for shape differences among regions. As a first step in the analysis, the shape data for each region were subjected to AOD permutation tests, to test for differences among the three classifications in Table 1 (males, females and juveniles). The random allocation to classes was performed according to the algorithm of Good (2000). Since the juveniles cannot be classified as male or female and not all regions showed negligible differences among the three classes, the juveniles were excluded from the remainder of the analysis.

Table 2. Variables measured on 109 *H. areolatus* tortoise shells.

Abbreviation	Variable description
CL	Straight carapace length
CW	Straight carapace width
CH	Shell height
V1, V2, V3, V4, V5	Straight length of vertebral scutes respectively
M6	Width of the sixth left marginal scute at the contact with the costal
PL	Straight maximum plastron length
MPL	Mid-plastron length from gular notch to anal notch
GL	Midline gular length
H	Midline humeral length
P	Midline pectoral length
AB	Midline abdominal length
F	Midline femoral length
A	Midline anal length

Similar AOD permutation tests were performed for each region separately to test for differences in shape between males and females only. From the results in Table 3 it is clear that there are no significant differences between the shapes of males and females. The permutation tests for FWC, FCM and EC were performed with 10 000 replicates while for FSC, GK and LK the exact permutation distributions were calculated.

Table 3. Achieved significance levels (ASL) in testing for differences between the shape of male and female tortoise shells for each geographical region.

Region	FWC	FCM	FSC	EC	GK	LK
ASL	0.1469	0.3076	1.0000	0.1174	0.5429	0.6000

Since the shape does not differ significantly between male and female tortoises, all adult tortoise shells were used in an AOD permutation test for testing differences in shape between the geographical regions. The achieved significance level based on 10 000 replicates was found to be ASL = 0.0002. The shape does indeed differ statistically significantly among the regions.

To visually represent the shape differences among the regions, an AOD biplot is shown in Figure 1. Both the individual sample points and the class means, as well as the variables are represented in the biplot.

Comparing the samples from the different geographical regions, it is clear that the tortoises from the Great Karoo (GK) differ most noticeably from the norm (point of concurrency of biplot axes corresponding with zero on each variable). Orthogonally projecting the GK class mean onto the biplot axes reveals that shape differences manifest in lower values for height and width.

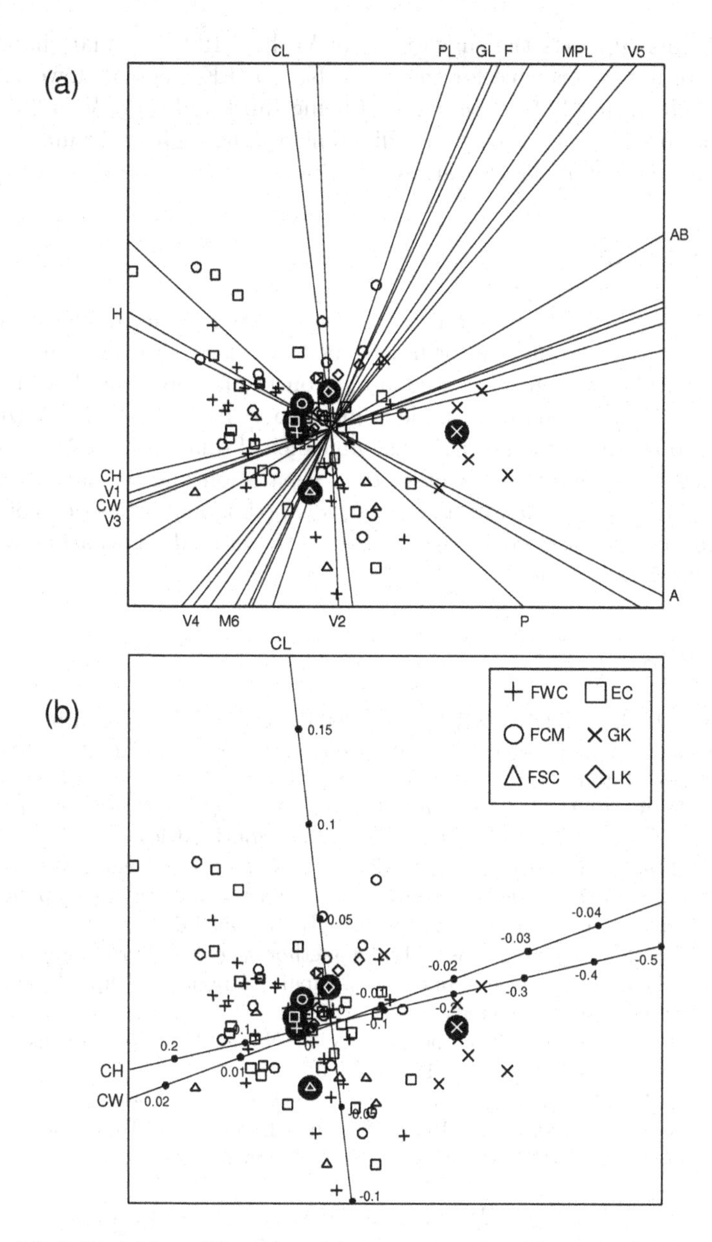

Fig. 1. (a) AOD biplot. All 17 variables are represented by biplot axes. Due to the large number of axes, no calibration of axes is shown but the axis direction is indicated by the convention of naming biplot axes on the positive side. (b) The biplot is identical to top panel but the representation of axes 4 to 17 is suppressed.

Although this supports the suggestions of Varhol (1998), similar shape differences are not perceptible for the Little Karoo (LK) region. However, the tortoise shells from FSC exhibit signs of being short and wide. Visually comparing the spread of observations for different regions confirm the uncertainty about homogeneity of class covariance matrices.

6 Conclusion

In this paper testing for differences among classes was accomplished without making any distributional or homogeneity of class covariance matrix assumptions. AOD inference is based simply on the assumption of being able to calculate distances among sample points. Apart from employing permutation tests to establish the statistical significance of difference an AOD biplot is proposed for visually representing class separation and overlap. Utilising the AOD biplot as a multivariate scatterplot, orthogonal projections of class means onto biplot axes provide researchers with detailed information regarding characteristics of each class.

References

BOYCOTT, R.C. and BOURQUIN, O. (2000): *The Southern African Tortoise Book: A Guide to Southern African Tortoises, Terrapins and Turtles.* Privately published, Hilton, South Africa.

BRANCH, W.R. (1998): *Field Guide to the Snakes and Other Reptiles of Southern Africa.* 3rd ed. Struik Publishers, Cape Town, South Africa.

FLURY, B. (1997): *A first course in multivariate statistics.* Springer, New York.

GABRIEL, K.R. (1971): The biplot graphical display of matrices with application to principal component analysis. *Biometrika*, 58, 453467.

GARDNER, S (2001): Extensions of biplot methodology to discriminant analysis with applications of non-parametric principal components. Unpublished PhD thesis, Stellenbosch University, South Africa.

GERMANO, D.J. (1993): Shell morphology of North American tortoises. *The American Midland Naturalist*, 129, 319335.

GOOD, P. (2000): *Permutation tests.* 2nd ed. Springer, New York.

GOWER, J.C. and HAND, D.J. (1996): *Biplots.* Chapman and Hall, London.

GOWER, J.C. and HARDING, D.J. (1988): Nonlinear biplots. *Biometrika*, 75, 445455.

GOWER, J.C. and KRZANOWSKI, W.J. (1999): Analysis of distance for structured multivariate data and extensions to multivariate analysis of variance. *Applied Statistics*, 48, 505519.

VARHOL, R. (1998): The molecular systematics of Southern African testudinidae. Unpublished MSc thesis, University of Cape Town, South Africa.

Revised Boxplot Based Discretization as the Kernel of Automatic Interpretation of Classes Using Numerical Variables

Karina Gibert and Alejandra Pérez-Bonilla

Department Statistics and Operations Research.
Technical University of Catalonia.
Campus Nord, Edif. C5, C - Jordi Girona 1-3, 08034 Barcelona, SPAIN.
karina.gibert@upc.edu; alejandra.perez@upc.edu

Abstract. In this paper the impact of improving *Boxplot based discretization (BbD)* on the methodology of *Boxplot based induction rules (BbIR)*, oriented to the automatic generation of conceptual descriptions of classifications that can support later decision-making is presented.

1 Introduction

In automatic classification where the classes composing a certain domain are to be discovered, one of the most important required processes and one of the less standardized ones, probably, is the interpretation of classes, closely related to *validation* and critical for usefulness of the discovered knowledge. The interpretation of the classes, so important to understand the meaning of the obtained classification as well as the structure of the domain, used to be done in an artistic-like way. But this process becomes more and more complicated as the number of classes grows. This work is involved with the automatic generation of useful interpretations of classes in such a way that decisions about the treatment or action associated to a new object can be modelled and it is oriented to develop, in the long term, decision support system. Such methods are especially needed in the context of decision-making with number of variables or classes too big for human interpretation. In this work we focus on first step for automatic generation of the interpretation of classes.

This is different from what is pursued by other inductive learning techniques as association rules algorithms, see Dietterich and Michalski (1993), where the set of produced association rules use to be huge in Data Mining context and the greater is the number of variables or/and classes in involved in the analysis, the more complex are the generated rules and the more difficult the interpretation of classes from those rules.

Decision trees (COBWEB, Michalsky in Michalski (1980)) are also an alternative for assigning concepts to groups of objects, but there is no guarantee that terminal nodes remain pure, i.e. composed by elements of a single class and many different terminal nodes may correspond to the same class, which produces long description of the classes. On the other hand, algorithms such as COBWEB are very expensive and it takes a lot of time to get the decision tree when the number of objects and variables are really big as in Data Mining.

Boxplot based induction rules (BbIR), see Gibert and Pérez-Bonilla (2006), is a proposal to produce compact concepts associated to the classes, oriented to express the differential characteristic of every class in such a way that the user can easily understand which is the underlying classification criterion and can easily decide the treatment or action to be assigned to each class. Given a classification, the idea is to provide an automatic interpretation for it that supports the construction of intelligent decision support systems. The core of this process is a method for discretizing numerical variables in such a way that particularities of the classes are elicited called *Boxplot based discretization (BbD)*. In this work special details of *BbD* are presented.

A particular application to Waste Water Treatment Plants (WWTP) is in progress and results appear to be very promising. Examples used in this paper come for this real application. The presented proposal integrates different findings from a series of previous works, see Gibert (1996), Gibert and Roda (2000), in a single methodological tool which takes advantage of the hierarchical structure of the classification to overcome some of the limitations observed in Gibert (1996) and Gibert et al. (1998).

This paper is organized as follows: After the introduction, previous work of this research is in Section 2. The Section 3 presents Revising Boxplot based discretization. Finally in Section 4 conclusions and future work are discussed.

2 Previous work

The present research is based on previous works in which the automatic process of characterization of classes has been analyzed. The main idea was to automatically analyze conditional distributions through *Multiple boxplot*[1], (see Figure 1) in order to identify *characterizing variables*, introduced in Gibert et al. (1998) as main concepts which are in the kernel of this work:

> Let us consider $\mathcal{I}$ as the set of n objects to be analyzed. They are described by K numerical variables $X_k, (k = 1 : K); x_{ik}$ is the value taken by variable X_k for object i.

- Given a variable X_k and a partition $\mathcal{P}$ of $\mathcal{I}$, x is a *characterizing value* of class $C \in \mathcal{P}$ if $\exists i \in C$ tq $x_{ik} = x$ and $\forall i \notin C, x_{ik} \neq x$.

[1] It is a graphical tool introduced by Tukey (1977). For each class the range of the variable is visualized and rare observations (outliers) are marked as "∘" or "*". A box is displayed from Q1 (first quartile) to Q3 (third quartile) and the Median, usually inside the box, is marked with a vertical sign. Boxes include, then, the 50% of the elements of the class and the whiskers extend until the minimum and maximum.

- Variable X_k is *Totally characterizing* class $C \in \mathcal{P}$, if either one or more of the values taken by X_k in class C are *characterizing values* of C.

The concepts defined in Gibert (1996) are formal and general. There is no guarantee that characterizing values exist in a class, which immediately requires a proposal for dealing with their lack. A first procedure of characterization is to detect minimum sets of variables that distinguish a class from another one only using qualitative variables.

Paralelly, extension to numerical variables was faced in Gibert and Roda (2000). Numerical variables are very relevant in many domains, as is the case of WWTP. Analysis of high-order interactions in this context is not easy and in this case experts required graphical representations for interpreting the classes. Thus, the starting point of the presented proposal was to study *multiple boxplots*, Tukey (1977); Gibert and Pérez-Bonilla (2006), (Fig. 1).

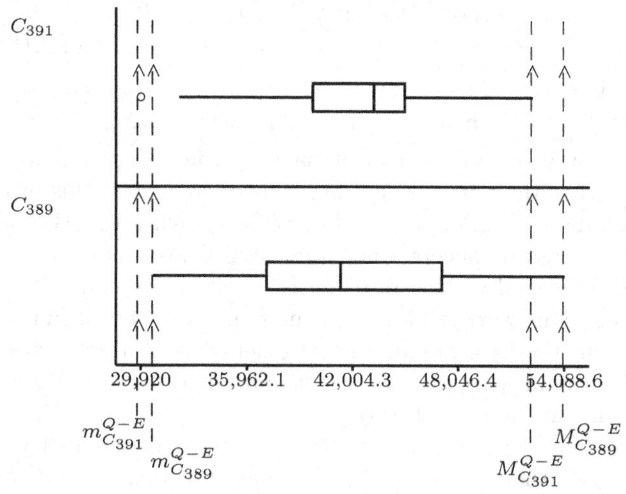

Fig. 1. Boxplot of Q-E (Inflow wastewater in daily m^3 of water) in WWTP *vs* a certain partition $\mathcal{P} = \{C_{391}, C_{389}\}$; see Gibert and Pérez-Bonilla (2006).

The *Boxplot based discretization (BbD)* is presented in Gibert and Pérez-Bonilla (2006) as an efficient way of transforming a numerical variable into a qualitative one in such a way that the cut points for discretizing identify where the set of classes with non-null intersection of X_k changes and it consists:

1. Calculate de *minimum* (m^k_C) and *maximum* (M^k_C) of X_k inside any class. Built $\mathcal{M}^k = \{m^k_{C_1}, \ldots, m^k_{C_\xi}, M^k_{C_1}, \ldots, M^k_{C_\xi}\}$, where $card(\mathcal{M}^k) = 2\xi$
2. Built the *set of cutpoints* $\mathcal{Z}^k$ by sorting $\mathcal{M}^k$ in increasing way into $\mathcal{Z}^k = \{z^k_i \; ; \; i = 1, \ldots, 2\xi\}$. At every z^k_i the set of intersecting classes changes. In Fig.1, for example, both C_{391} and C_{389} take values between 30,592.1 and 52,255.8 but only C_{391}, takes values between 29,920.0 and 30,592.1 while only C_{389} between 30,592.1 and 54,088.6 .

3. Built the *set of intervals I^k induced by $\mathcal{P}$ on X_k* by defining an interval I_s^k between every pair of consecutive values of $\mathcal{Z}^k$. $I^k = \{I_1^k, \ldots, I_{2\xi-1}^k\}$ is the *BbD* of X_k. The I_s^k intervals have variable length and the set of intersecting classes is constant all along the interval and changes from one to another.

In Vázquez and Gibert (2001) there is a proposal of building all the I_s^k following a single pattern: $I_s^k = (z_s^k, z_{s+1}^k]$ $\forall s > 1$ being $I_1^k = [z_1^k, z_2^k]$.

In Gibert and Pérez-Bonilla (2005) a deeper discussion about situations in which closed or open intervals are more convenient is presented.

In Gibert (2004) the formulation of the methodology *boxplot based induction rules (BbiR)* is presented. It is a method for generating probabilistic concepts with a minimum number of attributes on the basis of the *boxplot based discretization (BbD)* of X_k.

1. Use the *Boxplot based discretization* to build $\mathcal{I}^k = \{I_1^k, I_2^k, I_3^k, \ldots, I_{2\xi-1}^k\}$.
2. For every interval produce the rules: r_s : *If* $x_{ik} \in I_s^k \xrightarrow{p_{sC}} i \in C$
   where, $p_{sC} = P(C|I^k = I_s^k) = P(i \in C|x_{ik} \in I_s^k) = \frac{card\{i\,:\,x_{ik} \in I_s^k \wedge i\,\in C\}}{card\{i \in \mathcal{I} : x_{ik} \in I_s^k\}}$
   If $p_{sC} = 1$, I_s^k is a set of *characterizing values* of X_k. If $\forall s = 1 : 2\xi - 1$, $\exists C \in \mathcal{P}$ tq $p_{sC} = 1$ then X_k is a *totally characterizing variable.*

Although obtained results were satisfactory from an applied point of view, it is not clear that they are optimal in terms of coverage. This proposal was improved in Gibert and Pérez- Bonilla (2005) in such a way that the probability of the generated concepts increases, and yields more certain interpretations. The final goal of this research is to elaborate a new proposal, on the basis of previous work, which overcomes all the observed limitations and consolidates a methodology of automatic generation of interpretations from a given classification, giving support to the construction of intelligent decision support systems in the case of WWTP.

In Comas *et al* (2001) a comparison between a very primary version of the method and other inductive methods has shown that *BbIR* appears as a good *imitation* of the real process used by experts to manually interpret the classes. It also confirmed that modifying the method to provide more flexibility would sensibly improve its performance, (Gibert (2004)).

In Gibert and Pérez- Bonilla (2006), the *Boxplot based discretization (BbD)* was used in the *Methodology of conceptual characterization by embedded conditioning (CCEC)* which is a methodology for generating automatic interpretations of a given partition $\mathcal{P} \in \tau$, where $\tau = \{\mathcal{P}_1, \mathcal{P}_2, \mathcal{P}_3, \mathcal{P}_4, ..., \mathcal{P}_n\}$ is an indexed hierarchy of $\mathcal{I}$. Taking advantage of the hierarchy it is reduced to iteratively distinguish pairs of classes, what justifies that in this work only binary partitions are considered from now on.

3 Revising boxplot based discretization

For a binary partition $\mathcal{P}_2 = \{C_1, C_2\}$, Z^k always contains 4 elements which are minimum and maximum values of $C_1$and C_2 conveniently ordered. That is the reason why I^k will always have 3 intervals built upon Z^k values. In the particular case of binary partitions, the previous proposal (Vazquez and Gibert (2002)), established the following structure for I^k:

$$I_1^k = [z_1^k, z_2^k], I_2^k = (z_2^k, z_3^k], I_3^k = (z_3^k, z_4^k]$$

In Gibert and Pérez-Bonilla (2005) is evidenced that the rules generated from I^k following *BbIR* are sensitive to the form of the limits of every I_s^k.

Let us analyse the simple example in figure 1, where the multiple boxplot of a variable Q-E vs a binary partition called $\mathcal{P}$ is displayed. Table 1*(left)* shows minimum and maximum of variable Q-E in classes C_{391} and C_{389}, while Table 1*(right)* shows $\mathcal{M}^{Q-E}$, set of extreme values of Q-E$|\mathcal{P}$ and $\mathcal{Z}^{Q-E}$, the corresponding sorting in increasing order. Following the previous proposal of Vázquez and Gibert (2002), the $I^{Q-E} = \{I_1^{Q-E}, I_2^{Q-E}, I_3^{Q-E}\}$ is build in the following way:

$I_1^{Q-E} = [29920.0, 30592.2]$, $I_2^{Q-E} = (30592.2, 52255.8]$, $I_3^{Q-E} = (52255.8, 54088.6]$.

Table 1. (*left*): Minimum and Maximum of $Q-E$; (*right*): $\mathcal{M}^{Q-E}$ and $\mathcal{Z}^{Q-E}$.

classes	$min\ (m_C^{Q-E})$	$max\ (M_C^{Q-E})$
C_{391}^3	29,920.0	52,255.8
C_{389}^3	30,592.1	54,088.6

$\mathcal{M}^{Q-E}$	$\mathcal{Z}^{Q-E}$
29,920.0	29,920.0
52,255.8	30,592.1
30,592.1	52,255.8
54,088.6	54,088.6

From this, the *BbIR* this produces the following set of rules:

$$r_1: x_{i,Q-E} \in [29920.0, 30592.2] \xrightarrow{0.5} i \in C_{389}$$
$$r_2: x_{i,Q-E} \in [29920.0, 30592.2] \xrightarrow{0.5} i \in C_{391}$$
$$r_3: x_{i,Q-E} \in (30592.2, 52255.8] \xrightarrow{0.83} i \in C_{391}$$
$$r_4: x_{i,Q-E} \in (30592.2, 52255.8] \xrightarrow{0.17} i \in C_{389}$$
$$r_5: x_{i,Q-E} \in (52255.8, 54088.6] \xrightarrow{1.0} i \in C_{389}$$

From the Figure 1 it is clear that z_s^k, and in consequence M_C^k, are identifying the points where intersecting classes change; the area delimited by I_1^{Q-E} should certainly be assigned to C_{391} what is not according to r_1 and r_2. The reason why the probability of r_1 is 0.5 instead of 1 is that right limit of I_1^{Q-E} should be open instead of closed. Doing this redefinition, a new I^{Q-E} is defined as:

$I_1^{Q-E} = [29920.0, 30592.2)$, $I_2^{Q-E} = [30592.2, 52255.8]$, $I_3^{Q-E} = (52255.8, 54088.6]$.

and the new set of rules is:

$$r_1: x_{i,Q-E} \in [29920.0, 30592.2] \xrightarrow{1.0} i \in C_{391}$$
$$r_2: x_{i,Q-E} \in (30592.2, 52255.8] \xrightarrow{0.827} i \in C_{391}$$
$$r_3: x_{i,Q-E} \in (30592.2, 52255.8] \xrightarrow{0.173} i \in C_{389}$$
$$r_4: x_{i,Q-E} \in (52255.8, 54088.6] \xrightarrow{1.0} i \in C_{389}$$

Making a similar analysis on all different situations that can be found, Fig.2 shows the more convenient way of redefining the *Boxplot based discretization (BbD)* in each case, see Gibert and Pérez-Bonilla (2005) for details. Observing the column with header *pattern*, see Fig.2, it is easily seen that there are only 2 patterns of building I^k from Z^k according to the limits of this intervals. Both will generate I^k with 3 intervals:

1. **Open-*Center***: In the case 1 and case 2, 3 intervals are defined in such a way that the center (I_2^k) is an *open* interval by both sides. The pattern is the following: $I_1^k = [z_1^k, z_2^k]$,$I_2^k = (z_2^k, z_3^k)$ and $I_3^k = [z_3^k, z_4^k]$.
2. **Closed-*Center***: In the other cases (4 to 13), 3 intervals are defined in such a way that the center $[I_2^k]$ is a *closed* interval by both sides. The pattern is the following:$I_1^k = [z_1^k, z_2^k)$, $I_2^k = [z_2^k, z_3^k]$ and $I_3^k = (z_3^k, z_4^k]$.

These would represent a more realistic model for all different situations that can be found in the multiple Boxplot, detailed in Gibert and Pérez-Bonilla (2005).None of these patterns coincides with the proposal in Vazquez and Gibert (2001). From this analysis it was also seen that the condition to generate an *open center* pattern is: $M_{C2}^k < m_{C1}^k$ and $M_{C1}^k < m_{C2}^k$, all other cases should be treated as *closed center*. Using this new way of intervals generation it was seen that more certain rules can be induced from the classes which directly leads on more reliable interpretations. Table 2 shows the comparison between both proposals for some variables taken from the previously referred real application on WasteWater Treatment Plants (WWTP) Gibert and Roda (2000), which is in progress. The following systems of probabilistic rules then can be induced for the variables DBO-E (Biodegradable organic matter (m/l)) and SS-S (Suspended Solids (mg/l)) :

Table 2. Comparison between both proposal.

Original Boxplot based discretization	Revised Boxplot based discretization
$r_1 : x_{DBO-E,i} \in [90.00, 220.0] \xrightarrow{0.96} C_{390}$	$r_1 : x_{DBO-E,i} \in [90.00, 220.0) \xrightarrow{1.0} C_{390}$
$r_2 : x_{DBO-E,i} \in [90.00, 220.0] \xrightarrow{0.04} C_{383}$	$r_2 : x_{DBO-E,i} \in [220.0, 382.0] \xrightarrow{0.78} C_{390}$
$r_3 : x_{DBO-E,i} \in (220.0, 382.0] \xrightarrow{0.78} C_{390}$	$r_3 : x_{DBO-E,i} \in [220.0, 382.0] \xrightarrow{0.22} C_{383}$
$r_4 : x_{DBO-E,i} \in (220.0, 382.0] \xrightarrow{0.22} C_{383}$	$r_4 : x_{DBO-E,i} \in (382.0, 987.0] \xrightarrow{1.0} C_{383}$
$r_5 : x_{DBO-E,i} \in (382.0, 987.0] \xrightarrow{1.0} C_{383}$	
$r_1 : x_{SS-S,i} \in [2.80, 3.200] \xrightarrow{0.5} C_{389}$	$r_1 : x_{SS-S,i} \in [2.80, 3.200) \xrightarrow{1.0} C_{391}$
$r_2 : x_{SS-S,i} \in [2.80, 3.200] \xrightarrow{0.5} C_{391}$	$r_2 : x_{SS-S,i} \in [3.20, 20.00] \xrightarrow{0.74} C_{391}$
$r_3 : x_{SS-S,i} \in (3.20, 20.00] \xrightarrow{0.75} C_{391}$	$r_3 : x_{SS-S,i} \in [3.20, 20.00] \xrightarrow{0.26} C_{389}$
$r_4 : x_{SS-S,i} \in (3.20, 20.00] \xrightarrow{0.25} C_{389}$	$r_4 : x_{SS-S,i} \in (20.0, 174.8] \xrightarrow{1.0} C_{391}$
$r_5 : x_{SS-S,i} \in (20.0, 174.8] \xrightarrow{1.0} C_{391}$	

In most of the cases the number of rules with probability 1 produced by *Revised BbD* increases (rules with null probability can be eliminated).

Case N	Characteristic	Multiple Boxplot	Pattern CCEC
1	$M^k_{C2} < m^k_{C1}$	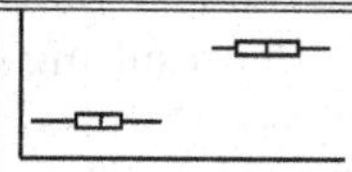	[ ],( ), [ ]
2	$M^k_{C1} < m^k_{C2}$	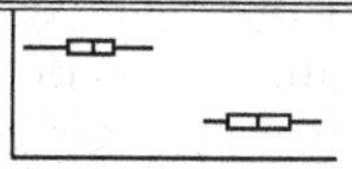	[ ],( ), [ ]
3	$m^k_{C1} = m^k_{C2} \wedge M^k_{C1} = M^k_{C2}$	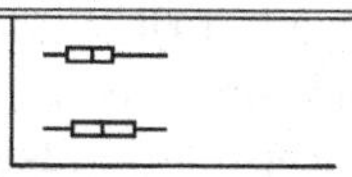	[ ), [ ], ( ]
4	$m^k_{C1} > m^k_{C2} \wedge M^k_{C1} > M^k_{C2} \wedge m^k_{C1} < M^k_{C2}$	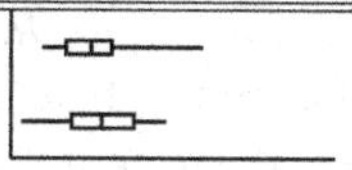	[ ), [ ], ( ]
5	$m^k_{C1} < m^k_{C2} \wedge M^k_{C1} < M^k_{C2} \wedge m^k_{C2} > M^k_{C1}$	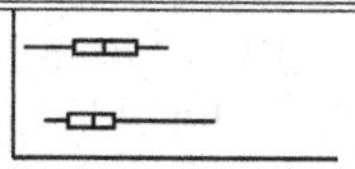	[ ), [ ], ( ]
6	$m^k_{C1} < m^k_{C2} \wedge M^k_{C1} > M^k_{C2} \wedge m^k_{C1} < M^k_{C2}$	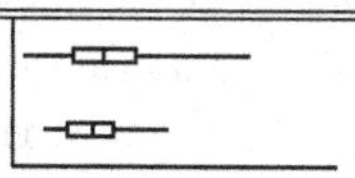	[ ), [ ], ( ]
7	$m^k_{C1} > m^k_{C2} \wedge M^k_{C1} < M^k_{C2} \wedge m^k_{C1} < M^k_{C2}$	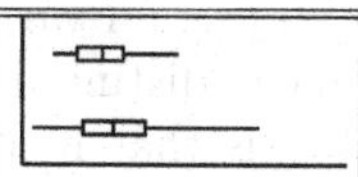	[ ), [ ], ( ]
8	$m^k_{C1} = m^k_{C2} \wedge M^k_{C1} < M^k_{C2}$	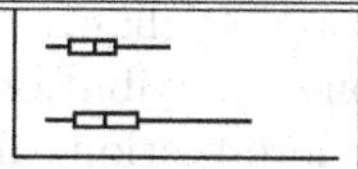	[ ), [ ], ( ]
9	$m^k_{C1} = m^k_{C2} \wedge M^k_{C1} > M^k_{C2}$	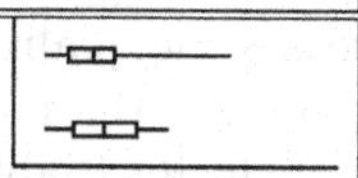	[ ), [ ], ( ]
10	$m^k_{C1} > m^k_{C2} \wedge M^k_{C1} = M^k_{C2}$	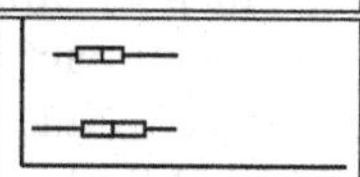	[ ), [ ], ( ]
11	$m^k_{C1} < m^k_{C2} \wedge M^k_{C1} = M^k_{C2}$	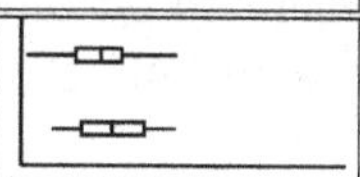	[ ), [ ], ( ]
12	$M^k_{C1} = m^k_{C2}$	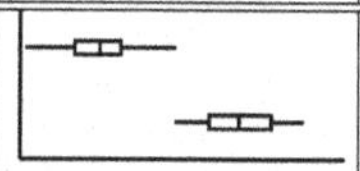	[ ), [ ], ( ]
13	$M^k_{C2} = m^k_{C1}$	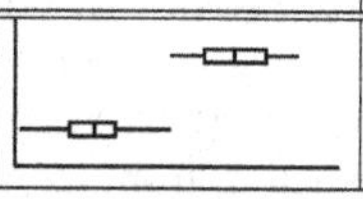	[ ), [ ], ( ]

Fig. 2. Way of redefining the *BbD* in each case.

4 Conclusions and future work

In this paper a revision of *Boxplot based discretization (BbD)* is presented in such a way that the resulting discretization of a numerical variable allows induction of more certain rules. The *BbD* is a step of a wider methodology called *CCEC* and presented in Gibert and Pérez-Bonilla (2006) which is oriented to generate automatic interpretations from a group of classes in such a way that concepts associated to classes are built taking advantage of hierarchical structure of the underlying clustering.
The *BbD*, is a quick and effective method for discretizing numerical variables for generating conceptual model of the domain, which will greatly support the posterior decision-making. *Revised BbD* has been included in *CCEC* and the whole methodology has been successfully applied this to real data coming from a Wastewater Treatment Plant. Benefits of this proposal are of special interest in the interpretation of partitions with great number of classes. The induced model can be included as a part of an Intelligent Decision Support System to recommend decisions to be taken in a certain new situation, see R.-Roda *et al.* (2002). The main requirement of *CCEC* is to have an efficient way of discretizing numerical variables according to the subsets of classes that can share values of a certain variable, so the generated concepts can express particularities than can distinguish one class from the others. The main property of *Revised BbD* is that it allows finding the cut points of the variable where class overlapping changes by a simple and cheap sorting of extreme values of conditioned distributions, which is extremely cheap compared with directly analysing intersections among classes with continuous variables. The *Revised BbD* increases the quality of produced knowledge and decision making support incrementing the number of certain rules. At present, different criteria for deciding which variable is to be kept in the final interpretation are being analysed to see the impact of *BbD* on final interpretation. As Fayyad *et al.* (1996)point out, Data Mining should also be involved with "interpretation of the patterns generated by Data Mining algorithms". *CCEC* (and *Revised BbD*) is trying to contribute to this particular issue.

References

COMAS, J., DZEROSKI, S. & GIBERT, K. (2001): KD by means of inductive methods in wastewater treatment plant data. *AI Communications.* **14**, 45-62.

DIETTERICH, T.G. & MICHALSKI, R.S. (1983): A comparative review of selected methods for learning from examples. *Mach. Learn. New York.* **2**, 41-81.

FAYYAD, U. *et al.*(1996): From Data Mining to Knowledge Discovery: An overview. *Advances in Knowledge Discovery and Data Mining.* AAAI/MIT. 1-34.

GIBERT, K. (2004): Técnicas híbridas de inteligencia artificial y estadística para el descubrimiento de conocimiento y la minería de datos. *Tendencias de la minería de datos en España.* Riquelme, J. C. 119-130.

GIBERT, K. (1996): The use of symbolic information in automation of statistical treatment for ill-structured domains. *AI Communications.* **9**, 36-37.

GIBERT, K., ALUJA, T. & CORTÉS, U. (1998): Knowledge Discovery with Clustering Based on Rules. Interpreting results. In: *LNAI.* **1510**, Springer. 83-92.

GIBERT, K & PÉREZ-BONILLA, A. (2005): Análisis y propiedades de la metodología Caracterización Conceptual por Condicionamientos Sucesivos (CCCS). Research DR 2005/14, EIO. UPC, Barcelona.

GIBERT, K. & PÉREZ-BONILLA, A. (2006): Automatic generation of interpretation as a tool for modelling decisions. In: Springer. *III International Conference on Modeling Decisions for Artificial Intelligence*, Tarragona, in press.

GIBERT, K. & RODA, I. (2000): Identifying characteristic situations in Waste Water Treatmet Plants. In: *Workshop on Binding Environmental Sciences and Artificial Intelligence*, **1**, 1-9.

MICHALSKI, R.S. (1980): Knowledge acquisition through conceptual clustering: A theoretical framework and algorithm for partitioning data. In: *International Journal of Policy Analysis and Information Systems*, **4**, 219-243.

RODA, -R.*et al.*, I. (2002): A hybrid supervisory system to support WWTP operation: implementation and validation. In: *Water Science and Tech.* **45**, 289-297.

TUKEY, J.W. (1977): *Exploratory Data Analysis.* Cambridge, MA: Ad-Wesley.

VÁZQUEZ, F. & GIBERT, K. (2001): Generación automática de reglas difusas en dominios poco estructurados con variables numéricas. In: *IXth Conferencia de la Asociación Española para la Inteligencia Artificial*, **1**, 143-152.

Part VI

Data and Web Mining

Comparison of Two Methods for Detecting and Correcting Systematic Error in High-throughput Screening Data

Andrei Gagarin[1], Dmytro Kevorkov[1], Vladimir Makarenkov[2], and Pablo Zentilli[2]

[1] Laboratoire LaCIM, Université du Québec à Montréal, C.P. 8888, Succ. Centre-Ville, Montréal (Québec), Canada, H3C 3P8
[2] Département d'Informatique, Université du Québec à Montréal, C.P. 8888, Succ. Centre-Ville, Montréal (Québec), Canada, H3C 3P8

Abstract. High-throughput screening (HTS) is an efficient technological tool for drug discovery in the modern pharmaceutical industry. It consists of testing thousands of chemical compounds per day to select active ones. This process has many drawbacks that may result in missing a potential drug candidate or in selecting inactive compounds. We describe and compare two statistical methods for correcting systematic errors that may occur during HTS experiments. Namely, the collected HTS measurements and the hit selection procedure are corrected.

1 Introduction

High-throughput screening (HTS) is an effective technology that allows for screening thousands of chemical compounds a day. HTS provides a huge amount of experimental data and requires effective automatic procedures to select active compounds. At this stage, active compounds are called hits; they are preliminary candidates for future drugs. Hits obtained during primary screening are initial elements for the determination of activity, specificity, physiological and toxicological properties (secondary screening), and for the verification of structure-activity hypotheses (tertiary screening) (Heyse (2002)).

However, the presence of random and systematic errors has been recognized as one of the major hurdles for successful implementing HTS technologies (Kaul (2005)). HTS needs reliable data classification and quality control procedures. Several methods for quality control and correction of HTS data have been recently proposed in the scientific literature. See for example the papers of Zhang et al. (1999), Heyse (2002), Heuer et al. (2003), and Brideau et al. (2003).

There are several well-known sources of systematic error (Heuer et al. (2003)). They include reagents evaporation or decay of cells which usually show up as smooth trends in the plate mean or median values. Another typical error can be caused by the liquid handling or malfunctioning of pipettes. Usually this generates a localized deviation of expected values. A variation in the incubation time, a time drift in measuring different wells or different plates, and reader effects may appear as smooth attenuations of measurements over an assay. This kind of effects may have a significant influence on the selection process of active compounds. They can result in an underestimation (false negative hits) or overestimation (false positive hits) of the number of potential drug targets.

We have developed two methods to minimize the impact of systematic errors when analyzing HTS data. A systematic error can be defined as a systematic variability of the measured values along all plates of an assay. It can be detected, and its effect can be removed from raw data, by analyzing the background pattern of plates of the same assay (Kevorkov and Makarenkov (2005)). On the other hand, one can adjust the data variation at each well along the whole HTS assay to correct the traditional hit selection procedure (Makarenkov et al. (2006)). Methods described in Sections 3 and 4 originate from the two above-mentioned articles.

2 HTS procedure and classical hit selection

An HTS procedure consists of running samples (i.e. chemical compounds) arranged in 2-dimensional plates through an automated screening system that makes experimental measurements. Samples are located in wells. The plates are operated in sequence. Screened samples can be divided into active (i.e. hits) and inactive ones. Most of the samples are inactive, and the measured values for the active samples are significantly different from the inactive ones. In general, samples are assumed to be located in a random order, but it is not always the case in practice.

The mean values and standard deviations are calculated separately for each plate. To select hits in a particular plate, one usually takes the plate mean value μ and its standard deviation σ to identify samples whose values differ from the mean μ by at least $c\sigma$, where c is a preliminary chosen constant. For example, in the case of an inhibition assay, by choosing $c = 3$, we would select samples with the values lower than $\mu - 3\sigma$. This is the simplest and most widely-known method of hit selection. This method is applied on a plate-by-plate basis.

3 Correction by removing the evaluated background

This correction method is a short overview of the corresponding procedure of Kevorkov and Makarenkov (2005). To use it properly, we have to assume that all samples are randomly distributed over the plates and systematic error causes a repeatable influence on the measurements in all plates. Also, we have to assume that the majority of samples are inactive and that their average values measured for a large number of plates are similar. Therefore, the average variability of inactive samples is caused mainly by systematic error, and we can use them to compute the assay background. In the ideal case, the measurements background surface is a plane, but systematic errors can introduce local fluctuations in it. The background surface and hit distribution surface of an assay represent a collection of scalar values which are defined per well and are plotted as a function of the well coordinates in a 3-dimentional diagram.

An appropriate statistical analysis of experimental HTS data requires a preprocessing. This will ensure the meaningfulness and correctness of the background evaluation and hit selection procedures. Therefore, we use normalization by plate and exclude outliers from the computations. Keeping in mind the assumptions and pre-procession requirements, the main steps of this method can be outlined as follows:

- Normalization of experimental HTS data by plate,
- Elimination of outliers from the computation (optional),
- Topological analysis of the evaluated background,
- Elimination of systematic errors by subtracting the evaluated background surface from normalized raw data,
- Selection of hits in the corrected data.

3.1 Normalization

Plate mean values and standard deviations may vary from plate to plate. To compare and analyze the experimental data from different plates, we need first to normalize all measurements within each plate.

To do this, we use classical *mean centering and unit variance standardization* of the data. Specifically, to normalize the input measurements, we apply the following formula:

$$x'_i = \frac{x_i - \mu}{\sigma}, \tag{1}$$

where x_i, $i = 1, 2, \ldots, n$, is the input element value, x'_i, $i = 1, 2, \ldots, n$, is the normalized output element value, μ is the plate mean value, σ is the plate standard deviation, and n is the total number of elements (i.e. number of wells) in each plate. The output data will have the plate mean value $\mu' = 0$ and the plate standard deviation $\sigma' = 1$.

Another possibility discussed by Kevorkov and Makarenkov (2005) is to normalize all the plate values to a given interval. This normalization generally produces results similar to the described one.

3.2 Evaluated background

Systematic error is assumed to appear as a mean fluctuation over all plates. Therefore, an assay background can be defined as the mean of normalized plate measurements, i.e.:

$$z_i = \frac{1}{N} \sum_{j=1}^{N} x'_{i,j}, \tag{2}$$

where $x'_{i,j}$, $i = 1, 2, \ldots, n$, $j = 1, 2, \ldots, N$, is the normalized value at well i of plate j, z_i is the background value at well i, and N is the total number of plates in the assay.

Clearly, Formula 2 is more meaningful for a large number of plates: in this case the values of inactive samples will compensate the outstanding values of hits. To make Formula 2 useful and more accurate for an assay with a small number of plates, one can exclude hits and outliers from the computations. Thus, the evaluated background will not be influenced by the outstanding values and will better depict systematic errors.

3.3 Subtraction of evaluated background

Analyzing the distribution of selected hits, we can tell whether any systematic error is present or not in the assay: hits should be more or less evenly distributed over all wells. Otherwise, the hit amounts vary substantially from one well to another indicating the presence of systematic errors.

Deviations of the evaluated background surface from the zero plane indicate an influence of systematic errors on the measured values. Therefore, it is possible to correct raw HTS data by subtracting the evaluated background, defined by Formula 2, from the normalized values of each plate, given in Formula 1. After that, we can reassess the background surface and hit distribution again.

4 Well correction method

This section is a concise description of the well correction approach presented in detail in Makarenkov et al. (2006). We have to make the assumptions stated in the previous section about input HTS data and positions of samples in wells. The main steps of the well correction method are the following:

- Normalization of all sample values by plate,
- Analysis of hit distribution in the raw data,
- Hit and outlier elimination (optional),
- Correction and normalization of samples by well,
- Normalization of all samples by plate,
- Selection of hits in the corrected data.

Similarly to the evaluated background approach, the normalization of all samples by plate is done here using the mean centering and unit variance standardization procedure described above. The hit distribution surface can be computed as a sum of selected hits by well along the whole assay. If this surface is significantly different from a plane, it implies the presence of systematic errors in the assay measurements. Excluding hits and outliers from the computation, we obtain the non-biased estimates for the mean values and standard deviations of inactive samples in plates.

4.1 Well correction technique

Once the data are plate-normalized, we can analyze their values at each particular well along the entire assay. The distribution of inactive measurements (i.e. excluding hits and outliers) along wells should be zero-mean centered if systematic error is absent in the dataset.

However, a real distribution of values by well can be substantially different from the ideal one. Such an example is shown in the article by Makarenkov et al. (2006). A deviation of the well mean values from zero indicates the presence of systematic errors. Experimental values along each well can have ascending and descending trends (Makarenkov et al. (2006)). These trends can be discovered using the linear least-squares approximation (e.g. the trends can be approximated by a straight line).

In the case of approximation by a straight line ($y = ax + b$), the line-trend is subtracted from or added to the initial values bringing the well mean value to zero (x denotes the plate number, and y is the plate-normalized value of the corresponding sample). For the analysis of large industrial assays, one can also use some non-linear functions for the approximation. On the other hand, an assay can be divided into intervals and a particular trend function characterizing each interval can be determined via an approximation. After that, the well normalization using the mean centering and unit variance standardization procedure is carried out. Finally, we normalize the well-corrected measurements in plates and reexamine the hit distribution surface.

5 Results and conclusion

To compare the performances of the two methods described above, we have chosen an experimental assay of the HTS laboratory of McMaster University (`http://hts.mcmaster.ca/Competition_1.html`). These data consist of a screen of compounds that inhibit the *Escherichia coli* dihydrofolate reductase. The assay comprises 1250 plates. Each plate contains measurements for 80 compounds arranged in 8 rows and 10 columns. A description of the hit follow-up procedure for this HTS assay can be found in Elowe et al. (2005).

Table 1 shows that the proposed correction methods have slightly increased the number of selected hits. However, the standard deviation of selected hits by well and the χ-square values (obtained using the χ-square contingency test with α-parameter equal to 0.01; the null hypothesis, H_0, here is that the hit distribution surface is a constant plane surface) become smaller after the application of the correction procedures. Moreover, the well correction method allowed the corresponding hit distribution surface to pass the χ-square contingency test in both cases (using 2.5σ and 3σ thresholds for hit selection). Figure 1 shows that the hit distribution surfaces have become closer to planes after the application of the correction methods.

To demonstrate the effectiveness of the proposed correction procedures, we have also conducted simulations with random data. Thus, we have con-

Table 1. Results and statistics of the hit selection carried out for the raw, background removed (Rem. backgr.) and well-corrected (Well correct.) McMaster data.

	Raw data	Rem. backgr.	Well correct.	Raw data	Rem. backgr.	Well correct.
Hit selection threshold	3σ	3σ	3σ	2.5σ	2.5σ	2.5σ
Mean value of hits per well	3.06	3.13	3.08	6.93	6.93	7.03
Standard deviation	2.17	2.16	2.06	3.93	3.55	2.61
Min number of hits per well	0	0	0	1	2	2
Max number of hits per well	10	10	10	19	22	15
χ-square value	121.7	118	109.1	175.8	143.8	76.6
χ-square critical value	111.14	111.14	111.14	111.14	111.14	111.14
χ-square contingency H_0	No	No	Yes	No	No	Yes

sidered random measurements generated according to the standard normal distribution. The randomly generated dataset also consisted of 1250 plates having wells arranged in 8 rows and 10 columns. The initial data did not contain any hit. However, the traditional hit selection procedure has found 119 false positive hits in the random raw data using the 3σ threshold. The correction methods detected 117 (removed background) and 104 (well correction) false positive hits.

Then, we have randomly added 1% of hits to the raw random data. The hit values were randomly chosen from the range $[\mu - 3.5\sigma; \mu - 4.5\sigma]$, where μ denotes the mean value and σ denotes the standard deviation of the observed plate. After that, the data with hits were modified by adding the values $4c, 3c, 2c, c, 0, 0, -c, -2c, -3c$, and $-4c$ to the 1st, 2nd, ..., and 10th columns, respectively, thus simulating a systematic error in the assay, where the variable c was consequently taking values $0, \sigma/10, 2\sigma/10, \ldots$, and $5\sigma/10$. The value $c = 0$ does not create any systematic error, but bigger values of c increase systematic error proportionally to the standard deviation σ.

For each value of the noise coefficient c, hits were selected in the raw, background removed and well-corrected datasets using the 3σ threshold. The hit detection rate as well as the false positive and false negative rates were assessed. The hit detection rate was generally higher for both corrected datasets. Figure 2(a) shows that the background and well correction procedures successfully eliminated systematic error from the random data. Both methods were robust and showed similar results in terms of the hit detection rate. However, the well correction method systematically outperformed the background method in terms of the false positive hit rate (see Figure 2(b)).

In conclusion, we developed two statistical methods that can be used to refine the analysis of experimental HTS data and correct the hit selection procedure. Both methods are designed to minimize the impact of systematic error in raw HTS data and have been successfully tested on real and artificial datasets. Both methods allow one to bring the hit distribution surface closer to a plane surface. When systematic error was not present in the data,

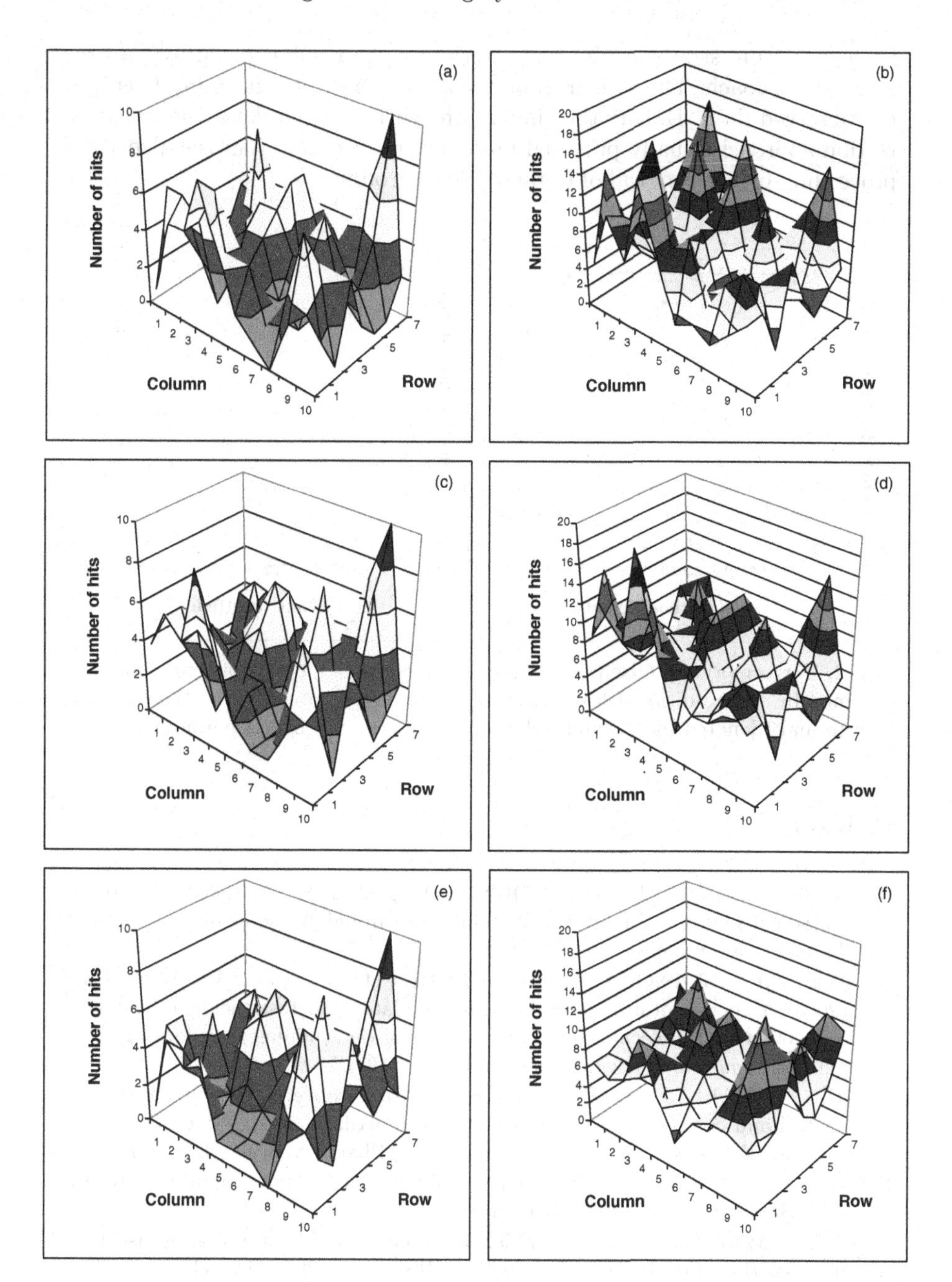

Fig. 1. Hit distribution surfaces computed for the 3σ and 2.5σ hit selection thresholds for the raw (a and b), background removed (c and d), and well-corrected (e and f) McMaster datasets.

both correcting strategies did not deteriorate the results shown by the traditional approach. Thus, their application does not introduce any bias into the observed data. During the simulations with random data, the well correction approach usually provided more accurate results than the algorithm proceeding by the removal of evaluated background.

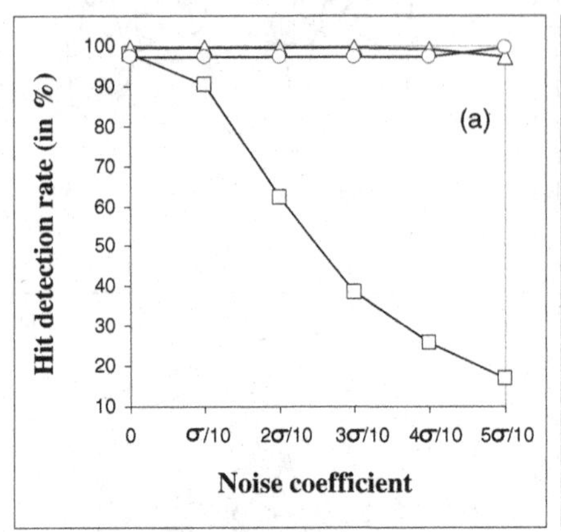

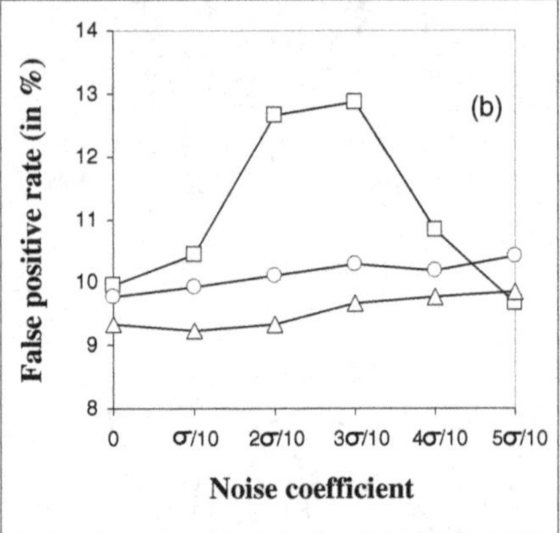

Fig. 2. Correct (a) and false positive (b) detection rates for the noisy random data obtained by the traditional hit selection procedure (denoted by □), the removed background (denoted by ○), and well correction (denoted by △) methods.

References

BRIDEAU, C., GUNTER, B., PIKOUNIS, W. and LIAW, A. (2003): Improved statistical methods for hit selection in high-throughput screening. *Journal of Biomolecular Screening, 8, 634-647.*

ELOWE, N.H., BLANCHARD, J.E., CECHETTO, J.D. and BROWN, E.D. (2005): Experimental screening of dihydrofolate reductase yields a "test set" of 50,000 small molecules for a computational data-mining and docking competition. *Journal of Biomolecular Screening, 10, 653-657.*

HEUER, C., HAENEL, T., PRAUSE, B. (2003): A novel approach for quality control and correction of HTS data based on artificial intelligence. *The Pharmaceutical Discovery & Development Report.* PharmaVentures Ltd. [Online].

HEYSE, S. (2002): Comprehensive analysis of high-throughput screening data. In: *Proceedings of SPIE 2002*, 4626, 535-547.

KAUL, A. (2005): The impact of sophisticated data analysis on the drug discovery process. *Business Briefing: Future Drug Discovery 2005.* [Online]

KEVORKOV, D. and MAKARENKOV, V. (2005): Statistical analysis of systematic errors in HTS. *Journal of Biomolecular Screening, 10, 557-567.*

MAKARENKOV, V., KEVORKOV, D., GAGARIN, A., ZENTILLI, P., MALO, N. and NADON, R. (2006): An efficient method for the detection and elimination of systematic error in high-throughput screening. Submitted.

ZHANG, J.H., CHUNG, T.D.Y. and OLDENBURG, K.R. (1999): A Simple Statistic Parameter for Use in Evaluation and Validation of High Throughput Screening Assays. *Journal of Biomolecular Screening, 4, 67-73.*

kNN Versus SVM in the Collaborative Filtering Framework

Miha Grčar, Blaž Fortuna, Dunja Mladenič, and Marko Grobelnik

Jožef Stefan Institute,
Jamova 39, SI-1000 Ljubljana, Slovenia

Abstract. We present experimental results of confronting the k-Nearest Neighbor (kNN) algorithm with Support Vector Machine (SVM) in the collaborative filtering framework using datasets with different properties. While k-Nearest Neighbor is usually used for the collaborative filtering tasks, Support Vector Machine is considered a state-of-the-art classification algorithm. Since collaborative filtering can also be interpreted as a classification/regression task, virtually any supervised learning algorithm (such as SVM) can also be applied. Experiments were performed on two standard, publicly available datasets and, on the other hand, on a real-life corporate dataset that does not fit the profile of ideal data for collaborative filtering. We conclude that the quality of collaborative filtering recommendations is highly dependent on the quality of the data. Furthermore, we can see that kNN is dominant over SVM on the two standard datasets. On the real-life corporate dataset with high level of sparsity, kNN fails as it is unable to form reliable neighborhoods. In this case SVM outperforms kNN.

1 Introduction and motivation

The goal of collaborative filtering is to explore a vast collection of items in order to detect those which might be of interest to the active user. In contrast to content-based recommender systems which focus on finding contents that best match the user's query, collaborative filtering is based on the assumption that similar users have similar preferences. It explores the database of users' preferences and searches for users that are similar to the active user. The active user's preferences are then inferred from preferences of the similar users. The content of items is usually ignored.

In this paper we explore how two different approaches to collaborative filtering – the memory-based k-Nearest Neighbor approach (kNN) and the model-based Support Vector Machine (SVM) approach – handle data with different properties. We used two publicly available datasets that are commonly used in collaborative filtering evaluation and, on the other hand, a dataset derived from real-life corporate Web logs. The latter does not fit the profile of ideal data for collaborative filtering. Namely, collaborative filtering is usually applied to research datasets with relatively low sparsity. Here we have included a real-life dataset of a company in need of providing collaborative filtering recommendations. It turned out that this dataset has much higher sparsity than usually handled in the collaborative filtering scenario.

The rest of this paper is arranged as follows. In Sections 2 and 3 we discuss collaborative filtering algorithms and data quality for collaborative filtering. The three datasets used in our experiments are described in Section 4. In Sections 5 and 6 the experimental setting and the evaluation results are presented. The paper concludes with the discussion and some ideas for future work (Section 7).

2 Collaborative filtering in general

There are basically two approaches to the implementation of a collaborative filtering algorithm. The first one is the so called "lazy learning" approach (also known as the memory-based approach) which skips the learning phase. Each time it is about to make a recommendation, it simply explores the database of user-item interactions. The model-based approach, on the other hand, first builds a model out of the user-item interaction database and then uses this model to make recommendations. "Making recommendations" is equivalent to predicting the user's preferences for unobserved items.

The data in the user-item interaction database can be collected either explicitly (explicit ratings) or implicitly (implicit preferences). In the first case the user's participation is required. The user is asked to explicitly submit his/her rating for the given item. In contrast to this, implicit preferences are inferred from the user's actions in the context of an item (that is why the term "user-item interaction" is used instead of the word "rating" when referring to users' preferences in this paper). Data can be collected implicitly either on the client side or on the server side. In the first case the user is bound to use modified client-side software that logs his/her actions. Since we do not want to enforce modified client-side software, this possibility is usually omitted. In the second case the logging is done by a server. In the context of the Web, implicit preferences can be determined from access logs that are automatically maintained by Web servers.

Collected data is first preprocessed and arranged into a user-item matrix. Rows represent users and columns represent items. Each matrix element is in general a set of actions that a specific user took in the context of a specific item. In most cases a matrix element is a single number representing either an explicit rating or a rating that was inferred from the user's actions.

Since a user usually does not access every item in the repository, the vector (i.e. the matrix row), representing the user, is missing some/many values. To emphasize this, we use the terms "sparse vector" and "sparse matrix".

The most intuitive and widely used algorithm for collaborative filtering is the so called k-Nearest Neighbor algorithm which is a memory-based approach. Technical details can be found, for example, in Grcar (2004). The algorithm is as follows:

1. Represent each user by a sparse vector of his/her ratings.

2. Define the similarity measure between two sparse vectors. In this paper, we consider two widely used measures: (i) the Pearson correlation coefficient which is used in statistics to measure the degree of correlation between two variables (Resnick et al. (1994)), and (ii) the Cosine similarity measure which is originally used in information retrieval to compare between two documents (introduced by Salton and McGill (1983)).
3. Find k users that have rated the item in question and are most similar to the active user (i.e. the user's neighborhood).
4. Predict the active user's rating for the item in question by calculating the weighted average of the ratings given to that item by other users from the neighborhood.

The collaborative filtering task can also be interpreted as a classification task, classes being different rating values (Billsus and Pazzani (1998)). Virtually any supervised learning algorithm can be applied to perform classification (i.e. prediction). For each user a separate classifier is trained (i.e. a model is built – hence now we are talking about a model-based approach). The training set consists of feature vectors representing items the user already rated, class labels being ratings from the user. Clearly the problem occurs if our training algorithm cannot handle missing values in the sparse feature vectors. It is suggested by Billsus and Pazzani (1998) to represent each user by several instances (optimally, one instance for each possible rating value). On a 1–5 rating scale, user A would be represented with 5 instances, namely {A, 1}, {A, 2}, ..., {A, 5}. The instance {A, 3}, for example, would hold ones ("1") for each item that user A rated 3 and zeros ("0") for all other items. This way, we fill in the missing values. We can now use such binary feature vectors for training. To predict a rating, we need to classify the item into one of the classes representing rating values. If we wanted to predict ratings on a continuous scale, we would have to use a regression approach instead of classification.

In our experiments we confronted the standard kNN algorithm (using Pearson and Cosine as the similarity measures) with SVM classifier and SVM regression (Vapnik (1998)). In the case of SVM we did not convert the user-item matrix into the dense binary representation, as SVM can handle sparse data directly.

3 Sparsity problem and data quality for collaborative filtering

The fact that we are dealing with a sparse matrix can result in the most concerning problem of collaborative filtering – the so called sparsity problem. In order to be able to compare two sparse vectors, similarity measures require some values to overlap. Furthermore, the lower the amount of overlapping values, the lower the reliability of these measures. If we are dealing with high level of sparsity, we are unable to form reliable neighborhoods.

Sparsity is not the only reason for the inaccuracy of recommendations provided by collaborative filtering. If we are dealing with implicit preferences, the ratings are usually inferred from the user-item interactions, as already mentioned earlier in the text (Section 2). Mapping implicit preferences into explicit ratings is a non-trivial task and can result in false mappings. The latter is even more true for server-side collected data in the context of the Web since Web logs contain very limited information. To determine how much time a user was reading a document, we need to compute the difference in time-stamps of two consecutive requests from that user. This, however, does not tell us wether the user was actually reading the document or he/she, for example, went out to lunch, leaving the browser opened. There are also other issues with monitoring the activities of Web users, which can be found in Rosenstein (2000).

From this brief description of data problems we can conclude that for applying collaborative filtering, explicitly given data with low sparsity are preferred to implicitly collected data with high sparsity. The worst case scenario is having highly sparse data derived from Web logs. However, collecting data in such a manner requires no effort from the users and also, the users are not obliged to use any kind of specialized Web browsing software. This "conflict of interests" is illustrated in Figure 1.

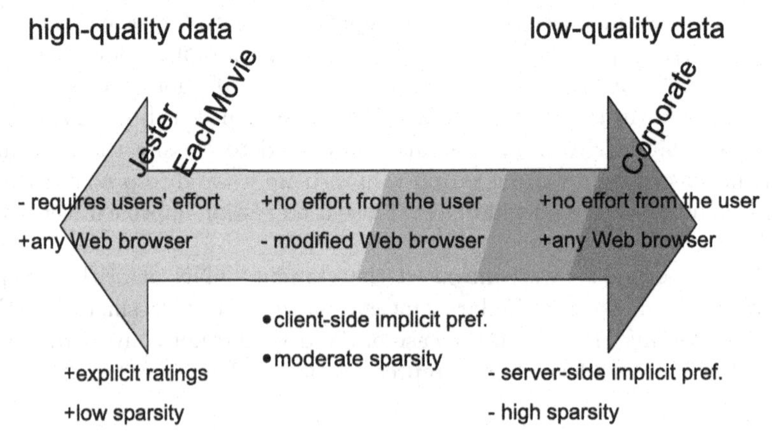

Fig. 1. Data characteristics that influence the data quality, and the positioning of the three datasets used in our experiments, according to their properties.

4 Data description

For our experiments we used three distinct datasets. The first dataset was EachMovie (provided by Digital Equipment Corporation) which contains explicit ratings for movies. The service was available for 18 months. The second dataset with explicit ratings was Jester (provided by Goldberg et al. (2001)) which contains ratings for jokes, collected over a 4-year period. The third dataset was derived from real-life corporate Web logs. The logs contain accesses to an internal digital library of a fairly large company. The time-span of acquired Web logs is 920 days. In this third case the user's preferences are implicit and collected on the server side, which implies the worst data quality for collaborative filtering.

In contrast to EachMovie and Jester, Web logs first needed to be extensively preprocessed. Raw logs contained over 9.3 million requests. After all the irrelevant requests (i.e. failed requests, non-GET requests, requests by anonymous users, requests for images, index pages, and other irrelevant pages) were removed we were left with only slightly over 20,500 useful requests, which is 0.22% of the initial database size. For detailed description of the preprocessing that was applied see Grčar et al. (2005). Note that only the dataset termed "Corporate 1/2/3/2" in Grčar et al. (2005) was considered in this paper.

Table 1 shows the comparison between the three datasets. It is evident that a low number of requests and somewhat ad-hoc mapping onto a discrete scale are not the biggest issues with our corporate dataset. The concerning fact is that the average number of ratings per item is only 1.22, which indicates extremely poor overlappingness. Sparsity is consequently very high, 99.93%. The other two datasets are much more promising. The most appropriate for collaborative filtering is the Jester dataset with very low sparsity, followed by EachMovie with higher sparsity but still relatively high average number of ratings per item. Also, the latter two contain explicit ratings, which means that they are more reliable than the corporate dataset (see also Figure 1).

Table 1. The data characteristics of the three datasets, showing the explicitness of ratings (explicit, implicit), size of the dataset and the level of sparsity.

	Ratings		Size			Sparsity		
	Explicit/ implicit	Scale	Num of users	Num of items	Num of ratings	%**	Avg # of r'tings/usr	Avg # of ratings/item
EachMovie	Explicit	Discrete 0–5	61,131	1,622	2,558,871	97.42	41.86	1,577.60
Jester	Explicit	Continuous −10 – +10	73,421	100	4,136,360	43.66	56.34	41,363.60
Corporate	Implicit	Discrete 1–3*	1,850	16,941	20,669	99.93	11.17	1.22

*after preprocessing
**computed as the number of missing values divided by the user-item matrix size (i.e. the number of rows times the number of columns)

5 Experimental setting

To be able to perform evaluation we built an evaluation platform (Grčar et al. (2005)). We ran a series of experiments to see how the accuracy of collaborative filtering recommendations differs between the two different approaches and the three different datasets (from EachMovie and Jester we considered only 10,000 randomly selected users to speed up the evaluation process). Ratings from each user were partitioned into "given" and "hidden" according to the "all-but-30%" evaluation protocol. The name of the protocol implies that 30% of all the ratings were hidden and the remaining 70% were used to form neighborhoods.

We applied three variants of memory-based collaborative filtering algorithms: (i) k-Nearest Neighbor using the Pearson correlation (kNN Pearson), (ii) k-Nearest Neighbor using the Cosine similarity measure (kNN Cosine), and (iii) the popularity predictor (Popularity). The latter predicts the user's ratings by simply averaging all the available ratings for the given item. It does not form neighborhoods or build models and it provides each user with the same recommendations. It serves merely as a baseline when evaluating collaborative filtering algorithms (termed "POP" in Breese et al. (1998)). For kNN variants we used a neighborhood of 120 users (i.e. k=120), as suggested in Goldberg et al. (2001).

In addition to the variants of the memory-based approach we also applied two variants of the model-based approach: SVM classifier, and SVM regression. In general, SVM classifier can classify a new example into one of the two classes: positive or negative. If we want to predict ratings, we need a multi-class variant of SVM classifier, classes being different rating values. The problem also occurs when dealing with continuous rating scales such as Jester's. To avoid this, we simply sampled the scale interval and thus transformed the continuous scale into a discrete one (in our setting we used $0.\overline{3}$ precision to sample the Jester's rating scale).

Although the work of Billsus and Pazzani (1998) suggests using items as examples, the task of collaborative filtering can equivalently be redefined to view users as examples. Our preliminary results showed that it is best to choose between these two representations with respect to the dataset properties. If the dataset is more sparse "horizontally" (i.e. the average number of ratings per user is lower than the average number of ratings per item), it is best to take users as examples. Otherwise it is best to take items as examples. Intuitively, this gives more training examples to build models which are consequently more reliable. With respect to the latter, we used users as examples when dealing with EachMovie (having on average 41.86 ratings per user vs. 1,577.60 ratings per item) and Jester datasets (having on average 56.34 ratings per user vs. 41,363.60 ratings per item) and items as examples when dealing with the corporate dataset (having on average 11.17 ratings per user vs. 1.22 ratings per item).

We combined several binary SVM classifiers in order to perform multi-class classification. Let us explain the method that was used on an example. We first transform the problem into a typical machine learning scenario with ordered class values as explained earlier in the previous paragraph. Now, let us consider a discrete rating scale from 1 to 5. We need to train 4 SVMs to be able to classify examples into 5 different classes (one SVM can only decide between positive and negative examples). We train the first SVM to be able to decide weather an example belongs to class 1 (positive) or to any of the classes 2–5 (negative). The second SVM is trained to distinguish between classes 1–2 (positive) and classes 3–5 (negative). The third SVM distinguishes between classes 1–3 (positive) and classes 4–5 (negative), and the last SVM distinguishes between classes 1–4 (positive) and class 5 (negative). In order to classify an example into one of the 5 classes, we query these SVMs in the given order. If the first one proves positive, we classify the example into class 1, if the second one proves positive, we classify the example into class 2, and so on in that same manner. If all of the queries prove negative, we classify the example into class 5. We used SVM classifier as implemented in Text-Garden (`http://www.textmining.net`). We built a model only if there were at least 7 positive and at least 7 negative examples available (because our preliminary experiments showed that this is a reasonable value to avoid building unreliable models).

SVM regression is much more suitable for our task than SVM classifier. It can directly handle continuous and thus also ordered discrete class values. This means we only need to train one model as opposed to SVM classifier where several models need to be trained. We used SVM regression as implemented in Text-Garden. As in the case of SVM classifier, we built a model only if there were at least 15 examples available.

Altogether we ran 5 experiments for each dataset-algorithm pair, each time with a different random seed (we also selected a different set of 10,000 users from EachMovie and Jester each time). When applying collaborative filtering to the corporate dataset, we made 10 repetitions (instead of 5) since this dataset is smaller and highly sparse, which resulted in less reliable evaluation results. Thus, we ran 100 experiments altogether.

We decided to use normalized mean absolute error (NMAE) as the accuracy evaluation metric. We first computed NMAE for each user and then we averaged it over all the users (termed "per-user NMAE") (Herlocker et al. (2004)). MAE is extensively used for evaluating collaborative filtering accuracy and was normalized in our experiments to enable us to compare evaluation results from different datasets.

6 Evaluation results

We present the results of experiments performed on the three datasets using the described experimental setting (see Section 5). We used two-tailed paired

Student's t-Test with significance 0.05 to determine if the differences in results are statistically significant.

We need to point out that in some cases the algorithms are unable to predict the ratings, as the given ratings do not provide enough information for the prediction. For instance, Popularity is not able to predict the rating if there are no ratings in the given data for a particular item. When calculating the overall performance we exclude such ratings from the evaluation, as we are mainly interested in the quality of prediction when available, even if the percentage of available predictions is low. We prefer the system to provide no recommendation if there is not enough data for a reasonably reliable prediction.

As mentioned earlier in Section 4, the three datasets have different characteristics that influence the accuracy of predictions. Jester is the dataset with the lowest sparsity and thus the most suitable for the application of collaborative filtering of the three tested datasets. We see that the kNN methods significantly outperform the other three methods. kNN Pearson slightly yet significantly outperforms kNN Cosine. SVM classifier also performs well, significantly outperforming SVM regression and Popularity. Interestingly, SVM regression performs significantly worse than Popularity.

EachMovie is sparser than Jester yet much more suitable for the application of collaborative filtering than the corporate dataset. Here kNN Cosine performs significantly better than kNN Pearson, followed by SVM classifier and Popularity. kNN Pearson slightly yet significantly outperforms Popularity. Again, SVM regression performs significantly worse than Popularity.

The corporate dataset is the worst of the three – it is extremely sparse and collected implicitly on the server side. It reveals the weakness of the kNN approach – lack of overlapping values results in unreliable neighborhoods. Notice that we do not provide the results of applying SVM classifier on this dataset, as the quality of the corporate dataset is too low for the classification setting. We see that SVM regression and Popularity perform best in this domain. The difference between them is not significant but they both significantly outperform the kNN approach. In this paper we are not concerned with the inability to predict but it is still worth mentioning that SVM regression can predict 72% of the hidden ratings, Popularity 23.7%, and the kNN approach only around 8% of the hidden ratings.

7 Discussion and future work

In our experimental setting we confronted the k-Nearest Neighbor algorithm with Support Vector Machine in the collaborative filtering framework. We can see that on our datasets, kNN is dominant on datasets with relatively low sparsity (Jester). On the two datasets with high to extremely high level of sparsity (EachMovie, the corporate dataset), kNN starts failing as it is unable to form reliable neighborhoods. In such case it is best to use a model-

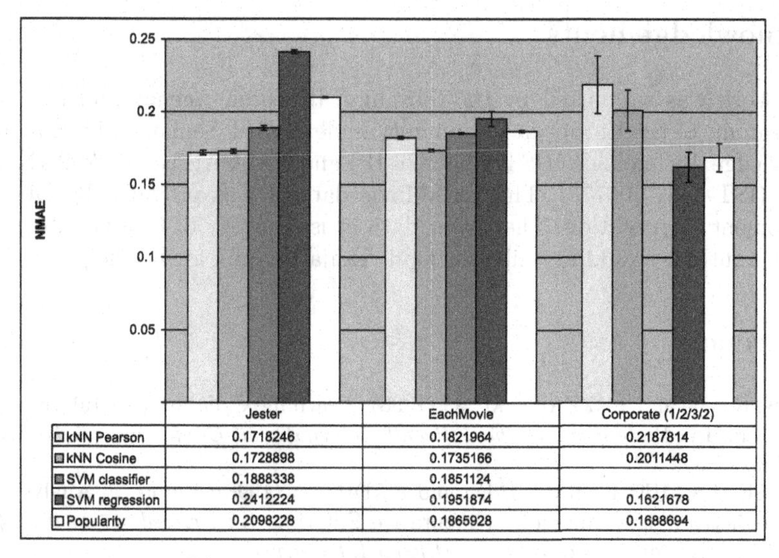

	Jester	EachMovie	Corporate (1/2/3/2)
kNN Pearson	0.1718246	0.1821964	0.2187814
kNN Cosine	0.1728898	0.1735166	0.2011448
SVM classifier	0.1888338	0.1851124	
SVM regression	0.2412224	0.1951874	0.1621678
Popularity	0.2098228	0.1865928	0.1688694

Fig. 2. The results of the experiments.

based approach, such as SVM classifier or SVM regression. Another strong argument for using the SVM approaches on highly sparse data is the ability to predict more ratings than with the variants of the memory-based approach.

Interestingly, Popularity performs extremely well on all domains. It fails, however, when recommending items to eccentrics. We noticed that the true value of collaborative filtering (in general) is shown yet when computing NMAE over some top percentage of eccentric users. We defined eccentricity intuitively as MAE (mean absolute error) over the overlapping ratings between "the average user" and the user in question (greater MAE yields greater eccentricity). The average user was defined by averaging ratings for each particular item. This is based on the intuition that the ideal average user would rate every item with the item's average rating. Our preliminary results show that the incorporation of the notion of eccentricity can give the more sophisticated algorithms a fairer trial. We computed average per-user NMAE only over the top 5% of eccentric users. The power of the kNN algorithms over Popularity became even more evident. In the near future we will define an accuracy measure that will weight per-user NMAE according to the user's eccentricity, and include it into our evaluation platform.

In future work we also plan to investigate if the observed behaviour – that SVM regression outperforms the kNN approaches at a certain level of sparsity – holds in general or only for the three datasets used in our evaluation.

Also interesting, the Cosine similarity works just as well as Pearson on EachMovie and Jester. Early researches show much poorer performance of the Cosine similarity measure (Breese et al. (1998)).

Acknowledgements

This work was supported by the Slovenian Research Agency and the IST Programme of the European Community under SEKT Semantically Enabled Knowledge Technologies (IST-1-506826-IP) and PASCAL Network of Excellence (IST-2002-506778). The EachMovie dataset was provided by Digital Equipment Corporation. The Jester dataset is courtesy of Ken Goldberg et al. The authors would also like to thank Tanja Brajnik for her help.

References

BILLSUS, D., and PAZZANI, M. J. (1998): Learning Collaborative Information Filers. In: *Proceedings of the Fifteenth International Conference on Machine Learning.*

BREESE, J.S., HECKERMAN, D., and KADIE, C. (1998): Empirical Analysis of Predictive Algorithms for Collaborative Filtering. In: *Proceedings of the 14th Conference on Uncertainty in Artificial Intelligence.*

CLAYPOOL, M., LE, P., WASEDA, M., and BROWN, D. (2001): Implicit Interest Indicators. In: *Proceedings of IUI'01.*

DEERWESTER, S., DUMAIS, S.T., and HARSHMAN, R. (1990): Indexing by Latent Semantic Analysis. In: *Journal of the Society for Information Science, Vol. 41, No. 6, 391–407.*

GOLDBERG, K., ROEDER, T., GUPTA, D., and PERKINS, C. (2001): Eigentaste: A Constant Time Collaborative Filtering Algorithm. In: *Information Retrieval, No. 4, 133–151.*

GRCAR, M. (2004): User Profiling: Collaborative Filtering. In: *Proceedings of SIKDD 2004 at Multiconference IS 2004, 75–78.*

GRCAR, M., MLADENIC D., GROBELNIK, M. (2005): Applying Collaborative Filtering to Real-life Corporate Data. In: *Proceedings of the 29th Annual Conference of the German Classification Society (GfKl 2005)*, Springer, 2005.

HERLOCKER, J.L., KONSTAN, J.A., TERVEEN, L.G., and RIEDL, J.T. (2004): Evaluating Collaborative Filtering Recommender Systems. In: *ACM Transactions on Information Systems, Vol. 22, No. 1, 5–53.*

HOFMANN, T. (1999): Probabilistic Latent Semantic Analysis. In: *Proceedings of the 15th Conference on Uncertainty in Artificial Intelligence.*

MELVILLE, P., MOONEY, R.J., and NAGARAJAN, R. (2002): Content-boosted Collaborative Filtering for Improved Recommendations. In: *Proceedings of the 18th National Conference on Artificial Intelligence, 187–192.*

RESNICK, P., IACOVOU, N., SUCHAK, M., BERGSTROM, P., and RIEDL, J. (1994): GroupLens: An Open Architecture for Collaborative Filtering for Netnews. In: *Proceedings of CSCW'94, 175–186.*

ROSENSTEIN, M. (2000): What is Actually Taking Place on Web Sites: E-Commerce Lessions from Web Server Logs. In: *Proceedings of EC'00.*

SALTON, G., McGILL, M.J. (1983): Introduction to Modern Information Retrieval. McGraw-Hill, New York.

VAPNIK, V. (1998): Statistical Learning Theory. Wiley, New York.

Mining Association Rules in Folksonomies

Christoph Schmitz[1], Andreas Hotho[1], Robert Jäschke[1,2], Gerd Stumme[1,2]

[1] Knowledge & Data Engineering Group, Department of Mathematics and Computer Science, University of Kassel, Wilhelmshher Allee 73, D–34121 Kassel, Germany, http://www.kde.cs.uni-kassel.de

[2] Research Center L3S, Expo Plaza 1, D–30539 Hannover, Germany, http://www.l3s.de

Abstract. Social bookmark tools are rapidly emerging on the Web. In such systems users are setting up lightweight conceptual structures called folksonomies. These systems provide currently relatively few structure. We discuss in this paper, how association rule mining can be adopted to analyze and structure folksonomies, and how the results can be used for ontology learning and supporting emergent semantics. We demonstrate our approach on a large scale dataset stemming from an online system.

1 Introduction

A new family of so-called "Web 2.0" applications is currently emerging on the Web. These include user-centric publishing and knowledge management platforms like Wikis, Blogs, and social resource sharing systems. In this paper, we focus on resource sharing systems, which all use the same kind of lightweight knowledge representation, called *folksonomy*. The word 'folksonomy' is a blend of the words 'taxonomy' and 'folk', and stands for conceptual structures created by the people.

Resource sharing systems, such as Flickr[1] or del.icio.us,[2] have acquired large numbers of users (from discussions on the del.icio.us mailing list, one can approximate the number of users on del.icio.us to be more than one hundred thousand) within less than two years. The reason for their immediate success is the fact that no specific skills are needed for participating, and that these tools yield immediate benefit for each individual user (e.g. organizing ones bookmarks in a browser-independent, persistent fashion) without too much overhead. Large numbers of users have created huge amounts of information within a very short period of time. As these systems grow larger, however, the users feel the need for more structure for better organizing their resources. For instance, approaches for tagging tags, or for bundling them, are currently discussed on the corresponding news groups. Currently, however, there is a lack of theoretical foundations adapted to the new opportunities which has to be overcome.

[1] http://www.flickr.com/

[2] http://del.icio.us

A first step towards more structure within such systems is to discover knowledge that is already implicitly present by the way different users assign tags to resources. This knowledge may be used for recommending both a hierarchy on the already existing tags, and additional tags, ultimately leading towards *emergent semantics* (Staab et al. (2002), Steels (1998)) by converging use of the same vocabulary. In this sense, knowledge discovery (KDD) techniques are a promising tool for bottom-up building of conceptual structures.

In this paper, we will focus on a selected KDD technique, namely association rules. Since folksonomies provide a three-dimensional dataset (users, tags, and resources) instead of a usual two-dimensional one (items and transactions), we present first a systematic overview of projecting a folksonomy onto a two-dimensional structure. Then we will show the results of mining rules from two selected projections on the del.icio.us system.

This paper is organized as follows. Section 2 reviews recent developments in the area of social bookmark systems, and presents a formal model. In Section 3, we briefly recall the notions of association rules, before providing a systematic overview over the projections of a folksonomy onto a two-dimensional dataset in Section 4. In Section 5, we present the results of mining association rules on data of the del.icio.us system. Section 6 concludes the paper with a discussion of further research topics on knowledge discovery within folksonomies.

2 Social resource sharing and folksonomies

Social resource sharing systems are web-based systems that allow users to upload their resources, and to label them with names. The systems can be distinguished according to what kind of resources are supported. Flickr,[3] for instance, allows the sharing of photos, del.icio.us[4] the sharing of bookmarks, CiteULike[5] and Connotea[6] the sharing of bibliographic references, and 43Things[7] even the sharing of goals in private life. Our own upcoming system, called *BibSonomy*,[8] will allow to share simultaneously bookmarks and BibTeX entries (see Fig. 1).

In their core, these systems are all very similar. Once a user is logged in, he can add a resource to the system, and assign arbitrary labels, so-called *tags*, to it. We call the collection of all his assignments his *personomy*, and the collection of all personomies is called *folksonomy*. The user can also explore the personomies of other users in all dimensions: for a given user he can see

[3] http://www.flickr.com/
[4] http://del.icio.us/
[5] http://www.citeulike.org/
[6] http://www.connotea.org/
[7] http://www.43things.com/
[8] http://www.bibsonomy.org

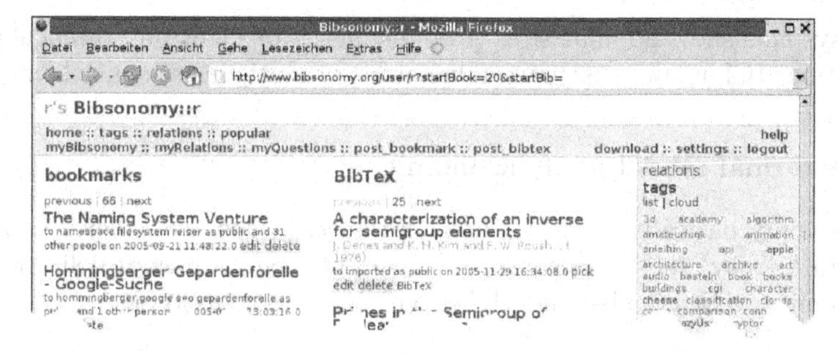

Fig. 1. Bibsonomy displays bookmarks and (BIBTEXbased) bibliographic references simultaneously.

the resources that user has uploaded, together with the tags he has assigned to them (see Fig. 1); when clicking on a resource he sees which other users have uploaded this resource and how they tagged it; and when clicking on a tag he sees who assigned it to which resources.

The systems allow for additional functionality. For instance, one can copy a resource from another user, and label it with one owns tags. Overall, these systems provide a very intuitive navigation through the data.

2.1 State of the art

There are currently virtually no scientific publications about folksonomy-based web collaboration systems. Among the rare exceptions are Hammond et al. (2005) and Lund et al. (2005) who provide good overviews of social bookmarking tools with special emphasis on folksonomies, and Mathes (2004) who discusses strengths and limitations of folksonomies. The main discussion on folksonomies and related topics is currently only going on mailing lists, e.g. Connotea (2005). To the best of our knowledge, the ideas presented in this paper have not been explored before, but there is a lot of recent work dealing with folksonomies.

Mika (2005) defines a model of semantic-social networks for extracting lightweight ontologies from del.icio.us. Besides calculating measures like the clustering coefficient, (local) betweenness centrality or the network constraint on the extracted one-mode network, Mika uses co-occurence techniques for clustering the concept network.

There are several systems working on top of del.icio.us to explore the underlying folksonomy. CollaborativeRank[9] provides ranked search results on top of del.icio.us bookmarks. The ranking takes into account, how early someone bookmarked an URL and how many people followed him or her.

[9] http://collabrank.org/

Other systems show popular sites (Populicious[10]) or focus on graphical representations (Cloudalicious[11], Grafolicious[12]) of statistics about del.icio.us.

2.2 A formal model for folksonomies

A folksonomy basically describes users, resources, tags, and allows users to assign (arbitrary) tags to resources. We present here a formal definition of folksonomies, which is also underlying our BibSonomy system.

Definition 1. A *folksonomy* is a tuple $\mathbb{F} := (U, T, R, Y, \prec)$ where

- U, T, and R are finite sets, whose elements are called *users*, *tags* and *resources*, resp.,
- Y is a ternary relation between them, i.e., $Y \subseteq U \times T \times R$, called assignments, and
- $\prec$ is a user-specific *subtag/supertag-relation*, i.e., $\prec \subseteq U \times ((T \times T) \setminus \{(t,t) \mid t \in T\})$.

The *personomy* $\mathbb{P}_u$ of a given user $u \in U$ is the restriction of $\mathbb{F}$ to u, i.e., $\mathbb{P}_u := (T_u, R_u, I_u, \prec_u)$ with $I_u := \{(t,r) \in T \times R \mid (u,t,r) \in Y\}$, $T_u := \pi_1(I_u)$, $R_u := \pi_2(I_u)$, and $\prec_u := \{(t_1, t_2) \in T \times T \mid (u, t_1, t_2) \in \prec\}$.

Users are typically described by their user ID, and tags may be arbitrary strings. What is considered as a resource depends on the type of system. In del.icio.us, for instance, the resources are URLs, and in Flickr, the resources are pictures. In our BibSonomy system, we have two types of resources, bookmarks and BibTeXentries. From an implementation point of view, resources are internally represented by some ID.

In this paper, we do not make use of the subtag/supertag relation for sake of simplicity. I.e., $\prec = \emptyset$, and we will simply note a folksonomy as a quadruple $\mathbb{F} := (U, T, R, Y)$. This structure is known in Formal Concept Analysis (Wille (1982), Ganter and Wille (1999)) as a *triadic context* (Lehmann and Wille (1995), Stumme (2005)). An equivalent view on folksonomy data is that of a tripartite (undirected) hypergraph $G = (V, E)$, where $V = U \dot\cup T \dot\cup R$ is the set of nodes, and $E = \{\{u,t,r\} \mid (u,t,r) \in Y\}$ is the set of hyperedges.

2.3 Del.ico.us — a folksonomy-based social bookmark system

In order to evaluate our folksonomy mining approach, we have analyzed the popular social bookmarking sytem del.icio.us. Del.icio.us is a server-based system with a simple-to-use interface that allows users to organize and share bookmarks on the internet. It is able to store in addition to the URL a

[10] http://populicio.us/
[11] http://cloudalicio.us/
[12] http://www.neuroticweb.com/recursos/del.icio.us-graphs/

description, a note, and tags (i. e., arbitrary labels). We chose del.icio.us rather than our own system, Bibsonomy, as the latter is going online only after the time of writing of this article. For our experiments, we collected from the del.ico.us system $|U| = 75,242$ users, $|T| = 533,191$ tags and $|R| = 3,158,297$ resources, related by in total $|Y| = 17,362,212$ triples.

3 Association rule mining

We assume here, that the reader is familiar with the basics of association rule mining introduced by Agrawal et al. (1993). As the work presented in this paper is on the conceptual rather than on the computational level, we refrain in particular from describing the vast area of developing efficient algorithms. Many of the existing algorithms can be found at the Frequent Itemset Mining Implementations Repository.[13] Instead, we just recall the definition of the association rule mining problem, which was initially stated by Agrawal et al. (1993), in order to clarify the notations used in the following. We will not use the original terminology of Srikant et al, but rather exploit the vocabulary of Formal Concept Analysis (FCA) (Wille (1982)), as it better fits with the formal folksonomy model introduced in Definition 1.[14]

Definition 2. A *formal context* is a dataset $\mathbb{K} := (G, M, I)$ consisting of a set G of *objects*, a set M of *attributes*, and a binary relation $I \subseteq G \times M$, where $(g, m) \in I$ is read as "object g has attribute m".

In the usual basket analysis scenario, M is the set of items sold by a supermarket, G is the set of all transactions, and, for a given transaction $g \in G$, the set $g^I := \{m \in M | (g, m) \in I\}$ contains all items bought in that transaction.

Definition 3. For a set X of attributes, we define $A' := \{g \in G \mid \forall m \in X\colon (g, m) \in I\}$. The *support* of A is calculated by $\text{supp}(A) := \frac{|A'|}{|G|}$.

Definition 4 (Association Rule Mining Problem (Agrawal et al. (1993))). Let $\mathbb{K}$ be a formal context, and minsupp, minconf $\in [0, 1]$, called *minimum support* and *minimum confidence thresholds*, resp. The *association rule mining problem* consists now of determining all pairs $A \to B$ of subsets of M whose *support* $\text{supp}(A \to B) := \text{supp}(A \cup B)$ is above the threshold minsupp, and whose *confidence* $\text{conf}(A \to B) := \frac{\text{supp}(A \cup B)}{\text{supp}(A)}$ is above the threshold minconf.

As the rules $A \to B$ and $A \to B \setminus A$ carry the same information, and in particular have same support and same confidence, we will consider in this

[13] http://fimi.cs.helsinki.fi/

[14] For a detailed discussion about the role of FCA for association rule mining see (Stumme (2002)).

paper the additional constraint prevalent in the data mining community, that premise A and conclusion B are to be disjoint.[15]

When comparing Definitions 1 and 2, we observe that association rules cannot be mined directly on folksonomies, because of their triadic nature. One either has to define some kind of triadic association rules, or to transform the triadic folksonomy into a dyadic formal context. In this paper, we follow the latter approach.

4 Projecting the folksonomy onto two dimensions

As discussed in the previous section, we have to reduce the three-dimensional folksonomy to a two-dimensional formal context before we can apply any association rule mining technique. Several such projections have already been introduced in Lehmann and Wille (1995). In Stumme (2005), we provide a more complete approach, which we will slightly adapt to the association rule mining scenario.

As we want to analyze all facets of the folksonomy, we want to allow to use any of the three sets U, T, and R as the set of objects – on which the support is computed – at some point in time, depending on the task on hand. Therefore, we will not fix the roles of the three sets in advance. Instead, we consider a triadic context as symmetric structure, where all three sets are of equal importance. For easier handling, we will therefore denote the folksonomy $\mathbb{F} := (U, T, R, Y)$ alternatively by $\mathbb{F} := (X_1, X_2, X_3, Y)$ in the following.

We will define the set of objects – i. e., the set on which the support will be counted – by a permutation on the set $\{1, 2, 3\}$, i. e., by an element σ of the full symmetric group S_3. The choice of a permutation indicates, together with one of the aggregation modes 'Ↄ', 'ʍ', '$\exists n$' with $n \in \mathbb{N}$, and '$\forall$', on which formal context $\mathbb{K} := (G, M, I)$ the association rules are computed.

- $\mathbb{K}^{\sigma,\text{Ↄ}} := (X_{\sigma(1)} \times X_{\sigma(3)}, X_{\sigma(2)}, I)$ with $((x_{\sigma(1)}, x_{\sigma(3)}), x_{\sigma(2)}) \in I$ if and only if $(x_1, x_2, x_3) \in Y$.
- $\mathbb{K}^{\sigma,\text{ʍ}} := (X_{\sigma(1)}, X_{\sigma(2)} \times X_{\sigma(3)}, I)$ with $(x_{\sigma(1)}, (x_{\sigma(2)}, x_{\sigma(3)})) \in I$ if and only if $(x_1, x_2, x_3) \in Y$.
- $\mathbb{K}^{\sigma,\exists n} := (X_{\sigma(1)}, X_{\sigma(2)}, I)$ with $(x_{\sigma(1)}, x_{\sigma(2)}) \in I$ if and only if there exist n different $x_{\sigma(3)} \in X_{\sigma(3)}$ with $(x_1, x_2, x_3) \in Y$.
- $\mathbb{K}^{\sigma,\forall} := (X_{\sigma(1)}, X_{\sigma(2)}, I)$ with $(x_{\sigma(1)}, x_{\sigma(2)}) \in I$ if and only if for all $x_{\sigma(3)} \in X_{\sigma(3)}$ holds $(x_1, x_2, x_3) \in Y$. The mode '$\forall$' is thus equivalent to '$\exists n$' if $|X_{\sigma(3)}| = n$.

[15] In contrast, in FCA, one often requires A to be a subset of B, as this fits better with the notion of *closed itemsets* which arose of applying FCA to the association mining problem (Pasquier et al. (1999), Zaki and Hsiao (1999), Stumme (1999)).

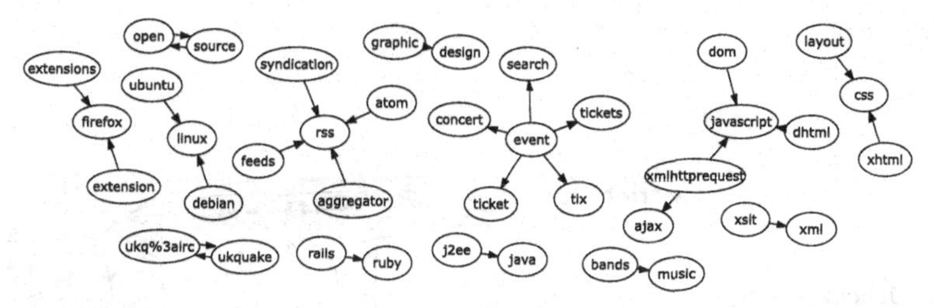

Fig. 2. All rules with two elements of $\mathbb{K}_1$ with .05 % support, 50 % confidence

These projections are complemented by the following way to 'cut slices' out of the folksonomy. A slice is obtained by selecting one dimension (out of user/tag/resource), and then fixing in this dimension one particular instance.

- Let $x := x_{\sigma(3)} \in X_{\sigma(3)}$. $\mathbb{K}^{\sigma,x} := (X_{\sigma(1)}, X_{\sigma(2)}, I)$ with $(x_{\sigma(1)}, x_{\sigma(2)}) \in I$ if and only if $(x_1, x_2, x_3) \in Y$.

In the next section, we will discuss for a selected subset of these projections the kind of rules one obtains from mining the formal context that is resulting from the projection.

5 Mining association rules on the projected folksonomy

After having performed one of the projections described in the previous section, one can now apply the standard association rule mining techniques as described in Section 3. Due to space restrictions, we have to focus on a subset of projections. In particular, we will address the two projections $\mathbb{K}^{\sigma_i,\mathfrak{I}}$ with $\sigma_1 :=$ id and $\sigma_2 := (1 \mapsto 1,\ 2 \mapsto 3,\ 3 \mapsto 2)$. We obtain the two dyadic contexts $\mathbb{K}_1 := (U \times R, T, I_1)$ with $I_1 := \{((u,r),t) | (u,t,r) \in Y\}$ and $\mathbb{K}_2 := (T \times U, R, I_2)$ with $I_2 := \{(t,u),r) | (u,t,r) \in Y\}$.

An association rule $A \to B$ in $\mathbb{K}_1$ is read as Users assigning the tags from A to some resources often also assign the tags from B to them. This type of rules may be used in a recommender system. If a user assigns all tags from A then the system suggests him to add also those from B.

Figure 2 shows all rules with one element in the premise and one element in the conclusion that we derived from $\mathbb{K}_1$ with a minimum support of 0.05 % and a minimum confidence of 50 %. In the diagram one can see that our interpretation of rules in $\mathbb{K}_1$ holds for these examples: users tagging some webpage with *debian* are likely to tag it with *linux* also, and pages about *bands* are probably also concerned with *music*. These results can be used in a recommender system, aiding the user in choosing the tags which are most helpful in retrieving the resource later.

Another view on these rules is to see them as subsumption relations, so that the rule mining can be used to learn a taxonomic structure. If many resources tagged with *xslt* are also tagged with *xml*, this indicates, for example,

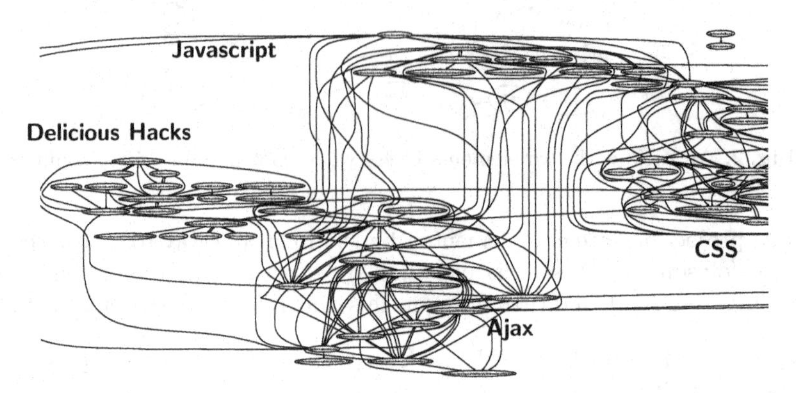

Fig. 3. Rules with two elements of $\mathbb{K}_2$ with 0.05 % support, and 10 % confidence

that *xml* can be considered a supertopic of *xslt* if one wants to automatically populate the $\prec$ relation. Figure 2 also shows two pairs of tags which occur together very frequently without any distinct direction in the rule: *open source* occurs as a phrase most of the time, while the other pair consists of two tags (*ukquake* and *ukq:irc*), which seem to be added automatically to any resource that is mentioned in a particular chat channel.

The second example are association rules $A \to B$ in $\mathbb{K}_2$ which are read as Users labelling the resources in A with some tags often also assign these tags to the resources in B. In essence both resources have to have something in common. Figure 3 shows parts of the resulting graph for applying association rules with 0.05 % support, and 10 % confidence on $\mathbb{K}_2$. Only associations rules with one element in premise and one element in conclusion are considered in the graph. In the figure 3 we identified four major areas in the graph which we labeled with the topics *delicious hacks*, *Javascript*, *Ajax*, and *CSS*. The topics can be derived by applying the FolkRank (Hotho et al. (2006)) on some of the resources of interest, which also yields relevant users and other resources for the respective area, such that communities of interest can be identified.

6 Conclusion

In this paper, we have presented a formal model of folksonomies as a set of triples – or, equivalently, a tripartite hypergraph. In order to apply association rule mining to folksonomies, we have systematically explored possible projections of the folksonomy structure into the standard notion of "shopping baskets" used in rule mining.

For two selected projections, we demonstrated the outcome of rule mining on a large-scale folksonomy dataset. The rules can be applied for different purposes, such as recommending tags, users, or resources, populating the supertag relation of the folksonomy, and community detection.

Future work includes the tighter integration of the various techniques we used here, namely, association rule mining, FolkRank ranking, and graph clustering, to further contribute to the abovementioned applications.

References

AGRAWAL, R., IMIELINSKI, T. and SWAMI, A. (1993): Mining association rules between sets of items in large databases. In: *Proc. of SIGMOD 1993*, pp. 207–216. ACM Press.

CONNOTEA (2005): Connotea Mailing List. https://lists.sourceforge.net/lists/listinfo/connotea-discuss.

GANTER, B. and WILLE, R. (1999): *Formal Concept Analysis : Mathematical foundations.* Springer.

HAMMOND, T., HANNAY, T., LUND, B. and SCOTT, J. (2005): Social Bookmarking Tools (I): A General Review. *D-Lib Magazine*, 11 (4).

HOTHO, A., JÄSCHKE, R., SCHMITZ, C. and STUMME, G. (2006): Information Retrieval in Folksonomies: Search and Ranking. In: *submitted for publication at ESWC 2006.*

LEHMANN, F. and WILLE, R. (1995): A triadic approach to Formal Concept Analysis. In: G. Ellis, R. Levinson, W. Rich and J. F. Sowa (Eds.), *Conceptual Structures: Applications, Implementation and Theory*, vol. 954 of *Lecture Notes in Computer Science*. Springer. ISBN 3-540-60161-9.

HANNAY, T. (2005): Social Bookmarking Tools (II): A Case Study - Connotea. *D-Lib Magazine*, 11 (4).

MATHES, A. (2004): Folksonomies – Cooperative Classification and Communication Through Shared Metadata. http://www.adammathes.com/academic/computer-mediated-communication/folksonomies.html.

MIKA, P. (2005): Ontologies Are Us: A Unified Model of Social Networks and Semantics. In: Y. Gil, E. Motta, V. R. Benjamins and M. A. Musen (Eds.), *ISWC 2005*, vol. 3729 of *LNCS*, pp. 522–536. Springer-Verlag, Berlin Heidelberg.

PASQUIER, N., BASTIDE, Y., TAOUIL, R. and LAKHAL, L. (1999): Closed set based discovery of small covers for association rules. In: *Actes des 15mes journes Bases de Donnes Avances (BDA'99)*, pp. 361–381.

STAAB, S., SANTINI, S., NACK, F., STEELS, L. and MAEDCHE, A. (2002): Emergent semantics. *Intelligent Systems, IEEE*, 17 (1):78.

STEELS, L. (1998): The Origins of Ontologies and Communication Conventions in Multi-Agent Systems. *Autonomous Agents and Multi-Agent Systems*, 1 (2):169.

STUMME, G. (1999): Conceptual Knowledge Discovery with Frequent Concept Lattices. FB4-Preprint 2043, TU Darmstadt.

STUMME, G. (2002): Efficient Data Mining Based on Formal Concept Analysis. In: A. Hameurlain, R. Cicchetti and R. Traunmller (Eds.), *Proc. DEXA 2002*, vol. 2453 of *LNCS*, pp. 534–546. Springer, Heidelberg.

STUMME, G. (2005): A Finite State Model for On-Line Analytical Processing in Triadic Contexts. In: B. Ganter and R. Godin (Eds.), *ICFCA*, vol. 3403 of *Lecture Notes in Computer Science*, pp. 315–328. Springer. ISBN 3-540-24525-1.

WILLE, R. (1982): Restructuring lattices theory : An approach based on hierarchies of concepts. In: I. Rival (Ed.), *Ordered Sets*, pp. 445–470. Reidel, Dordrecht-Boston.

ZAKI, M. J. and HSIAO, C.-J. (1999): ChARM: An efficient algorithm for closed association rule mining. Technical Report 99–10. Tech. rep., Computer Science Dept., Rensselaer Polytechnic.

Empirical Analysis of Attribute-Aware Recommendation Algorithms with Variable Synthetic Data

Karen H. L. Tso and Lars Schmidt-Thieme

Computer-based New Media Group (CGMN),
Department of Computer Science, University of Freiburg,
George-Köhler-Allee 51, 79110 Freiburg, Germany
{tso,lst}@informatik.uni-freiburg.de

Abstract. Recommender Systems (RS) have helped achieving success in E-commerce. Delving better RS algorithms has been an ongoing research. However, it has always been difficult to find adequate datasets to help evaluating RS algorithms. Public data suitable for such kind of evaluation is limited, especially for data containing content information (attributes). Previous researches have shown that the performance of RS rely on the characteristics and quality of datasets. Although, a few others have conducted studies on synthetically generated data to mimic the user-product datasets, datasets containing attributes information are rarely investigated. In this paper, we review synthetic datasets used in RS and present our synthetic data generator that considers attributes. Moreover, we conduct empirical evaluations on existing hybrid recommendation algorithms and other state-of-the-art algorithms using these synthetic data and observe the sensitivity of the algorithms when varying qualities of attribute data are applied to the them.

1 Introduction

Recommender systems(RS) have acted as an automated tool to assist users to find products accustomed to their tastes. RS algorithms generate recommendations that are expected to fit the users purchase preferences. The prevalent algorithm in practice uses the nearest-neighbor method, called collaborative filtering (CF ; Goldberg et al. (1992)). Methods that rely only on attributes and disregard the rating information of other users, are commonly called the Content-Based Filtering (CBF). They have shown to perform very poorly. Yet, attributes usually contain valuable information; hence it makes it desirable to include attribute information in CF models, so called hybrid CF/CBF filtering methods. There are several proposals on integrating attributes in CF for ratings. For instance, few others attempt linear combination of recommendation of CBF and CF predictions (Claypool et al. (1999), Good et al. (1999), Li and Kim (2003) and Pazzani (1999)). There also exists methods that apply a CBF and a CF model sequentially (Melville et al. (2002)) or view it as a classification problem (Basilico and Hofmann (2004), Basu et al (1998) and Schmidt-Thieme (2005)). As we lose the simplicity of CF, we do not consider those more complex methods here.

When evaluating these recommendation algorithms, suitable datasets of users and items have always been demanding, especially when diversity of public data is limited. It is not enough to compare only the recommendation quality of algorithms but also the sensitivity of the algorithms especially when varying qualities of data are provided to the algorithms. Thus, one should investigate the behavior of the algorithms as systematic changes are applied to the data. Although there are already a few attempts in generating synthetic data for the use in RS, to our best knowledge, there is no prior approach in generating synthetic data for evaluating recommender algorithms that incorporate attributes.

In this paper, (i) we will propose our Synthetic Data Generator which produces user-item and object attribute datasets and introduce the use of entropy to measure the randomness in the artificial data. (ii) we will survey some of the existing hybrid methods that consider attribute information in CF for predicting items. In addition, we will conduct empirical evaluations on three existing hybrid recommendation algorithms and other state-of-the-art algorithms and investigate their performances when synthetic datasets with varying qualities of synthetic attributes data are applied.

2 Related works

One of the most widely known Synthetic Data Generators (SDG) in data mining is the one provided by the IBM Quest group (Agrawl and Srikant (1994)). It generates data with a structure and was originally intended for evaluating association rule algorithms. Later on, Deshpande and Karypis (2004) used this SDG for evaluating their item-based top-N recommendation algorithm. Popescul et al. (2001) have proposed a simple approach by assigning a fixed number of users and items into clusters evenly and draw a uniform probability for each user and item in each cluster. A similar attempt has been done for Usenet News (Konstan et al. (1997) and Miller et al. (1997)) as well as Aggarwal et al. (1999) for their horting approach. Traupman and Wilensky (2004) tried to reproduce data by introducing skewed data to the synthetic data similar to a real dataset. Another approach is to produce datasets by first sampling a complete dataset and re-sample the data again by missing data effect (Marlin et al. (2005)).

The focus of this paper is to investigate SDG for CF algorithms which consider attributes. To the best of our knowledge, there is no prior attempts in examining SDGs for hybrid RS algorithms.

3 Synthetic data generator

The SDG can be divided into two phases: drawing distributions and sampling data. In the first phase, it draws distribution of User Cluster (UC) and Item Cluster (IC), next it affiliates UC or IC with object attribute respectively as well as to associate the UC and IC. Using these generated correlations, users, items, ratings and object attribute datasets can then be produced in the second phase.

3.1 Drawing distributions

To create the ratings and attributes datasets, we generate five random distributions models:

- $P(UC)$, how users are distributed in the UC.
- $P(IC)$, how items are distributed in the IC.
- $P(A \mid UC) \; \forall \; UC$, how user attributes (A) are distributed in UC.
- $P(B \mid IC) \; \forall \; IC$, how item attributes (B) are distributed in IC.
- $P(IC \mid UC) \; \forall \; UC$, how UC are distributed in IC.
- q be the probability that an item in IC_i is assigned to UC_j.

The SDG first draws $P(UC)$ and $P(IC)$ from a Dirichlet distribution (with parameters set to 1). This asserts that the sum of $P(UC)$ or $P(IC)$ forms to one. $P(B \mid IC)$ shows the affiliation of item attributes with the item clusters by drawing $|B|$ number of attributes from a Chi-square distribution rejecting values greater than 1 for each IC. Other types of distribution have also been examined, yet, Chi-square distribution has shown to give the most diverse entropy range. Likewise, the correlation between UC and IC is done by drawing the distribution $P(IC \mid UC)$, but in this case M is replaced with N and $|B|$ with M. User attributes can be drawn with similar manner, however, the attribute-aware CF algorithms we discuss in this paper do not take user-attributes into account.

By virtue of the randomness in those generated models, it is necessary to control or to measure the informativeness of these random data. Hence, we apply the Information Entropy and compute the average normalized entropy.

$$H(X) = - \sum_{x \in \mathrm{dom}(X)} \frac{P(x) \log_2 P(x)}{\log_2 |\mathrm{dom}(X)|}. \tag{1}$$

The conditional entropy for the item-attribute data therefore is:

$$H(B_i|IC) = - \sum_{b=0}^{1} \sum_{j \in \mathrm{dom}\, IC} \frac{P(B_i = b, IC = j) \cdot \log_2 P(B_i = b | IC = j)}{\log_2 |\mathrm{dom}\, IC|} \tag{2}$$

In our experiment, $H(B|IC)$ is sampled for 11 different entropy values by varying the degrees of freedom of the Chi-square distribution. By rejection sampling, $P(B \mid IC)$ is drawn iteratively with various Chi-square degrees of freedom until desired entropies, $H(B|IC)$, have been reached. We expect that as the entropy increases, which implies the data is less structured, the recommendation quality should decrease. The overall drawing distributions process is summarized in (Algo. 1).

Algorithm 1 Drawing distribution

Input: $|A|, |B|, N, M, H_A, H_B, H_{IC}, \epsilon_A, \epsilon_B, \epsilon_C$
Output: $P(UC), P(IC), P(A|UC), P(B|IC), P(IC|UC)$
$P(UC) \sim Dir_{a_1, a_2 \ldots, a_N}$
$P(IC) \sim Dir_{b_1, b_2 \ldots, b_M}$
$P(A|UC) = S\chi^2 ED(|A|, N, H_A, \epsilon_A)$
$P(B|IC) = S\chi^2 ED(|B|, M, H_B, \epsilon_B)$
$P(IC|UC) = S\chi^2 ED(N, M, H_{IC}, \epsilon_{IC})$

$S\chi^2 ED(N, M, H_{XY}, \epsilon_{XY})$:
d=1
repeat
$P(X_i|Y_j) \sim \chi^2_d|_{[0,1]} \quad \forall i = 1...N, \forall j = 1...M$
d=d+1
until $|H(X|Y) - H_{XY}| < \epsilon_{XY}$
return $P(X|Y)$

Algorithm 2 Sampling data

$uc_u \sim P(UC)$ *user class of user u*
$ic_i \sim P(IC)$ *item class of item i*
$oc_{k,l} \sim P(IC_k|UC_l)$ *items of class k prefered by user of class l*
$o_{u,i} \sim binom(q) \quad \forall u, i : oc_{ic_i, uc_u} = 1$
$o_{u,i} = 0$ *else*
$b_{i,t} \sim P(B_t|IC = ic_i)$ *item i contains attribute t*

3.2 Sampling data

Once these distributions have been drawn, users, items, ratings and item-attributes data are then sampled accordingly to those distributions. Firstly, users are assigned to user clusters by random sampling $P(UC)$. Similar procedure applies for sampling items. The user-item(ratings) data is generated by sampling $P(IC_k|UC_l)$ of users belonging to UC_l and then assigning q portion of items belonging to IC_k to the sampled users. The affiliation between items and attributes is done by sampling $P(B_t \mid IC = ic_i)$ of items which contain attribute B_t. The same procedure can be applied to generate the user-attributes datasets. The overall sampling data process is summarized in (Algo. 2).

4 Hybrid attribute-aware CF methods

These three existing hybrid methods (Tso and Schmidt-Thieme (2005)) are selected to be evaluated using the synthetic data.

Sequential CBF and CF is the adapted version of an existing hybrid approach, Content-Boosted CF, originally proposed by Melville et al. (2002) for predicting ratings. This method has been conformed to the predicting items problem here. It first uses CBF to predict ratings for unrated items and then filters out ratings with lower scores (and applies CF to recommend topN items.

Joint Weighting of CF and CBF (Joint-Weighting CF-CBF), first applies CBF on attribute-dependent data to infer the fondness of users for attributes. In parallel, user-based CF is used to predict topN items with ratings-dependent data. Both predictions are joint by computing their geometric mean.

Attribute-Aware Item-Based CF (Attr-Item-based CF) extends item-based CF (Deshpande and Karypis (2004)). It exploits the content/attribute information by computing the similarities between items using attributes thereupon combining it with the similarities between items using ratings-dependent data.

For the last two algorithms, λ is used as a weighting factor to vary the significance applied to CF or CBF.

5 Evaluation and experimental results

In this section, we present the evaluation of the selected attributes-aware CF algorithms using artificial data generated by SDG and compare their performances with their corresponding non-hybrid base models: user-based and item-based CF as well as to observe the behavior of the algorithms after supplement of attributes.

Metrics Our paper focuses on the item prediction problem, which is to predict a fixed number of top recommendations and not the ratings. Suitable evaluation metrics are Precision, Recall and F1.

Parameters Due to the nature of collaborative filtering, the size of neighborhood has significant impact on the recommendation quality (Herlocker et al. (1999)). Thus, each of the randomly generated data should have an assorted neighborhood sizes for each method. In our experiments, we have selected optimal neighborhood sizes and λ parameters for the hybrid methods by means of a grid search. See Fig. 1. Threshold and max, for the Sequential CBF-CF are set to 50 and 2 accordingly as chosen in the original model. For more detail explanation of the parameters used in those algorithms, please refer to (Tso and Schmidt-Thieme (2005)) and (Melville et al. (2002)).

Method	Neighborhood Size	λ
user-based CF	35-50	–
item-based CF	40-60	–
joint weighting CF–CBF	35-50	0.15
attr-aware item-based CF	40-60	0.15

Fig. 1. The parameters chosen for the respective algorithms.

Description	Symbol	Value
Number of users	n	250
Number of items	m	500
Number of User Clusters	N	5
Number of Item Clusters	M	10
Number of Item Attributes	$\|B\|$	50
Probability of i in IC assigned to a UC	q	0.2

Fig. 2. The parameters settings for the synthetic data generator.

Experimental Results In our experiments, we have generated five different trials. For each trial, we produce one dataset of user-item (ratings) and eleven different item-attributes datasets with increasing entropy from 0-1 with 0.1 interval, by rejection sampling. In addition, to reduce the complexity of the experiment, it is assumed that the correlation between the user and item clusters to be fairly well-structure and have a constant entropy of 0.05. The results of the average of five random trials where only item-attributes with entropy of 0.05 are presented in Fig. 3.

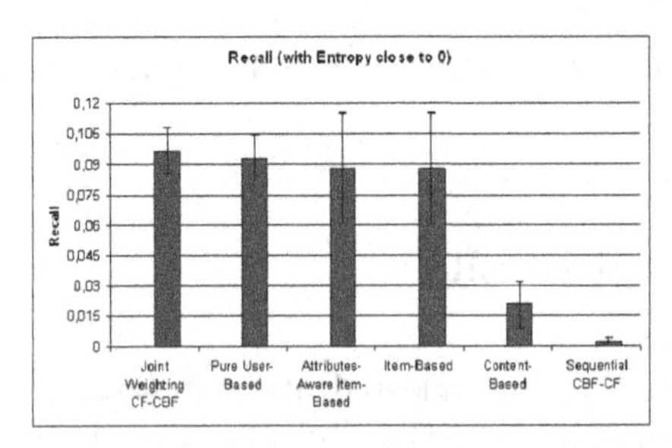

Fig. 3. Recall with entropy ≤ 0.05

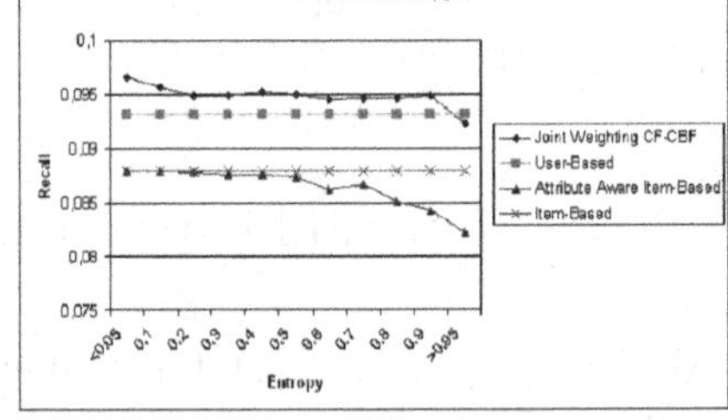

Fig. 4. Recall vs. Entropy from 0-1

As shown in Fig. 3, Joint-Weighting CF-CBF achieves the highest Recall value by around 4% difference w.r.t. its base method. On the other hand, Attr-Item-based CF does not seem to be effective at all as attributes are appended to its base model. It also has a very high standard deviation. This suggests that the algorithms to be rather unstable and unreliable. Although Melville et al. (2002) reported that CBCF performed better than user-based and pure CBF for ratings, it fails to provide quality top-N recommendations for items in our experiments. Therefore, we will focus our evaluation on the other two algorithms in the rest of the paper. As the aim of the paper is to examine the behavior of the models as the characteristic of attribute data varies, what is more important is to observe the performance as entropy varies. As anticipated, the recommendation quality increases, when there exists more structure in the data. The results of an average of five random trials of item-attribute datasets with eleven various entropies are presented in Fig. 4.

trials of item-attribute datasets with eleven various entropies are presented in Fig. 4.

We can see that for both Attr-Item-based CF and Joint-Weighting CF-CBF algorithms, the quality of recommendation reaches its peaks when the entropy approaches zero and it gradually decreases as entropy increases. As for Attr-Item-based CF, although it carries the right entropy trend, its peak does not surpass its base model and the quality drops gradually below its base model, which does not make use of attributes. On the other hand, Joint-Weighting CF-CBF, the value of recall descends gradually as the entropy raises, still the recall maintain above its base-model until entropy approaches 1 where recall plummets to below its base-line score. This shows that Joint-Weighting CF-CBF algorithm has a rather reliable performance when mixed qualities of attributes are applied to it.

6 Conclusions and future works

The aim of this paper is to conduct an empirical evaluation on three existing hybrid recommendation models and other state-of-the-art algorithms with data generated by the SDG presented in this paper. In particular, attribute data with varying qualities has been observed. Joint-Weighting CF-CBF, appears to enhance recommendations quality when reasonable amount of informative attributes are presented. This implies that the algorithm should have fairly consistent performances in realistic cases, where mixed attributes are presented most of the time. The other algorithms do not seem to be sensitive to attributes. Yet, we expect the outcomes could be ameliorated by adding more structural dependency between clusters. In addition, currently the data are only controlled by the entropy of item-attribute datasets; however, other distributions such as the user-item data should also be investigated when various entropies are considered. Furthermore, more extensive experiments should be done to examine the effect of varying other parameters settings.

References

AGGARWAL, C.C., WOLF, J.L., WU, K.L. and YU, P.S. (1999): Horting hatches an egg: A new graph-theoretic approach to collaborative filtering. In: ACM-SIGKDD International Conference on Knowledge Discovery and Data Mining. ACM, New York.

AGRAWL, R. and SRIKANT, R. (1994): Fast algorithms for mining association rules. In: VLDB Conference, Santiago, Chile. 487-499.

BASILICO, J. and HOFMANN, T. (2004): Unifying collaborative and content-based filtering. In: 21st International Conference on Machine Learning. Banff, Canada.

BASU, C., HIRSH, H., and COHEN, W. (1998): Recommendation as classification: Using social and content-based information in recommendation. In: Workshop on Recommender Systems. AAAI Press, Reston, Va. 11-15.

CLAYPOOL, M., GOKHALE, A. and MIRANDA, T. (1999): Combining content-based and collaborative filters in an online newspaper. In: SIGIR-99 Workshop on Recommender Systems: Algorithms and Evaluation.

DESHPANDE, M. and KARYPIS, G. (2004): Item-based top-N recommendation Algorithms. In: *ACM Transactions on Information Systems* **22/1**, *143-177.*

GOLDBERG, D., NICHOLS, D., OKI, B.M. and TERRY, D. (1992): Using collaborative filtering to weave an information tapestry. In: *Commun. ACM* **35**, *61-70.*

GOOD, N., SCHAFER, J.B., KONSTAN, J., BORCHERS, A., SARWAR, B., HERLOCKER, J., and RIEDL, J. (1999): Combining Collaborative Filtering with Personal Agents for Better Recommendations. In: Conference of the American Association of Artificial Intelligence (AAAI-99), pp 439-446.

HERLOCKER, J., KONSTAN, J., BORCHERS, A., and RIEDL, J. (1999): An Algorithmic Framework for Performing Collaborative Filtering. In: ACM SIGIR'99. ACM press.

KONSTAN, J. A., MILLER, B. N. , MALTZ D., HERLOCKER, J. L., GORDON, L. R. and RIEDL, J. (1997): Group-Lens: Applying collaborative filtering to usenet news. In: Commun. ACM 40, 77-87.

LI, Q. and KIM, M. (2003): An Approach for Combining Content-based and Collaborative Filters. In: Sixth International Workshop on Information Retrieval with Asian Languages (ACL-2003) pp. 17-24.

MARLIN, B., ROWEIS, S. and ZEMEL, R. (2005): Unsupervised Learning with Non-ignorable Missing Data. In: 10th International Workshop on Artificial Intelligence and Statistics, 222-229.

MELVILLE, P., MOONEY, R.J. and NAGARAJAN, R. (2002): Content-boosted Collaborative Filtering. In: Eighteenth National Conference on Artificial Intelligence(AAAI-2002), 187-192. Edmonton, Canada.

MILLER, B.N., RIEDL, J. and KONSTAN, J.A. (1997): Experiences with GroupLens: Making Usenet useful again. In: USENIX Technical Conference.

PAZZANI, M.J. (1999): A framework for collaborative, content-based and demographic filtering. In: *Artificial Intelligence Review 13(5-6):393-408.*

POPESCUL, A., L.H. UNGAR, D.M. PENNOCK, and S. LAWRENCE (2001): Probabilistic models for unified collaborative and content-based recommendation in sparse-data environments. In: Seventeenth Conference: *Uncertainty in Artificial Intelligence*, 437-444.

SARWAR, B.M., KARYPIS, G., KONSTAN, J.A. and RIEDL, J. (2000): Analysis of recommendation algorithms for E-commerce. In: 2nd ACM Conference on Electronic Commerce. ACM, New York. 285-295.

SCHMIDT-THIEME, L. (2005): Compound Classification Models for Recommender Systems. In: IEEE International Conference on Data Mining (ICDM'05), 378-385.

TRAUPMAN, J. and WILENSKY, R. (2004): Collaborative Quality Filtering: Establishing Consensus or Recovering Ground Truth?. In: WebKDD 2004, Seattle, WA.

TSO, H.L.K. and SCHMIDT-THIEME, L. (2005): Attribute-Aware Collaborative Filtering. In: 29th Annual Conference of the German Classification Society 2005, Magdeburg, Germany.

Patterns of Associations in Finite Sets of Items

Ralf Wagner

Business Administration and Marketing, Bielefeld University
Universitätsstraße 25, 33615 Bielefeld, Germany

Abstract. Mining association rules is well established in quantitative business research literature and makes up an up-and-coming topic in marketing practice. However, reducing the analysis to the assessment and interpretation of a few selected rules does not provide a complete picture of the data structure revealed by the rules.

This paper introduces a new approach of visualizing relations between items by assigning them to a rectangular grid with respect to their mutual association. The visualization task leads to a quadratic assignment problem and is tackled by means of a genetic algorithm. The methodology is demonstrated by evaluating a set of rules describing marketing practices in Russia.

1 Introduction

Data mining applications frequently focus on the relations between item of finite set of similar items. For instance, recommender systems often need to predict the clients' interest in a particular item from the prior knowledge of their interest in other items of the systems' domain. Another example is the optimization of retail assortments. Here, the importance of each item has to be assessed by means of it's impact on the sales of other items of the assortment. Moreover, in web usage mining we are interested in the identifying a visitor's probability of traversing from any precursor page of an Internet portal to one of the target pages which enable financial transactions. The computation of association rules provides us with detailed knowledge about the relation between two items or–more generally–two item sets. But, even from small sets of items a large number of rules is likely to be computed. Reducing this number by increasing the minimum support and confidence, a rule has to meet to be considered, restricts our focus to "strong" relations only. Therefore, rules which might be interesting for the management because of their novelty will possibly be ignored. Using measures of interestingness as discussed by Hilderman and Hamilton (2001) and Wagner (2005a) does not provide us with a remedy, if we are interested in grasping the complete structure between the items. Accepting the old marketing wisdom that "a picture says a thousand words" to humans, this paper aims to introduce an innovative procedure displaying the structure of items by assigning them to a rectangular grid. This method avoids the difficulties arising from the transformation of co-occurrences to similarities for applying conventional projection methods, such as multidimensional scaling. The problems mentioned above might be summarized as "market basket" problems. Applications of cluster analysis in high dimensional feature spaces face the same problem at the stage of characterizing the clusters of a solution. The combination of some features gives reason for assigning an observation to a cluster. But, this combination has to be revealed and communicated to the analyst.

The remainder of this paper is structured as follows. In the next section the methodology is introduced. The subsequent section comprises details of a data set describing contemporary marketing practices in Russia and an outline of the pattern revealed by applying the methodology. The paper concludes with a final discussion of the procedure, its' limitations, and promising venues of further improvements.

2 Methodology

An association rule describes the relation between the set $\mathbb{X}$ of one or more items in the antecedent, which implies the items of a set $\mathbb{Y}$ are likely to be combined in one entry of a data base, e.g. in one market basket or one session recorded in a web log file. Let $\mathbb{K}$ be a finite set of items with $|\mathbb{K}| = K$ and $\mathbb{T}$ the set of entries in a data set. Already the simple enumeration of all rules has a time complexity much higher than $O(2^K)$ (Zaki and Ogihara (1999)). Since the user has to interpret the rules rather than simply to enumerate in order to grasp the structure some aid is needed. For our purpose we are interested in bilateral relations only, thus $|\mathbb{X}| = |\mathbb{Y}| = 1$ and $\mathbb{X} \cap \mathbb{Y} = \emptyset$ hold. The strength of the relation of two items is assessed by the support of the rule.

$$sup(\mathbb{X} \Rightarrow \mathbb{Y}) = \frac{|\{\tilde{\mathbb{T}} \in \mathbb{T} | (\mathbb{X} \cup \mathbb{Y}) \subseteq \tilde{\mathbb{T}}\}|}{|\mathbb{T}|} \quad (1)$$

Let $\mathcal{G}$ be a $I \times J = K$ grid in the two-dimensional plane. The grid should at least be as large as the number of items. The items are assigned to the grid so that if item k is frequently combined with item l in the entries of our database, both items are next to each other on the grid. The extent to which item k and item l are combined is quantified by the support $sup(\mathbb{X}_k \Rightarrow \mathbb{Y}_l)$ where the antecedent set consists only of item k and the consequent set consists only of item l. As is evident from equation 1, in this case $sup(\mathbb{X}_k \Rightarrow \mathbb{Y}_l) = sup(\mathbb{Y}_l \Rightarrow \mathbb{X}_k)$ holds and ensures a unique weight for each edge of the grid. In a two-dimensional grid, each item has no more than 4 neighbours, but the support for $K - 1$ relations has to be considered for each of the items in order to assign them on the grid. Let us consider a simple example of 6 items $\{\mathcal{A}, \mathcal{B}, \mathcal{C}, \mathcal{D}, \mathcal{E}, \mathcal{F}\}$ with a $sup(\mathbb{X} \Rightarrow \mathbb{Y}) = 0.2$, if the characters succeed each other in the alphabetical order and 0.1 otherwise. Figure 1 illustrates two possible configurations of assigning the six items on a 2×3 grid.

Although item $\mathcal{A}$ is connected to $\mathcal{B}$, $\mathcal{C}$ to $\mathcal{D}$, and $\mathcal{E}$ to $\mathcal{F}$ in the left configuration the alphabetical order is distorted and, thus, the structure underlying the data does not become obvious by this assignment. In the right

$$\begin{array}{ccccc} \mathcal{A} & \frac{0.1}{} & \mathcal{C} & \frac{0.1}{} & \mathcal{E} \\ |0.2 & & |0.2 & & |0.2 \\ \mathcal{B} & \overline{0.1} & \mathcal{D} & \overline{0.1} & \mathcal{F} \end{array} \qquad \begin{array}{ccccc} \mathcal{A} & \frac{0.2}{} & \mathcal{B} & \frac{0.2}{} & \mathcal{C} \\ |0.1 & & |0.1 & & |0.2 \\ \mathcal{F} & \overline{0.2} & \mathcal{E} & \overline{0.2} & \mathcal{D} \end{array}$$

Fig. 1. Examples of assigning items to a two-dimensional grid

configuration the alphabetical order is easily seen by considering the items in a clockwise manner. The edges in the left configuration capture a sum of support equal to one. In the right configuration the edges capture a sum of support equal 1.2. More generally, patterns in the data become emergent by maximizing the support captured by the grid. The maximization problem is:

$$\max \sum_{k=1}^{K} \sum_{i=1}^{I} \sum_{j=1}^{J} \left[\sum_{\substack{l=1 \\ l \neq k}}^{L} sup(\mathbb{X}_k \Rightarrow \mathbb{Y}_l) a_{ijk} a_{(i-1)jl} + \sum_{\substack{l=1 \\ l \neq k}}^{L} sup(\mathbb{X}_k \Rightarrow \mathbb{Y}_l) a_{ijk} a_{(i+1)jl} \right.$$
$$\left. + \sum_{\substack{l=1 \\ l \neq k}}^{L} sup(\mathbb{X}_k \Rightarrow \mathbb{Y}_l) a_{ijk} a_{i(j-1)l} + \sum_{\substack{l=1 \\ l \neq k}}^{L} sup(\mathbb{X}_k \Rightarrow \mathbb{Y}_l) a_{ijk} a_{i(j+1)l} \right] \tag{2}$$

$$\begin{aligned} \text{s.t.} \ \sum_{i=1}^{I} \sum_{j=1}^{J} a_{ijk} &= 1 \\ \sum_{k=1}^{K} a_{ijk} &\leq 1 \ \forall (i,j) \in \mathcal{G} \\ a_{ijk} &\in \{0,1\} \end{aligned}$$

The indicator variable a_{ijk} is equal to 1, if item k is assigned to knot (i,j) in the grid, otherwise it is equal to zero. The first constraint assures that each item is assigned exactly once. The second constraint ensures that each knot of the grid is covered by a maximum of one item. Limiting a_{ijk} to be equal to one or to zero in the third constraint guarantees a precise assignment. The boundary of the grid is defined by $a_{0jl} = a_{i0l} = a_{(I+1)jl} = a_{i(J+1)l} = 0 \ \forall l$. This problem can be rewritten in a linear form:

$$\max \sum_{k=1}^{K} \sum_{i=1}^{I} \sum_{j=1}^{J} \left[\sum_{\substack{l=1 \\ l \neq k}}^{L} sup(\mathbb{X}_k \Rightarrow \mathbb{Y}_l) b_{i(i-1)jkl} + \sum_{\substack{l=1 \\ l \neq k}}^{L} sup(\mathbb{X}_k \Rightarrow \mathbb{Y}_l) b_{i(i+1)jk} \right.$$
$$\left. + \sum_{\substack{l=1 \\ l \neq k}}^{L} sup(\mathbb{X}_k \Rightarrow \mathbb{Y}_l) b_{ij(j-1)kl} + \sum_{\substack{l=1 \\ l \neq k}}^{L} sup(\mathbb{X}_k \Rightarrow \mathbb{Y}_l) b_{ij(j+1)kl} \right] \tag{3}$$

$$
\begin{aligned}
\text{s.t.} \sum_{i=1}^{I}\sum_{j=1}^{J} a_{ijk} &= 1 \\
\sum_{k=1}^{K} a_{ijk} &\leq 1 \ \forall (i,j) \in \mathcal{G} \\
a_{ijk} + a_{(i-1)jl} - b_{i(i-1)jkl} &\leq 1 \\
a_{ijk} + a_{i(j-1)l} - b_{ij(j+1)kl} &\leq 1 \\
0 \leq b_{i(i-1)jkl} &\leq 1 \\
0 \leq b_{ij(j+1)kl} &\leq 1 \\
a_{ijk} &\in \{0,1\}
\end{aligned}
$$

$b_{i(i-1)jkl}$ and $b_{ij(j+1)kl}$ are additional variables enabling a linear formulation of the problem which is widely known as QAP (Quadratic Assignment Problem) and has been proven to be NP-hard (Sahni and Gonzalez (1976)). Cela (1998) outlines algorithms of solving the problem exactly with no more than 22 items. In order to overcome this restriction, we consider a meta-heuristic approach for the assignment. In a first step the items are assigned randomly on the grid. We use a genetic algorithm with a simple one point cross over and mutations with a repair function (Zbigniew and Fogel (2000)) to tackle the maximization problem. In each generation, new combinations of assignments are created by crossover and mutation using the roulette wheel method. The fitness function is given by the sum of support reflected by the grid. Moreover, the ten best solutions of the previous generation were added to the new population.

One of the major difficulties in solving the problem described in equation 3 with heuristic procedures is the danger of premature convergence toward sub-optimal solutions (Blum and Roli (2003)). For the approach presented herein the diversity is ensured by the mutation operator. It's simple form just performs a small random perturbation of an individual, introducing some kind of noise. Since each mutated string has to be repaired in order to satisfy the constrains that each item is assigned once and only once to a position of the grid, every mutation introduces at least two changes of the solution string and therefore, maintains the population diversity.

3 Empirical application

3.1 Domain and data

The data set used to demonstrate the methodology describes marketing practices in Russia. It comprises $|\mathbb{T}| = 72$ observations which were obtained using the standardized questionnaire developed within the *Contemporary Marketing Practices* project (CMP)[1]. The respondents are managers or directors

[1] The CMP group is an international research team investigating aspects of marketing practice. It has developed a classification scheme to describe current marketing practices and has published extensively on the empirical evidence and the conceptual framework. See http://cmpnetwork.webexone.com.

concerned with marketing activities of their companies. They answered questions describing the marketing practices of their companies or organizations. The items of interest are different aspects of modern marketing.

While the framework distinguishes between five aspects of marketing practices - transactional marketing (TM), data-base marketing (DM), e-marketing (EM), interactive marketing (IM), and network marketing (NM) - it does not assume that these practices are mutually exclusive. Thus companies can practice assorted combinations of TM, DM, IM, NM and EM. The question of interest is the extent to which they practice these different aspects and more important, how they combine the different activities. For example, a consumer goods firm may be expected to practice higher levels of TM and DM and lower levels of IM, EM, and NM. The nine dimensions used to define these five aspects of marketing practice are as follows:

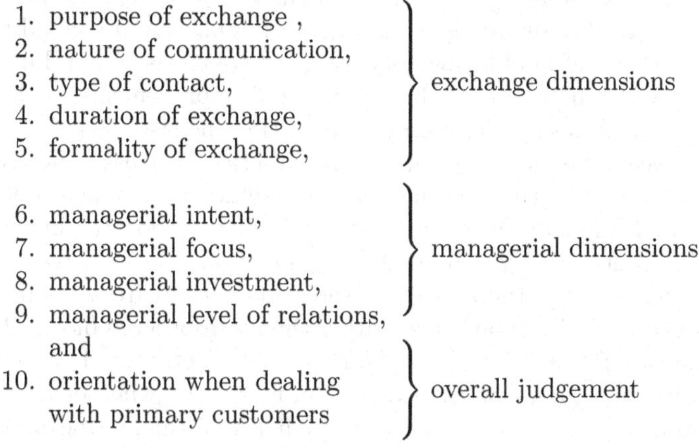

1. purpose of exchange ,
2. nature of communication,
3. type of contact,
4. duration of exchange,
5. formality of exchange,

} exchange dimensions

6. managerial intent,
7. managerial focus,
8. managerial investment,
9. managerial level of relations,

} managerial dimensions

and

10. orientation when dealing with primary customers

} overall judgement

Clearly, for each of the aspects the dimensions capture different meanings. The "managerial investment " (i.e. people, time, and money) might be a good example to clarify this.

- Transactional Marketing is characterized by investments in the classical marketing mix by means of product, promotion, price, and distribution activities or some combination of these (item: TM_8).
- Database Marketing requires ongoing investments in database technology to maintain and improve communication with the customers (item: DM_8).
- e-Marketing requires investments in operational assets (IT, website, logistics) and functional systems integration, e.g., marketing with IT (item: EM_8).
- For Interaction Marketing the investments need to be devoted to establishing and building personal relationships with individual customers (item: IM_8).
- Network Marketing is characterized by strategic investments in developing the organization's network relationships within their market(s) or wider marketing system (item: NM_8).

Each observation $\tilde{\mathbb{T}} \in \mathbb{T}$ of the data set is described by the activities which are judged to be important for the marketing activities of the company or organization. One item (TM_1) is missing in the data set due to a printing error in the questionnaires. Thus 49 items have to be assigned.

3.2 Results

In order to create a start population, the items reflecting the marketing activities were randomly assigned 200 times to the grid describing the contemporary marketing practices. Computing the cumulative support for each of these assignments yields the initial fitness for applying the roulette wheel procedure to select individuals for crossover and mutation. Each new population is made up of 100 individuals using the one point cross over operator and a repair function. An additional 100 individuals are generated by mutation. Moreover, the best 10 individuals of the former population are added to the new population as a of kind of memory for good solutions. A total of 1,300 populations have been computed, but after rapid improvements in the first populations, the best results did not improve during the last 450 populations.

Figure 2 reveals the pattern of marketing practices in Russia. For the interpretation it should be kept in mind that combinations of items, which are shared by many companies in this market are likely to be located in the center of the grid. Contrastingly, combinations holding only for few observations are likely to be located at the borders of the grid. Evident from the figure is the impression of mixed combinations rather than well-ordered blocks. In the lower left quadrant EM activities are adjoining the IM activities. For instance, IM_10 (overall orientation towards interaction marketing when dealing with primary customers) is directly connected with EM_7 (marketing activities are carried out by marketing specialists with technology specialists, and possibly senior managers). Moreover, the overall orientation towards the DM activities (DM_10) is directly connected with the items DM_2 (marketing planning is focused on issues related to the customers in the market(s) in addition to the offers) and DM_7 (marketing activities are carried out by specialist marketers, e.g., customer service manager, loyalty manager). The area of technology-driven marketing activities fades out to the upper left quadrant. The two fields are interconnected by EM items on the middle column. The upper field is characterized by adjoining transactional marketing activities. Overall, the transactional marketing as well as the network marketing activities seem to spread over the grid. However, the items of these marketing orientations are frequently adjoining. This contradicts some priors about the Russian marketing practices. As Russia has experienced a strong economic growth in the very last few years after the collapse of the Soviet Union, the vendors in these markets are expected to have a clear orientation towards the Transactional Marketing (cf. Wagner (2005b) for a detailed discussion). Surprisingly, the overall orientation towards Transactional Marketing (TM_10) is assigned to the border of the grid. This clearly indicates

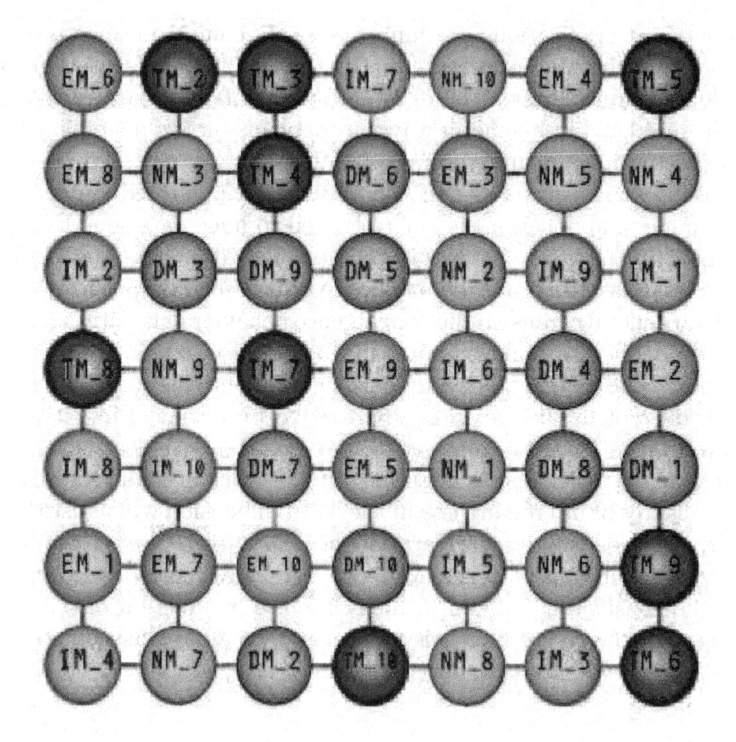

Fig. 2. Patterns of Contemporary Marketing Practices in Russia

that the Russian marketers are already adopting the modern sophisticated marketing techniques and assemble these to meet the requirements of their customers better than by simply adopting textbook knowledge.

4 Discussion

The methodology presented in this paper aims at a summary of the information that are revealed by analyzing the associations between items. By restricting to bilateral relations between items the mining task is simplified. A selection of rules by means of a reduction of the information to be revealed by the data analysis is evitable. The visualization provides an impression of the strengths of the relation between the items. In a general view the method can be considered as a mapping to a two-dimensional space similar to principal component analysis or correspondence analysis. But, in difference to many eigenvalue-decomposition based procedures this method provides a nonlinear projection based on co-occurrences instead of correlations. Moreover, the distances and the neighborhood relations might be easier to interpret than χ^2-distances in correspondence analysis.

The scope of applications are finite item sets comprising not too many items, because otherwise understanding the graphical representation would become rather cumbersome. The question of how many items are too many can not be answered in a general way, but depends on the distortion of previously expected blocks, such as brand families in market basked analysis or the marketing types considered in the application presented herein. Moreover, the range of applications is not restricted to huge data sets. Even from smaller sets the unknown patterns of relations between the items can be visualized. Scalability offers another challenge for the method proposed herein. Although no general claims on the convergence behavior of genetic algorithms can be proved, the assignment appears to be not that sensitive (in terms of computational effort) to an increase of the number of items as exact methods.

Two venues of further research seem to be promising. First, the items do not need to be assigned to a rectangular grid, but – as learned from self organizing maps – other shapes might be useful as well. Secondly, a projection of some measures of interestingness in the third dimension will assist the user to identify the most interesting parts of the grid without losing the context of the whole underlying patterns. Moreover, an assessment of the quality of fit has to be considered. The improvement of the fit function of the genetic algorithm might be a suitable basis fit measure for all applications that aim to reveal a priori unknown patterns.

References

BLUM, C. and ROLI, A. (2003): Metaheuristics in Combinatorial Optimazation: Overview and Conceptual Comparision, *ACM Computing Survey, 35/3, 268-308.*

CELA, E. (1997): *The Quadratic Assignment Problem: Theory and Algorithms*, Kluwer, Dordrecht.

HILDERMAN, R.J. and HAMILTON H.J. (2001): Evaluation of Interestingness Measures for Ranking Discovered Knowledge. In: D. Cheung, G.J. Williams, and Q. Li (Eds.): Advances in Knowledge Discovery and Data Mining, Springer, Berlin, 247–259.

SAHNI, S.K. and GRONZALEZ, T. (1976): P-Complete Approximation Problems, *Journal of Association of Computing Machinery, 23/3, 555–565.*

WAGNER, R. (2005a): Mining Promising Qualification Patterns. In: D. Baier and K.-D. Wernecke (Eds.): *Innovations in Classification, Data Science, and Information Systems.* Springer, Berlin, 249–256.

WAGNER, R. (2005b): Contemporary Marketing Practices in Russia, *European Journal of Marketing, Vol. 39/1-2, 199–215.*

ZAKI, M.J. and OGIHARA, M. (1998): Theoretical Foundations of Association Rules. In: Proceedings of 3rd ACM SIGMOD Workshop on Research Issues in Data Mining and Knowledge Discovery.

ZBIGNIEW, M. and FOGEL, D.B. (2000): *How to Solve it: Modern Heuristics*, Springer, Berlin.

Part VII

Analysis of Music Data

Generalized N-gram Measures for Melodic Similarity

Klaus Frieler

Institute for Systematic Musicology, University of Hamburg,
Neue Rabenstr. 13, D-20354 Hamburg, Germany

Abstract. In this paper we propose three generalizations of well-known N-gram approaches for measuring similarity of single-line melodies. In a former paper we compared around 50 similarity measures for melodies with empirical data from music psychological experiments. Similarity measures based on edit distances and N-grams always showed the best results for different contexts. This paper aims at a generalization of N-gram measures that can combine N-gram and other similarity measures in a fairly general way.

1 Introduction

For similarity comparisons melodies are often viewed as sequences (strings) of pitch symbols. This is a quite natural approach in the light of common practice music notation and indeed proves to be quite sufficient and adequate for many applications. However, some important aspects of melodies such as the order of tones regarding to pitch height or the dimension of rhythm are often left out. In our former work (Müllensiefen & Frieler (2004)) we acheived optimized similarity measures as linear combinations of similarity measures coming from different musical dimensions such as pitch, contour, rhythm, and implicit tonality. We now aim at more complex combinations with a generalization of the N-gram approach. To do this we first to set out some basic concepts, viewing melodies as (time) series of arbitrary length in an arbitrary event space. We go on with the definition of melodic similarity measures and present three common N-gram-based approaches for text similarity and their application to melodies. We will then develop a generalization of these N-gram measures using the concept of similarity on a lower level whereby we come close to some concepts of fuzzy logic.

2 Abstract melodies

The fundamental elements of symbolic music processing are discrete finite sequences. We will state some basic definitions at the beginning.

2.1 Discrete sequences

Definition 5 (sequences). Let $\mathbb{E}$ be a set and $N \leq M$ two integer numbers. A **sequence** over $\mathbb{E}$ is a discrete map

$$\phi : [N : M] \rightarrow \mathbb{E}$$
$$k \mapsto \phi(k)$$

We write $|\phi| = M - N + 1$ for the length of a sequence. A sequence is said to have **normal form**, if $N = 0$. For sequences of length N we write $F_N(\mathbb{E})$.

The N-gram approaches for similarity are based on the concept of subsequences length N (Downie (1999), Uitdenbogerd (2002)).

Definition 6 (N-gram). Let $s \in F_N(\mathbb{E})$ and $0 \leq i \leq j < N$. Then

$$\phi_j^i : [i : j] \rightarrow \mathbb{E}$$
$$k \mapsto \phi(k)$$

is called a **N-gram** or **subsequence** of ϕ of length $n = j - i + 1$. The set of all N-grams of ϕ of length n is notated with $\mathcal{S}_n(\phi)$.

Music can be abstracted and symbolically represented in many ways. Two main classes can be differentiated: audio (signal) oriented and notation (symbolical) oriented representations. For our purposes of measuring melodic similarity we use solely symbolically coded music on the abstraction level of common practice music notation. Most music theorist and music psychologists agree that onset and pitch are mostly sufficient to capture the 'essence' of a melody. We share this viewpoint and will give now the definition of an abstract melody, considered as a finite, discrete sequence in some event space.

Definition 7 (Melody). Let $\mathbb{E}$ be an event space. A finite, discrete map

$$\mu : [0 : N-1] \rightarrow \mathbb{R} \times \mathbb{E}$$
$$n \mapsto (t_n, p_n),$$

is called **melody** if

$$t_n < t_m \Leftrightarrow n < m.$$

The vaues p_n are called **generalized pitch**.

3 Similarity measures

After this introduction of the basic notions, we will now discuss similarity measures for melodies. At the beginning we had to make the basic choice between the complementary concepts of similarity and dissimilarity (i.e. distance) measures. In the statistical practice the use of distance measures is far more common, e.g. for MDS or cluster analysis. However, our aim was to compare many different (dis-)similarity measures on a common ground and moreover with experimental data. Because the task to judge similarity is much easier and more familiar for musical experts than to judge dissimilarity, we choose similarity as our basic concept. Because of the complementary nature of the two approaches every similarity measure can be formulated as dissimilarity measure and vice versa, but not uniquely, i.e. with some degree of freedom due to choice of normalization and transformation function. We state our definition of general similarity maps and of similarity measures for melodies.

Definition 8 (Similarity map). Let $\mathcal{M}$ be an arbitray set. A **similarity map** is a map

$$\begin{aligned} \sigma : \mathcal{M} \times \mathcal{M} &\to [0,1] \\ (\mu, \mu') &\mapsto \sigma(\mu, \mu') \end{aligned}$$

with the following properties

1. Symmetry: $\sigma(\mu, \mu') = \sigma(\mu, \mu')$
2. Self-identity: $\sigma(\mu, \mu) = 1$ and $\sigma(\emptyset, \mu) = 0 \ \forall \mu \neq \emptyset$

The similarity map $\sigma(\mu, \mu') = 1$ is called the **trivial** similarity map. A similarity map with

$$\sigma(\mu, \mu') = 1 \Leftrightarrow \mu = \mu'$$

is called **definite**. The value of $\sigma(\mu, \mu')$ is the **degree of similarity** of μ and μ'.

A similarity map can be viewed as a generalization of Kroneckers δ-operator. Between the two distinct cases *identic* and *non-identic* a non-trivial similarity map provides a whole continuum of similarity. This is related to concepts of fuzzy logic, where one has degrees of *belongingness* of elements to sets and degrees of *truth* for logical statements.

For similarity measures for melodies we demand additionally the invariance under pitch transposition, time shift and tempo change. For a more detailed discussion of this properties we refer the interested reader to Müllensiefen & Frieler (2004).

3.1 N-gram measures

N-gram based approaches form a set of standard techniques for measuring similarity of strings and texts and for abstract melodies as well (Downie (1999), Uitdenbogerd (2002)). All of them are more or less based on counting common N-grams in two strings. We will discuss here three basic forms, the

Count-Distinct, the Sum-Common and the Ukkonen measure. For this and for later purposes we first define the frequency $f_s(r)$ of a N-gram r with respect to a sequence s:

$$f_s(r) = \sum_{u \in S_n(s)} \delta(u, r)$$

(The δ-operator for sequences is obviously given through the identity of corresponding sequence elements.) The set of all distinct N-grams of a sequence s will be written as $\mathbf{n}(s)$

Definition 9 (Count-Distinct, Sum-Common and Ukkonen measure). Let s and t be two sequences over a event space $\mathbb{E}$, and let $0 < n \leq \min(|s|, |t|)$ be a positive integer.

1. The **Count-Distinct measure (CDM)** is the count of N-grams common to both sequences:

$$S_d(s,t) = \sum_{r \in \mathbf{n}(s) \cap \mathbf{n}(t)} 1 = |\mathbf{n}(s) \cap \mathbf{n}(t)|$$

2. The **Sum-Common measure (SCM)** is the sum of frequencies of N-grams common to both sequences

$$S_c(s,t) = \sum_{r \in \mathbf{n}(s) \cap \mathbf{n}(t)} f_s(r) + f_t(r),$$

3. The **Ukkonen measure (UM)** counts the absolute differences of frequencies of all distinct N-grams of both sequences:

$$S_u(s,t) = \sum_{r \in \mathbf{n}(s) \cup \mathbf{n}(t)} |f_s(r) - f_t(r)|$$

The Count-Distinct and the Sum-Common measures are (unnormalized) similarity measures, the Ukkonnen measures is a distance measure. To fulfill our definition of a similarity map we need to apply a normalization to the first and second, and for the UM we additionally need a suitable transformation to a similarity measure.

$$\sigma_d(s,t) = \frac{S_d(s,t)}{f(|\mathbf{n}(s)|, |\mathbf{n}(t)|)}$$

$$\sigma_c(s,t) = \frac{S_c(s,t)}{|s| + |t| - 2n + 2}$$

$$\sigma_u(s,t) = 1 - \frac{S_u(s,t)}{|s| + |t| - 2n + 2}$$

The choice of this normalizations come mainly from the self-identity property and from the desirable property of being definite. Natural choices for $f(x, y)$

are $\max(x, y)$ and $\frac{1}{2}(x + y)$. The normalization for the SCM comes from the cases of two identic constant sequences, and for the UM from the case of two constant sequences with no common N-grams. The choice of the functional form $1 - x$ to transform the UM into a similarity map after normalization is arbitrary but quite natural and simple. Of course any monotonic decreasing function with $f(0) = 1$ and $\lim_{x \to \infty} f(x) = 0$ like e^{-x} could have been used as well.

Example 1. We will give an short example with 4-grams. We consider the beginning 6 notes of a major and a minor scale, e.g. in **C** we have $s = \{\mathbf{C}, \mathbf{D}, \mathbf{E}, \mathbf{F}, \mathbf{G}, \mathbf{A}\}$ and $t = \{\mathbf{C}, \mathbf{D}, \mathbf{Eb}, \mathbf{F}, \mathbf{G}, \mathbf{Ab}\}$. We neglect the rhythm component. We will transform this to a presentation using semitone intervals to acheive transposition invariance:

$$s = \{2, 2, 1, 2, 2\}, t = \{2, 1, 2, 2, 1\}$$

We have two 4-grams for each melody

$$s_1 = \{2, 2, 1, 2\}, s_2 = \{2, 1, 2, 2\}$$

$$t_1 = \{2, 1, 2, 2\}, s_2 = \{1, 2, 2, 1\}$$

with only one common 4-gram s_2 resp. t_1. Thus the normalized CD similarity is

$$\sigma_d = \frac{1}{2}$$

the normalized SCM is likewise

$$\sigma_c = \frac{1 + 1}{|s| + |t| - 2 \cdot 4 + 2} = \frac{1}{2}$$

and the normalized UM is

$$\sigma_u = 1 - \frac{1 + |1 - 1| + 1}{5 + 5 - 8 + 2} = \frac{1}{2}$$

Incidentally, all three measures give the same value for this example.

4 Generalized N-grams

4.1 Reformulation

We now come to the generalization procedure for the three N-gram measures. The basic idea is to generalize identity of N-grams to similarity. A common N-gram is a N-gram present in both sequences, or, stated in other words, the frequencies with respect to both sequences of a common N-gram is greater than zero. We will use this idea to restate the N-gram measures in a way more suitable for generalization.

For the following let be s and t be sequences over $\mathbb{E}$ of length $|s| = N$ and $|t| = M$, $0 < n \leq \min(N, M)$, and $0 \leq \epsilon < 1$ be an arbitrary real constant. Moreover let σ be a similarity map for $F_n(\mathbb{E})$. A helpful function is the step function Θ which is defined as

$$\Theta(t) = \begin{cases} 1 & t > 0 \\ 0, & t \leq 0 \end{cases}$$

The **presence function** Θ_s of a N-gram r with respect to s can be defined as

$$\Theta_s(r) = \Theta(f_s(r) - \epsilon) = \begin{cases} 1 & r \in \mathcal{S}_n(s) \\ 0, & r \notin \mathcal{S}_n(s) \end{cases}$$

For compactness of presentation it is moreover useful to define the frequency ratio of a N-gram r with respect to s and t:

$$g_{st}(r) = \frac{f_s(r)}{f_t(r)}$$

We are now able to restate the basic definitions of the three N-gram measures.

Lemma 2. *The CDM, SCM and UM can be written as follows:*

1. *(CDM)*

$$S_d(s,t) = \sum_{r \in \mathcal{S}_n(s)} \frac{\Theta_t(r)}{f_s(r)}$$

2. *(SCM)*

$$S_c(s,t) = \sum_{r \in \mathcal{S}_n(s)} \Theta_t(r)(1 + g_{ts}(r))$$

3. *(UM)*

$$S_u(s,t) = \sum_{r \in \mathcal{S}_n(s)} 1 + \Theta_t(r)(\frac{1}{2}|1 - g_{ts}(r)| - 1) + (s \leftrightarrow t) \tag{1}$$

Proof. First, we note that a sum over distinct N-grams can be rewritten as sum over all N-grams:

$$\sum_{r \in \mathbf{n}(s)} F(r) = \sum_{r \in \mathcal{S}_n(s)} \frac{F(r)}{f_s(r)} \tag{2}$$

We will only shortly proof the formula for the Sum-Common-Measure, the other proofs follow similar lines of argumenation.

$$\begin{aligned} \sum_{r \in \mathcal{S}_n(s)} \Theta_t(r)(1 + g_{ts}(r)) &= \sum_{r \in \mathcal{S}_n(s)} \frac{\Theta_t(r)(f_s(r) + f_t(r))}{f_s(r)} \\ &= \sum_{r \in \mathbf{n}(s)} \Theta_t(r)(f_s(r) + f_t(r)) \\ &= \sum_{r \in \mathbf{n}(s) \cup \mathbf{n}(t)} (f_s(r) + f_t(r)) \square \end{aligned}$$

After this reformulation we again have to apply a normalization to fulfill our similarity map definition. We will state this without proof and define some auxiliary functions

$$\beta(s) = \sum_{r \in \mathcal{S}_n(s)} \frac{1}{f_s(r)},$$

$$\lambda(s,t) = \sum_{r \in \mathcal{S}_n(s)} 1 + g_{ts}(r) = \sum_{r \in \mathcal{S}_n(s)} \frac{f_s(r) + f_t(r)}{f_s(r)}$$

and

$$\eta(s,t) = \sum_{r \in \mathcal{S}_n(s)} 2 + |1 - g_{ts}(r)|$$

With this functions we can write down the normalized similarity measures in a compact way:

$$\sigma_d(s,t) = \frac{S_d(s,t)}{f(\beta(s), \beta(t))}$$
$$\sigma_c(s,t) = \frac{S_c(s,t)}{f(\lambda(s,t), \lambda(t,s))}$$
$$\sigma_u(s,t) = 1 - \frac{S_u(s,t)}{f(\eta(s,t), \eta(t,s))}$$

Again the function $f(x,y)$ can be $\max(x,y)$ or $\frac{1}{2}(x+y)$. For the following we fix f to the second form of the arithmetic mean.

4.2 Generalization

This new form of the N-gram measures can now be generalized. The main step is the introduction of generalized frequencies:

$$\nu_s(r) = \sum_{u \in \mathcal{S}_n(s)} \sigma(u,r) \geq f_s(r)$$

and generalized frequency ratios:

$$\omega_{st}(r) = \frac{\nu_s(r)}{\nu_t(r)}$$

By substituting frequencies with generalized frequencies the desired generalization is achieved. Now arbitrary similarity maps (and measures) can be used to judge the degree of similarity between N-grams. The old definition is contained as a special case of this generalized N-gram measure with Kronecker's δ as similarity map.

Example 2. We take the same two melodies as in example 1 with 4-grams and use a similarity measure based on the Levensthein-Distance $d(u,v)$ (see e.g. Müllensiefen & Frieler (2004) for a definition and discussion):

$$\sigma(u,v) = 1 - \frac{d(u,v)}{\max(|u|,|v|)}$$

First of all we need the similarities between all 4-grams:

$$\sigma(s_1,s_2) = \sigma(s_1,t_1) = \sigma(s_2,t_2) = \frac{1}{2}$$
$$\sigma(s_1,t_2) = 0$$

We can now calculate the generalized frequencies:

$$\nu_s(s_1) = \nu_s(s_2) = \nu_t(t_1) = \nu_t(t_2) = \nu_t(s_2) = \nu_s(t_1) = \frac{3}{2}$$
$$\nu_t(s_1) = \nu_s(t_2) = \frac{1}{2}$$

with generalized frequency ratios:

$$\omega_{st}(t_1) = \omega_{ts}(s_2) = 1$$
$$\omega_{st}(t_2) = \omega_{ts}(s_1) = \frac{1}{3}$$

Now we determine the presence functions of all 4-grams with $\epsilon = \frac{1}{2}$.

$$\Theta_t(s_1) = \Theta_s(t_2) = \Theta(\frac{1}{2} - \frac{1}{2}) = 0$$
$$\Theta_t(s_2) = \Theta_s(t_1) = \Theta(\frac{3}{2} - \frac{1}{2}) = 1$$

As a last step we determine values of the auxiliary functions:

$$\beta(s) = \beta(t) = \frac{4}{3}$$
$$\lambda(s,t) = \lambda(t,s) = \frac{10}{3}$$
$$\eta(s,t) = \eta(t,s) = \frac{14}{3}$$

With this preleminaries we can now calculate the three generalized N-grams measures. First the GCDM:

$$\begin{aligned}\sigma_d(s,t) &= \frac{2}{\beta(s)+\beta(t)} \sum_{r \in \mathcal{S}_n(s)} \frac{\Theta_t(r)}{\nu_s(r)} \\ &= \frac{2}{4/3+4/3} \frac{1}{\nu_s(s_2)} = \frac{1}{2}\end{aligned}$$

Then the GSCM:

$$\sigma_c(s,t) = \frac{2}{\lambda(s,t)+\lambda(t,s)} \sum_{r\in S_n(s)} \Theta_t(r)(1+\omega_{ts}(r))$$
$$= \frac{3}{10}(1+\omega_{ts}(s_2)) = \frac{3}{5},$$

and at last the GUM:

$$\sigma_u(s,t) = 1 - \frac{2}{\eta(s,t)+\eta(t,s)} \sum_{r\in S_n(s)} 1 + \Theta_t(r)(\frac{1}{2}|1-\omega_{ts}(r)|-1) + (s\leftrightarrow t)$$
$$= 1 - \frac{2}{14/3+14/3}(1+\frac{1}{2}|1-\omega_{ts}(s_2)|+1+\frac{1}{2}|1-\omega_{st}(t_1)|)$$
$$= 1 - \frac{3}{14}(1+1) = \frac{4}{7}$$

We see from this example that the generalized measures usually raise the similarity values compared to the orginal version.

There is some more possibility for further generalization. For this purpose we define a ramp function:

$$\rho(t) = \begin{cases} 0, \; t \le 0 \\ t, \; 0 \le t < 1 \\ 1, \; t \ge 1 \end{cases}$$

and can now generalize the presence function (with some real constant $a > 0$):

$$\theta_s(r) = \rho(\frac{\nu_s(r)}{a})$$

Example 3. We consider again the above example and will calculate the GCDM with this new generalized presence function ($a = 1$).

$$\theta_t(s_1) = \rho(\nu_t(s_1)) = \rho(\frac{1}{2}) = \frac{1}{2}$$
$$\theta_t(s_2) = \rho(\nu_t(s_2)) = \rho(\frac{3}{2}) = 1$$
$$\theta_s(t_1) = \rho(\nu_s(t_1)) = \rho(\frac{3}{2}) = 1$$
$$\theta_s(t_2) = \rho(\nu_s(t_2)) = \rho(\frac{1}{2}) = \frac{1}{2}$$

Applying this to the GCDM gives the following similarity value for our example.

$$\sigma_d(s,t) = \frac{2}{\beta(s)+\beta(t)} \sum_{r \in \mathcal{S}_n(s)} \frac{\theta_t(r)}{\nu_s(r)}$$
$$= \frac{2}{4/3+4/3}\left(\frac{\theta_t(s_1)}{\nu_s(s_1)} + \frac{\theta_t(s_2)}{\nu_s(s_2)}\right)$$
$$= \frac{3}{4}\left(\frac{1/2}{3/2} + \frac{1}{3/2}\right) = \frac{3}{4}$$

5 Conclusion

We proposed a generalization of well-known similarity measures based on N-grams. The application of these techniques to melodies made a generalization desirable because of the cognitively multidimensional nature of melodies. They can be viewed to some extent as string of symbols but this already neglects such important dimension as rhythm and pitch order. Besides the possibility of using linear combination of similarity measures that focus on different musical dimensions, it could be fruitful to combine this measure in a more compact way. This, however, waits for further research, particularly an implementation of the generalized N-gram measures proposed here (which is currently under development), and a comparision with existing empirical data and other similarity measures. Thus, this paper should be viewed as a first sketch of ideas in this direction.

References

DOWNIE, J. S. (1999): *Evaluating a Simple Approach to Musical Information retrieval: Conceiving Melodic N-grams as Text.* PhD thesis, University of Western Ontario .

MÜLLENSIEFEN, D. & FRIELER, K.(2004): Cognitive Adequacy in the Measurement of Melodic Similarity: Algorithmic vs. Human Judgments. *Computing in Musicology,* Vol. 13.

UITDENBOGERD, A. L. (2002): *Music Information Retrieval Technology.* PhD thesis, RMIT University Melbourne Victoria, Australia.

Evaluating Different Approaches to Measuring the Similarity of Melodies

Daniel Müllensiefen and Klaus Frieler

Institute of Musicology, University of Hamburg,
Neue Rabenstr. 13, D-20354 Hamburg, Germany

Abstract. This paper describes an empirical approach to evaluating similarity measures for the comparision of two note sequences or melodies. In the first sections the experimental approach and the empirical results of previous studies on melodic similarity are reported. In the discussion section several questions are raised that concern the nature of similarity or distance measures for melodies and musical material in general. The approach taken here is based on an empirical comparision of a variety of similarity measures with experimentally gathered rating data from human music experts. An optimal measure is constructed on the basis of a linear model.

1 Introduction

While working on an empirical project on human memory for melodies at Hamburg University (Müllensiefen (2004), Müllensiefen and Hennig (2005)) it soon became very clear that measuring the similarity of two given melodies is an important analytical tool in setting up a prediction model for what people can remember of melodies that they just heard once. But melodic similarity is not only a key concept in memory research or music psychology, but also several of musicology's subdisciplines have a strong need for valid and reliable similarity measures for melodies. Some of these subdisciplines are ethnomusicology, music analysis, copyright issues in music, and music information retrieval. So it is not surprising that many different approaches and concepts for measuring the similarity of melodies have been proposed in the literature in the last two decades. Several techniques for computing melodic similarity have been defined that cover distinct aspects or elements of melodies. Among these aspects are intervals, contour, rhythm, and tonality, each with several options to transform the musical information into numerical datasets. In combination with each of these aspects different approaches for constructing distance or similarity measures have been used with music data in the past. Some important algorithms are the edit distance, n-grams, geometric measures and hidden Markov models. In the literature there many examples of successful applications of specific similarity measures that combine an abstraction technique for a special musical aspect of melodies with a specific approach to computing similarities or distances. In the past, we dedicated several studies to the comparison of different approaches to the similarity measurement for melodies (Müllensiefen and Frieler (2004a, 2004b, 2006)) and its applications for example in folk music research (Müllensiefen and Frieler (2004c)). In this earlier work it was shown that these differently constructed similarity measures may generate very different similarity values for the same pair of melodies. And from the mathematical or algorithmical construction of the similarity measures it is by no means clear which one is the most adequate to be used in a specific research situation, like a memory model. The answer to this confusing situation was to compare the measurement values of different similarity measures to ratings from human judges that rate the similarity of melody pairs after listening to the two melodies. This paper first resumes our previous comparative work and in the discussion section we are able to adress some issues of melodic similarity from a meta-perspective.

2 Different approaches to measuring the similarity of melodies

To introduce a general framework for comparing different algorithms for similarity measurement it seems useful to first get a clear idea what a melody is on an abstract level. In a useful working definition that was pointed out earlier (Müllensiefen and Frieler (2004a)), a melody will be simply viewed as a time series, i.e., as a series of pairs of onsets and pitches (t_n, p_n), whereby pitch is represented by a number, ususally a MIDI number, and an onset is a real number representing a point in time. A similarity measure $\sigma(m_1, m_2)$ is then a symmetric map on the space of abstract melodies $\mathcal{M}$, mapping two melodies to a value between 0 and 1, where 1 means identity. The similarity measure should meet the constraints of symmetry, self-identity, and invariance under transposition in pitch, translation in time, and tempo changes. The construction of most algorithms for measuring melodic similarity involves the following processing stages:

1. Basic transformations (representations)
2. Main transformations
3. Computation

The most common basic transformations are projection, restriction composition and differentiation. Projections can be either on the time or pitch coordinate, (with a clear preference for pitch projections). Differentiation means using coordinate differences instead of absolute coordinates, i.e. intervals and durations instead of pitch and onsets.

Among the main transformations rhythmical weighting, *Gaussification* (see Frieler (2004a)), classifications and contourization are the most important. Rhythmical weighting can be done for quantized melodies, i.e. melodies where the durations are integer multiples of a smallest time unit T. Then each pitch of duration nT can be substituted by a sequence of n equal tones with duration T. After a pitch projection the weighted sequence will still reflect the rhythmical structure. The concept of rhythmical weighting has been widely used in other studies (e.g. Steinbeck (1982), Juhasz (2000)). Classification is mainly used to assign a difference between pitch or time coordinates to a class of musical intervals or rhythmic durations. Other studies used this idea of classification in very similar ways (e.g. Pauws (2002)). Contourization is based on the idea that the perceptionally important notes are the extrema, the turning points of a melody. One takes this extrema (which to take, depends on the model) and substitutes the pitches in between with linear interpolated values, for example. We used linear interpolation exclusively for all of the tested contour models. The contourization idea was employed, for example, in the similarity measures by Steinbeck (1982) and Zhou and Kankanhalli (2003).

For computing the similarity of melodies several basic techniques have been described in the literature. Most of these techniques have their origin in application areas other than music, e.g. text retrieval and comparing gene sequences. But for most of them it has been shown that an adaption for musical data is possible. It is impossible to explain these techniques here in detail, so the reader should refer to the following publications or may find a summary in Müllensiefen and Frieler (2004). Among the most prominent techniques for computing melodic similarity are the edit distance algorithm (McNab et al (1996) Uitdenbogerd (2002)), n-grams (Downie (1999)), correlation and difference coefficients (O'Maidin (1998), Schmuckler (1999)), hidden Markov models (Meek and Birmingham (2002)), and the so-called earth mover distance (Typke et al (2003)).

As is described in Müllensiefen and Frieler (2004a) we implemented 48 similarity measures into a common software framework. These 48 similarity measures were constructed as meaningful combinations of basic and main transformations plus a specific computing technique.

3 Experimental evaluation of melodic similarity measures

3.1 Experimental design

We conducted three rating experiments in a test-retest design. The subjects were musicology students with longtime practical musical experience. In the first experiment the subjects had to judge the similarity of 84 melody pairs taken from western popular music on a 7-point scale. For each original melody six comparison variants with errors were constructed, resulting in 84 variants of the 14 original melodies. The error types and their distribution were constructed according to the literature on memory errors for melodies (Sloboda and Parker (1985), Oura and Hatano (1988), Zielinska and Miklaszewski (1992)). Five error types with differing probabilities were defined: rhythm errors, pitch errors leaving the contour intact, pitch errors changing the contour, errors in phrase order, modulation errors (pitch errors that result in a transition into a new tonality). For the construction of the individual variants, error types and degrees were randomly combined.

The second and third experiment served as control experiments. In the second experiment two melodies from the first experiment were chosen and presented along with the original six variants plus six resp. five variants, which had their origin in completely different melodies. The third experiment used the same design as the first one, but tested a different error distribution for the variants and looked for the effects of transposition of the variants.

3.2 Stability and correlation of human ratings

Only subjects who showed stable and reliable judgments were taken into account for further analysis. From 82 participants of the first experiment 23 were chosen, which met two stability criteria: They rated the same pairs of reference melody and variant highly similar in two consecutive weeks, and they gave very high similarity ratings to identical variants. For the second experiment 12 out of 16 subjects stayed in the analysis. 5 out of 10 subjects stayed in the data analysis of the third experiment. We assessed the between-subject similarity of the judgements in the three experiments using two different, i.e. Cronbach's alpha and the Kaiser-Meyer-Olkin measure (Kaiser (1974)). The inter-personal jugdments of the selected subjects showed very high correlations:

- As an indicator of the coherence of the estimations of the latent magnitude 'true melodic similarity' Cronbach's alpha reached values of 0.962, 0.978, and 0.948 for subjects' ratings of the three experiments respectively.
- The Kaiser-Meyer-Olkin measure reached values as high as 0.89, 0.811, and 0.851 for the three experiments respectively.

This high correlation between the subjects' ratings led us to assume, that there is something like an objective similarity at least for the group of 'western musical experts', from which we took a sample.

3.3 Optimisation of similarity measures

It is an old assumption in music research that for the perception and mental computation of melodies all musical aspects play a role to a certain degree. We therefore considered melodic similarity to work on five musical dimensions: contour information, interval structure, harmonic content, rhythm and characteristic motives. For each dimension the euclidean distances of the included measures to the mean subjects' ratings were computed, and the best measure for each dimension was pre-selected to serve as an input for a linear regression. This regression was done for the data of all three experiments separately and used the step-wise variable selection procedure. The best five similarity measures for experiment 1 were (ordered according to their euclidean distances, minimum first):

- *coned* (edit distance of contourized melodies)
- *rawEdw* (edit distance of rhythmically weighted raw pitch sequences)
- *nGrCoord* (coordinate matching based on count of distinct n-grams of melodies)
- *harmCorE* (edit distance of harmonic symbols per bar, obtained with the help of Carol Krumhansl's tonality vectors (Krumhansl (1990))
- *rhytFuzz* (edit distance of classified length of melody tones)

From this input we obtained a linear combination of the two measures *rawEdw* and *nGrCoord* for the data from experiment 1, which was 28.5% better than the best single measure for that experiment in terms of the euclidean distance from the subjects ratings over all 84 melody pairs. The model reached an corrected R^2 value of 0.826 and a standard error of 0.662. Given these results the optimisation within the linear model can be seen as successful. As the experimental task and the constructed melodic variants to be rated differed systematically in experiment 2 and 3 different similarity measures were preselected for the five musical dimensions and linear regression lead to weighted combinations of similarity measures that were different for each experiment.

3.4 Applications

We used our similarity measures in several analytical tasks on a folk song collection that was investigated thoroughly by an expert ethnomusicologist (Sagrillo (1999)). For example we filtered out successfully variants and exact copies of melodies in a catalogue of about 600 melodies from Luxembourg using the optimised similarity measure from our experiment 3 (Müllensiefen and Frieler (2004c)). This specific linear combination of similarity measures was chosen because the experimental taks the subjects had to fullfill in experiment 3 came closest to the duty of finding highly similar melodies. A second application within this folk song research (Müllensiefen and Frieler (2004c)) was to predict if two given melodic phrases from the total of 3312 phrases in the catalogue belong to the same group as classified by Sagrillo. For this task, we again pre-selected the best five out of 48 similarity measures (this time according to their *area under curve* values after drawing a ROC curve for each similarity measure) and we subsequently used logistic regression to predict for each melody pair if the two melodies belonged to the same group or not. Further applications that we tested so far, are the measurement of melodic similarity in cases of plagiarism in pop songs where one melody is assumed to be an illegal copy of a previously existing melody, and the ordering of short melodic phrases from classical music (incipits) according to similarity criteria.

4 Discussion

Having worked very intensely for three years on the measurement of similarity between melodies, we came across several conceptual issues that are hardly discussed in the literature. We would like to pose the respective questions here and answer with some tentative hypotheses, but the field is still very open to discussion.

- **Homogeneity of human similarity judgements** For all experiments we conducted so far, we found very high correlations of similarity judgements between subjects (Cronbach's α with values > 0.9). This is not to forget that we always selected subjects on the basis of their within-subject reliability, i.e. subjects had to rate the same melody pair in two consequent weeks alike, and they should rate identical melodies as highly similar. The interesting fact is that subjects selected according to their within-subject reliability show a very high between-subjects correlation. The only bias that entered our selection procedure for subjects was the natural requirement that subjects should rate identical melodies as highly similar. But in their judgments of non-identical melodies, subjects were completely free to give their subjective evaluation of the rhythmical, pitch, and contour differences between the melodies. It could have turned out that some of the reliable subjects rated differences in the rhythmical structure as much more severe than others or that contour errors would have been of different importance to different subjects. This would have resulted in lower correlations as reflected by the between-subjects correlation measures, which we actually did not find. These high between-subject correlations could be interpreted as if there is a latent but clear inter-personal notion of melodic similarity that each subject tries to estimate in an experimental situation. This assumption of an inter-personal agreement on what melodic similarity actually is, lays the conceptual foundation for the statistical modelling of human similarity perception of melodies.
- **Human notion of melodic similarity may change** Although there seems to be a consensus on what is similar in melodies, this consensed notion may make different use of the information in the various musical dimensions. For example for melodies that are all very similar because they were constructed as variants from each other like in expriment 1 it was possible to model subjects ratings exclusively with similarity measures that exploit pitch information only. Whereas in experiment 2 where some of the to be compared melodies were drawn at random from a larger collection and were therefore very dissimilar, subjects' ratings could be modeled best including similarity measures that reflect rhythmical information and implicit harmonic content. So obviously, humans show an adaptive behaviour to different tasks, different stylistic repertoires, and different contexts of experimental materials. For modelling subjects' ratings there are two solutions to this adaptive behaviour:

1. Find a general similarity measure that works well in most situations, but be aware that it might not be the optimal measure to model a specific task with a specific repertoire of melodies.
2. Try to gather test data of that specific situation and run an optimisation on that test data before predicting similarity in that domain.

- **Distance vs. similarity measures for melodies** To our knowledge all studies in the literature that deal with the comparision of melody pairs make exclusive use of similarity measures to conceptualise the relationship between two given melodies. Distance measures are never used for example for clustering or ordering of melodies. This seems to reflect an intuitive cognitive approach towards the processsing of comparable melodies. Humans seem to make sense out of melodies that differ in only a few notes. Obviously music listeners are used to relate them to each other effortlessly. But unrelated melodies that differ strongly in most musical dimensions are hard to relate. From our data it was clear that the subjects were much better at differentiating small changes on the rating scale when the two melodies were quite similar as when they had little in common. This might be interpreted as a reflection of the distribution of similarity values in large melody collections. As was outlined in Müllensiefen and Frieler (2004b) the distribution of about 250.000 similarity values between about 700 folk song phrases show a gauss-like distribution, but the shape of the curve was much steeper. Almost all previous studies with the exception of Kluge (1974) use similarity measures that are bounded between 0 and 1. Kluge's special research interest lead him to consider negatively correlated melodies as well and his similarity measure was therefore bounded between -1 and 1. Among our own 48 similarity measures we used several measures based on vector correlations and we tried both variants: Measures between -1 and 1 and measures where we set all negative correlation values to 0. In comparison with our experimental rating data almost always the variants with limits of 0 and 1 showed a superior performance than their -1/1 analogues.

References

DOWNIE, J.S. (1999): *Evaluating a Simple Approach to Musical Information retrieval: Conceiving Melodic N-grams as Text.* PhD thesis, University of Western Ontario.

FRIELER, K. (2004). Beat and Meter Extraction Using Gaussified Onsets. In: *Proceedings of the 5th International Conference on Music Information Retrieval. Barcelona: Universitat Pompeu Fabra, 178-183.*

JUHASZ, Z. (2000): A Model of Variation in the Music of a Hungarian Ethnic Group. *Journal of New Music Research, 29/2, 159–172.*

KAISER, H.F. (1974): An Index of Factorial Simplicity. *Psychometrika, 39, 31-36.*

KLUGE, R. (1974): *Faktorenanalytische Typenbestimmung an Volksliedmelodien.* Leipzig: WEB Deutscher Verlag für Musik.

KRUMHANSL, C. (1990): *Cognitive Foundations of Musical Pitch.* New York: Oxford University Press.

MCNAB, R.J., SMITH, L. A., WITTEN, I.H., HENDERSON, C.L. and CUNNINGHAM, S.J. (1996). Towards the Digital Music Library: Tune Retrieval from Acoustic Input. In: *Proceedings ACM Digital Libraries, 1996.*

MEEK, C. and BIRMINGHAM, W. (2002): Johnny Can't Sing: A Comprehensive Error Model for Sung Music Queries. In: *ISMIR 2002 Conference Proceedings, IRCAM, 124–132.*

O'MAIDIN, D. (1998): A Geometrical Algorithm for Melodic Difference in Melodic Similarity. In: W.B. Hewlett & Eleanor Selfridge-Field. *Melodic Similarity: Concepts, Procedures, and Applications. Computing in Musicology 11.* MIT Press, Cambridge, 1998.

MÜLLENSIEFEN, D. (2004): *Variabilität und Konstanz von Melodien in der Erinnerung. Ein Beitrag zur musikpsychologischen Gedächtnisforschung.* PhD work, University of Hamburg.

MÜLLENSIEFEN, D. and FRIELER, K. (2004a): Cognitive Adequacy in the Measurement of Melodic Similarity: Algorithmic vs. Human Judgments. *Computing in Musicology, 13, 147–176.*

MÜLLENSIEFEN, D. and FRIELER, K. (2004b): Melodic Similarity: Approaches and Applications. In: S. Lipscomb, R. Ashley, R. Gjerdingen and P. Webster (Eds.). *Proceedings of the 8th International Conference on Music Perception and Cognition (CD-R).*

MÜLLENSIEFEN, D. and FRIELER, K. (2004c): Optimizing Measures of Melodic Similarity for the Exploration of a Large Folk-Song Database. In: *Proceedings of the 5th International Conference on Music Information Retrieval. Barcelona: Universitat Pompeu Fabra, 274–280.*

MÜLLENSIEFEN, D. and HENNIG, CH. (2005): Modeling Memory for Melodies. In: *Proceedings of the 29th Annual Conference of the German Society for Classification (GfKl).* Springer, Berlin.

OURA, Y. and HATANO, G. (1988): Memory for Melodies among Subjects Differing in Age and Experience in Music. *Psychology of Music 1988, 16, 91–109.*

SAGRILLO, D. (1999): *Melodiegestalten im luxemburgischen Volkslied: Zur Anwendung computergestützter Verfahren bei der Klassifikation von Volksliedabschnitten.* Bonn, Holos.

SCHMUCKLER, M.A. (1999): Testing Models of Melodic Contour Similarity. *Music Perception 16/3, 109–150.*

SLOBODA, J.A. and PARKER, D.H.H. (1985): Immediate Recall of Melodies. In: I. Cross, P. Howell and R. West. (Eds.). *Musical Structure and Cognition.* Academic Press, London, 143–167.

STEIBECK, W. (1982): *Struktur und Ähnlichkeit: Methoden automatisierter Melodieanalyse.* Bärenreiter, Kassel.

TYPKE, R., GIANNOPOULOS, P., VELTKAMP, R.C., WIERING, F., and VAN OOSTRUM, R. (2003): Using Transportation Distances for Measuring Melodic Similarity. In: *ISMIR 2003:Proceedings of the Fourth International Conference on Music Information Retrieval, 107–114.*

UITDENBOGERD, A.L. (2002): *Music Information Retrieval Technology.* PhD thesis, RMIT University Melbourne Victoria, Australia.

ZIELINSKA, H. and MIKLASZEWSKI, K. (1992): Memorising Two melodies of Different Style. *Psychology of Music 20, 95-111.*

Using MCMC as a Stochastic Optimization Procedure for Musical Time Series

Katrin Sommer and Claus Weihs

Department of Statistics, University of Dortmund,
D-44221 Dortmund, Germany

Abstract. Based on a model of Davy and Godsill (2002) we describe a general model for time series from monophonic musical sound to estimate the pitch. The model is a hierarchical Bayes Model which will be estimated with MCMC methods. All the parameters and their prior distributions are motivated individually. For parameter estimation an MCMC based stochastic optimization is introduced. In a simulation study it will be looked for the best implementation of the optimization procedure.

1 Introduction

Pitch estimation of monophonic sound can be supported by the joint estimation of overtones as demonstrated in (Weihs and Ligges (2006)). One important model including the fundamental frequency together with its overtones is formulated by Davy and Godsill (2002) with the number of relevant overtones to be estimated as well. Furthermore, the model can deal with time-variant amplitudes, i.e. with varying loudness of tones. Being defined as a hierarchical Bayes model, the model and the corresponding estimation procedure contain many parameters. MCMC estimation is computationally very expensive. As a short-cut, an MCMC based stochastic optimization is introduced for estimation. Moreover, in a simulation study we look for the best pre-fixings of parts of the parameters in order to restrict the parameter space.

2 Harmonic model

In this section a harmonic model is introduced and its components illustrated. The whole model which is based on the model of Davy and Godsill (2002) has the following structure:

$$y_t = \sum_{h=1}^{H} (a_{h,t} \cos(2\pi h f_0 t) + b_{h,t} \sin(2\pi h f_0 t)) + \varepsilon_t.$$

In this model one tone is composed out of harmonics from H partial tones. The first partial is the fundamental frequency f_0, the other $H-1$ partials are called overtones. The time-variant amplitudes of each partial tone are $a_{h,t}$ and $b_{h,t}$. Finally, ε_t is the model error.

In order to avoid this high complexity, the time-variant amplitudes are modelled with so-called basis functions:

$$y_t = \sum_{h=1}^{H} \sum_{i=0}^{I} \phi_{t,i} \left[a_{h,i} \cos(2\pi h f_0 t) + b_{h,i} \sin(2\pi h f_0 t) \right] + \varepsilon_t \qquad \text{with}$$

$$a_{h,t} = \sum_{i=0}^{I} a_{h,i}\phi_{t,i} \text{ and } b_{h,t} = \sum_{i=0}^{I} b_{h,i}\phi_{t,i}.$$

In our model the basis functions $\phi_{t,i}$ are Hanning windows with 50% overlap. Hanning windows are shifted stretched squared cosine functions defined as

$$\Phi_{t,i} := cos^2 \left[\pi(t - i\Delta)/(2\Delta) \right], \Delta = (N-1)/I, N = \text{no. of observations.}$$

In principle a basis function can be any non-oscillating function (Davy and Godsill (2002)).

Figure 1 shows the difference of modelling a tone with oder without time-variant amplitudes. In figure 1,left it can be seen, that the model assumes

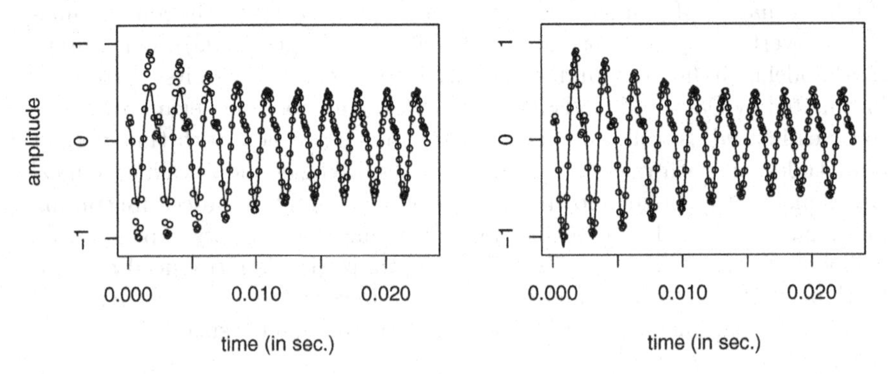

Fig. 1. Original oscillation (dots) and approximation (line) by a harmonic model with constant amplitudes (left) and with time-variant amplitudes (right), 4 basis functions

constant amplitudes over time. The model cannot depict the higher amplitudes at the beginning of the tone. Modelling with time-variant amplitudes (see figure 1,right) leads to better results.

3 Hierarchical Bayes-Model

In this section a hierarchical Bayes-Model is introduced for the parameters in the pitch model.

The likelihood of the tone is assumed to be normally distributed:

$$p(y \mid \theta, f_0, H, \sigma_\varepsilon^2) = \frac{1}{(2\pi\sigma_\varepsilon^2)^{N/2}} \exp\left[-\frac{1}{2\sigma_\varepsilon^2}(y - D\theta)^T(y - D\theta)\right],$$

where D is a matrix with sine and cosine entries multiplied by the values of the basis functions:

$$D_{t+1,2(Hi+h)-1} = \phi_{t,i} \cos(hf_0t), D_{t+1,2(Hi+h)} = \phi_{t,i} \sin(hf_0t).$$

For each time point and for each basis function the number of entries is 2 times the number of partial tones. So the matrix has the dimension $N \times 2H(I+1)$.

The priors in the Bayes-Model have a hierarchical structure:

$$p(\theta, f_0, H, \sigma_\varepsilon^2) = p(\theta \mid f_0, H, \sigma_\varepsilon^2)p(f_0 \mid H)p(H)p(\sigma_\varepsilon^2).$$

The amplitudes $a_{h,i}$ and $b_{h,i}$ are combined in one amplitude-vector θ:

$$\theta_{2(Hi+m)-1} = a_{h,i}, \theta_{2(Hi+m)} = b_{h,i}.$$

The following parameters determine the model: fundamental frequency f_0, number of partials H, parameter vector θ, and the predefined number of basis functions $I+1$. For these parameters we assume the following priors

- The amplitude vector θ is chosen normally distributed with expectation 0 and covariance matrix $\sigma_\varepsilon^2 \Sigma_\theta$: $\theta \mid f_0, H, \sigma_\varepsilon^2 \sim N(0, \sigma_\varepsilon^2 \Sigma_\theta)$, where Σ_θ is calculated with $\Sigma_\theta^{-1} = \frac{1}{\xi^2} D^T D$. ξ^2 is a inverse-gamma distributed hyperparameter with parameters $\alpha_\xi = 1$ and $\beta_\xi = 1$: $\xi \sim IG(1, 1)$.
- The prior density for the fundamental frequency f_0 is chosen to be uniformly distributed. One could propose a prior restricted to a discrete set of frequencies such as for the keys of the piano, but for simplicity the following uninformative prior is used $f_0 \mid H \sim U(0, \pi/H)$, where the upper limit π/H is chosen in order to be able to represent the highest involved partial.
- The number of partials H can be any positive integer depending on the instrument playing. Nevertheless, for generality one prior for all instruments has been chosen $H \sim \text{Poi}(\Lambda)$, where Λ is a so called hyperparameter of the Poisson distribution being Gamma distributed with parameters ε_1 and ε_2: $\Lambda \sim Gamma(0.5 + \varepsilon_1, \varepsilon_2)$. Here we set $\varepsilon_1 = 1$ and $\varepsilon_2 = 0.01$. Λ can be interpreted as the expected number of partials.
- The prior of the variance parameter σ_ε^2 is inverse gamma distributed with parameters α_ε and β_ε: $\sigma_\varepsilon^2 \sim IG(\alpha_\varepsilon, \beta_\varepsilon)$. In this case we chose $\alpha_\varepsilon = 1$ and $\beta_\varepsilon = 0.5$.

4 Stochastic optimization procedure

The basic idea of parameter estimation is optimization of likelihood by stochastic search for the best coefficients in given regions with given probability distributions. In standard MCMC methods the distributions are fully generated. This leads to a heavy computational burden. As a short cut, we used an optimal model fit criterion instead:

Every 50 MCMC iterations it is checked whether linear regression of the last 50 residuals against the iteration number delivers a slope significant at a previously specified level with a maximum number of iterations of 2000.

Figure 2 shows the decreasing error with an increasing number of iterations. Between iteration number 350 and iteration number 400 the slope is no more significant at a level of 10 %.

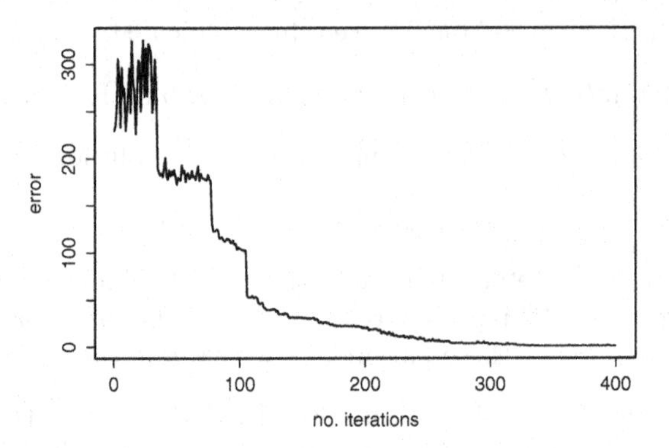

Fig. 2. Progression of model error

5 Simulation

5.1 Data

In a simulation study we aim at finding optimal levels of some of the unknown parameters of the hierarchical Bayes model and the estimation procedure. As a data base real audio data are used. We chose 5 instruments (flute, electric guitar, piano, trumpet and violin) each with 5 notes (220, 262, 440, 523 and 880 Hz) from the McGill database (McGill University Master Samples).

The instruments were chosen out of two groups, bowed instruments and wind instruments. There are three ways a bowed instrument can be played. They can be picked, bowed or stroke. We chose one instrument for each way. The two wind instruments can be distinguished as a woodwind instrument and a brass instrument.

The choice of the pitches was restricted by the availability of the data from the McGill database and the different ranges of the instruments.

For each tone $N = 512$ data points at a sampling rate 11025 Hz are used. Since the absolute overall loudness of the different recordings is not relevant, the waves are standardized to the interval $[-1, 1]$.

5.2 Model

In the simulation we estimated all the unknown parameters of the hierarchical Bayes model except the number of basis functions which was fixed to 1 to 5, where one basis function implies constant amplitudes over time.

5.3 Algorithm

There a some unknown parameter in the estimation algorithm optimized in our simulation. The main aim was to fix the stopping criterion. The questions are which significance level in the linear regression leads to best results and whether non-significance should be met more than once to avoid local optima. The significance level was varied from 0.05 to 0.55 in steps of 0.05. The procedure is stopped if non-significance is met once, twice or three times.

Moreover, 3 Markov chains with different frequency starting points are simulated. The frequency starting points are 175 Hz, 1230 Hz and the result $f\!f_{\text{Heur}}$ of a Heuristic fundamental frequency estimation (Ligges et al. (2002)):

$$f\!f_{\text{Heur}} = h + \frac{s-h}{2}\sqrt{ds/dh},$$

where h is the peaking Fourier frequency, s is the peaking neighbor frequency. The corresponding density values are dh and ds. The chain started at $f\!f_{\text{Heur}}$ is simulated with a Burn In of 200 iterations.

5.4 Design

Overall, the design leads to a full factorial design (5 instruments * 5 notes * 11 levels * 3 stops * 5 basis functions) with 4125 experiments each applied to 3 chains. In table 1 there is an overview of all components of the design. Instruments, notes and starting points define the environment of the simulation. For the three parameters level, stops and number of basis functions we looked for the best level.

Table 1. Factors and correspondending levels of the full factorial design

	factor	level
data	instruments	flute, electric guitar, piano, trumpet, violin
	notes	220, 262, 440, 523, 880 Hz
chains	starting points	175 Hz, 1230 Hz and $f\!f_{\text{Heur}}$
optimization	level	0.05, 0.1, ..., 0.55
	stops	1, 2, 3
	no. of basis functions	1, 2, ..., 5

5.5 Results

The deviations of the estimated frequencies from the real frequencies are measured in cents. 100 cent correspond to one halftone. The estimated frequencies are assigned to the correct note if the estimated values are enclosed by the interval of [-50,50] cents around the real frequency.

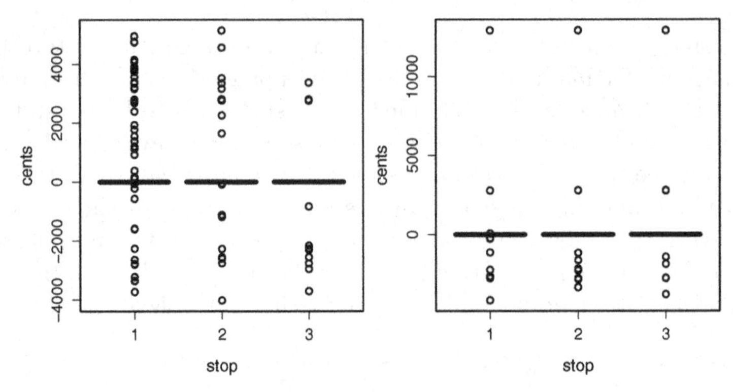

Fig. 3. Boxplots of deviations of estimated fundamental frequency, 3 basis functions, starting point = 175 Hz (left), and ff_{Heur} , BurnIn = 200 (right) iterations

Figure 3,left shows the deviation of the estimated frequencies from the fundamental frequency in cents. The estimates are from the chain with 3 basis functions and 175 Hz as starting point. It can be seen that there are less wrongly estimated frequencies if the stopping criterion is met more than once. The results improve by using the outcome of the Heuristics as the starting point and dismissing the first 200 iterations of the chain (see figure 3,right).

Now the three Markov chains with the different starting points are combined in that for each estimated frequency the chain with the minimal error is chosen. This way, there are only few estimates outside the bounds of [-50,50] cents. Actually almost all values are enclosed by the interval [-25,25] cents, see figure 4 for an outcome from three basis functions. Moreover, it can be seen that now the number of times non-significance has to be met is no more relevant. Hence it is sufficient so choose "meet non-significance once" as the stopping criterion.

Finally we looked for the best level for the stopping criterion and an optimal number of basis functions. A low level of 10 % for the stopping criterion appeared to be sufficient. With higher levels there is no improvement. Figure 5 shows the estimated deviation in cents for all levels. Here, the chain and the number of basis functions with the smallest error are chosen. As shown before, most values are inside the interval of [-25,25] cents. Only few frequen-

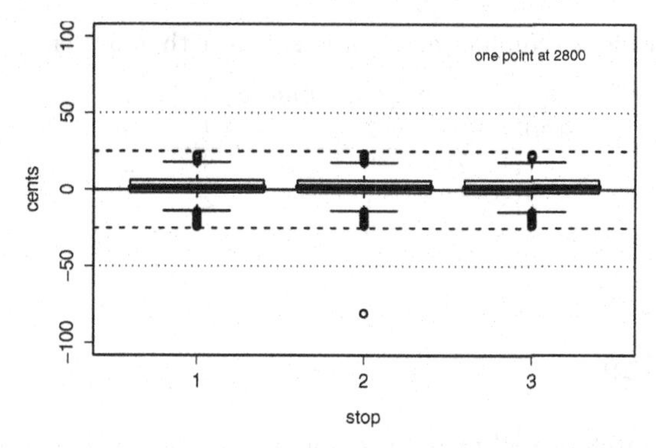

Fig. 4. Boxplots of deviations of estimated fundamental frequency, 3 basis functions, choosing chain with minimal error

cies are estimated incorrectly, i.e. are outside the interval of [-50,50] cents.

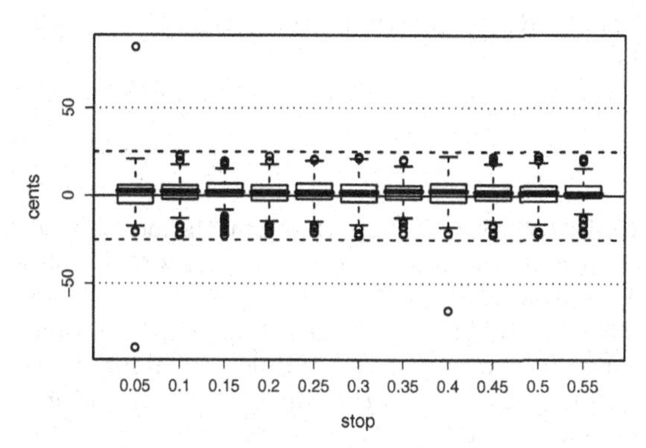

Fig. 5. Boxplots of deviations of estimated fundamental frequency, 3 basis functions, choosing chain with minimal error

Concerning the number of basis functions only one basis function, meaning constant amplitudes over time, lead to the highest number of incorrectly estimated fundamental frequencies. The modulation with time-variant amplitudes generally lead to far better results with an optimum for 3 basis functions. In Table 2 the number of deviations from the fundamental frequencies bigger than 50 cents are shown.

Table 2. Number of deviations bigger than 50 cents

	no. basis functions				
stop	1	2	3	4	5
1	21	3	0	8	2
2	11	1	1	2	1
3	8	2	1	0	1

6 Conclusion

In this paper a pitch model for monophonic sound has been introduced. The unknown parameters have been estimated with an MCMC method implemented as a stochastic optimization procedure. In a simulation study the optimal settings for some free optimization parameters have been determined. Next steps will include merging the vibrato model of Rossignol et al. (1999) with our model and the expansion to a polyphonic model.

Acknowledgements
This work has been supported by the Graduiertenkolleg "Statistical Modelling" of the German Research Foundation (DFG). We thank Uwe Ligges for his kind support.

References

DAVY, M. and GODSILL, S.J. (2002): Bayesian Harmonic Models for Musical Pitch Estimation and Analysis. *Technical Report 431*, Cambridge University Engineering Department.

LIGGES, U., WEIHS, C., HASSE-BECKER, P. (2002): Detection of Locally Stationary Segments in Time Series. In: Härdle, W., Rönz, B. (Hrsg.): *COMPSTAT 2002 - Proceedings in Computational Statistics - 15th Symposium held in Berlin, Germany.* Heidelberg: Physica, 285–290.

McGill University Master Samples. McGill University, Quebec, Canada. `http://www.music.mcgill.ca/resources/mums/html/index.htm`

ROSSIGNOL, S., RODET, X., DEPALLE, P., SOUMAGNE, J. and COLLETTE, J.-L.(1999): Vibrato: Detection, Estimation, Extraction, Modification. *Digital Audio Effects Workshop (DAFx'99).*

WEIHS, C. and LIGGES, U. (2006): Parameter Optimization in Automatic Transcription of Music. In Spiliopoulou, M., Kruse, R., Nürnberger, A., Borgelt, C. and Gaul, W. (eds.): *From Data and Information Analysis to Knowledge Engineering.* Springer, Berlin, 740 – 747.

Local Models in Register Classification by Timbre

Claus Weihs, Gero Szepannek, Uwe Ligges, Karsten Luebke, and Nils Raabe

Fachbereich Statistik, Universität Dortmund, 44221 Dortmund, Germany
e-mail: weihs@statistik.uni-dortmund.de

Abstract. Investigating a data set containing different sounds of several instruments suggests that local modelling may be a promising approach to take into account different timbre characteristics of different instruments. For this reason, some basic ideas towards such a local modelling are realized in this paper yielding a framework for further studies.

1 Introduction

Sound characteristics of orchestra instruments derived from spectra are currently a very important research topic (see, e.g., Reuter (1996, 2002)). The sound characterization of voices has, however, many more facets than for instruments because of the sound variation in dependence of technical level and emotional expression (see, e.g., Kleber (2002)).

During a former analysis of singing performances (cp. Weihs and Ligges (2003)) it appeared that register can be identified from the spectrum even after elimination of pitch information. In this paper this observation is assessed by means of a systematic analysis not only based on singing performances but also on corresponding tones of high and low pitched instruments.

For similar work on spectral analysis for discrimination of timbre of musical instruments see, for example, McAdams et al. (1999), Horner et al. (2004), and Röver et al. (2005).

The big aim of our work is to provide basic research on sound characteristics. The aim of this particular report is to investigate some basic ideas of local modelling towards achieving the goal of classification of the register of instrument or singer by timbre, i.e. by the spectrum after pitch information is eliminated. To this end, pitch independent characteristics of spectral densities of instruments and voices are generated.

Similar to the voice prints introduced in Weihs and Ligges (2003) we use masses and widths of peaks of the first 13 partials, i.e. of the fundamental and the first 12 overtones. These variables are computed for representatives of all tones involved in the classical christmas song "Tochter Zion" composed by G.F. Händel. For the singing performances the first representative of each note was chosen, for the instruments the representatives were chosen from the "McGill University Master Samples" (Opolko and Wapnick (1987), see also Section 2). In Section 3 existing models such as those used for global modelling by Weihs et al. (2005b) and Szepannek et al. (2005) are extended towards local modelling of the different instruments and singers.

2 Data

The analyses of this paper are based on time series data from an experiment with 17 singers performing the classical song "Tochter Zion" (Händel) to a standardized piano accompaniment played back by headphones (cp. Weihs et al. (2001)). The singers could choose between two accompaniment versions transposed by a third in order to take into account the different voice types (Soprano and Tenor vs. Alto and Bass). Voice and piano were recorded at different channels in CD quality, i.e. the amplitude of the corresponding vibrations was recorded with constant sampling rate 44100 hertz in 16-bit format. The audio data sets were transformed by means of a computer program into wave data sets. For time series analysis the waves were reduced to 11025 Hz (in order to restrict the number of data), and standardized to the interval $[-1, 1]$. Since the volume of recording was already controlled individually, a comparison of the absolute loudness of the different recordings was not sensible anyway. Therefore, by our standardization no additional information was lost.

Since our analyses are based on variables derived from tones corresponding to single notes, we used a suitable segmentation procedure (Ligges et al. (2002), Weihs and Ligges (2005)) in order to get data of segmented tones corresponding to notes. For further analysis the first representative of the notes with identical pitch in the song was chosen. This leads to 9 different representatives per voice in "Tochter Zion".

The notes involved in the analyzed song were also identified in the "McGill University Master Samples" in the Alto and in the Bass version for the following instruments:

Alto version (McGill notation, referred to as *'A'*): *aflute-vib, cello-bv, elecguitar1, elecguitar4, flute-flu, flute-vib, frehorn, frehorn-m, marimba, piano-ld, piano-pl, piano-sft, tromb-ten, viola-bv, viola-mv, violin-bv, violin-mv.*

Bass version (referred to as *'B'*): *bflute-flu, bflute-vib, cello-bv, elecguitar1, elecguitar4, frehorn, frehorn-m, marimba, piano-ld, piano-pl, piano-sft, tromb-ten, tromb-tenm, viola-mv.*

Thus, 17 high instruments and 14 low instruments were chosen together with 10 high female singers and 7 male. The McGill database contains almost all notes played by a huge collection of instruments, hence it was easily possible to reproduce a song at same pitch height if the instrument of interest can play the note in question.

From the periodogram corresponding to each tone corresponding to an identified note pitch independent voice characteristics are derived (cp. Weihs and Ligges (2003)). For our purpose we only use the size and the shape corresponding to the first 13 partials, i.e. to the fundamental frequency and the first 12 overtones, in a periodogram. To calculate the periodograms we used a window size of 2048 observations.

In order to measure the size of the peaks in the spectrum, the mass (weight) of the peaks of the partials are determined as the sum of the percentage shares of those parts of the corresponding peak in the spectrum which are higher than a pre-specified threshold. The shape of a peak cannot easily be described. Therefore, we only use one simple characteristic of the shape as a variable, namely the width of the peak of the partials. The width of a peak is measured by the range (i.e. maximum - minimum) of the Fourier frequencies (in Hertz) of the peak with a spectral height above a pre-specified threshold. For a discussion on choosing an appropriate measure for peak widths see Szepannek et al. (2005).

Overall, every tone is characterized by the above 26 variables which are used as a basis for classification. Mass is measured as a percentage (%), whereas width is measured in Hertz.

3 Towards local modelling

3.1 Global modelling

Global modelling of the data have been described by Weihs et al. (2005b) and Szepannek et al. (2005). In these papers, we have classified the register of different instruments and singers using common techniques, in particular the classical linear discriminant analysis (LDA). Resulting error rates based on 10-fold cross validation are 0.345 (using LDA with all 26 variables) and 0.340 (LDA with mass and width of the first 2 partials as variables after variable selection). Even using variable selection techniques and applying other classification methods, it was hardly possible to reduce the error rate below 0.34. Obviously, the resulting classification errors of a global classification are not satisfying at all.

3.2 Introduction to local modelling

In the following subsection, local models are built for all instruments separately (indexed by l, where $\{1, \ldots, L\}$ represent all instruments). For each instrument a local classification rule for the register is developed, returning posterior probabilities $p_l(k|x)$ given the instrument l and an observed tone with masses and widths x to belong to either high or low register, denoted as classes k.

Since the goal consists in classification of *any* observed note to the correct register, the instrument (or voice) playing this note can not assumed to be known in advance. For this reason, the problem consists in finding a classification rule out of all L local classification models.

How classification rules may be constructed out of all local (classification) models l, $l = 1, \ldots, L$, will be explained in the following subsections.

For each local model parameters are estimated and all observations of the complete data set were predicted with all local models. The instruments mentioned in Section 2 were collected into the following groups: *cello, elecguitar, flute, frenchhorn, marimba, piano, singer, trombone, violin-viola.*

Table 1. Best performing local models using LDA for each instrument group.

instrument	selected variables	# cases	L1o error
cello	mass02 + mass05	18	0.056
elecguitar	mass02 + mass03 + mass04 + width02	36	0.111
flute	mass06 + width04	45	0.200
frenchhorn	mass08	36	0.056
marimba	width01	18	0.111
piano	mass01 + mass07	54	0.148
singer	width01	153	0.209
trombone	mass01	27	0.074
violin, viola	mass13	45	0.111
weighted L1o error		432	0.150

All local models are generated on (local) variable subsets, found by the stepclass-algorithm using LDA (see Weihs et al. (2005a)). Variables are added stepwise to the data set until the *L1o* cross validated error rate can not be improved by more than 1%.

Table 1 shows the LDA based local models for the different instruments. Selected variables are shown in the table where *mass01* denotes the mass of the first partial (fundamental frequency) and *width02* the width of the second partial. Moreover, the number of cases used for L1o cross validation is shown as well as the L1o error rate. The L1o error weighted by the number of observations falling in the different local models is 0.15. This is a lower error bound for the error rate that would be possible in case of a perfect choice of the correct local models for a new observation of an unknown instrument.

In a second idealized step, let us assume that the correct instrument of a new observation is still known, but we do want a global variable selection used for all local models. In this case, *stepclass* variable selection results in a model including the variables *mass01*, *mass02*, *width01*, and *width02*, i.e. the four measures of the first two partials. The corresponding overall L1o error rate is 0.194.

3.3 Comparing posterior probabilities

A first straightforward intuition for the combination of local models consists in comparing the posterior probabilities of the different local models. This is possible by classifying an object to the class $\hat{k}$ with the highest posterior probability *(maximum-posterior rule)* of all local models l

$$\hat{k} = \arg\max_{k} \left(\max_{l} p_l(k|x) \right) \tag{1}$$

or to that class $\hat{k}$ possessing the largest average posterior probability averaged over all local models *(average-posterior rule)*

$$\hat{k} = \arg\max_k \sum_l p_l(k|x) \tag{2}$$

A third competitor is *majority voting* of the local classification rules

$$\hat{k} = \arg\max_k \sum_l I_{[p_l(k|x) > p_l(j|x)\, \forall\, j \neq k]}(k), \tag{3}$$

where I is the indicator function. The variable subset selection is performed for every local model separately, leading to possibly different variable subsets of the different local models (= instruments).

Using the 'maximum-posterior' classification rule (equation 1) results in an error rate of 0.479. Both averaging the local posterior probabilities and majority voting lead to an improvement of the total error rate to the same value of 0.391. The obtained error rate is worse compared to global modelling as described in the previous subsection.

Surprisingly, using the maximum-posterior classification rule (see equation 1) more than 50% of the chosen local models are of class 'marimba'. On the other hand, the singer's local model is never chosen. Ignoring the possibly misleading local marimba models for the analysis can slightly but not drastically improve the total error rates up to 0.42 (0.365 for averaging and 0.388 for majority voting).

3.4 Comparing densities

In some situation an observation may have a large posterior probability of class k close to one in some local model l even if its density $f_l(x|k)$ given this class (here: the register) and the instrument is very small.

Assume for simplicity equal prior probabilities $p_l(k) = p_l(1)\, \forall\, k, l$. Then,

$$p_l(k|x) = \frac{f_l(k|x)}{\sum_k f_l(k|x)} = \frac{f_l(x,k)}{f_l(x) p_l(1)} \frac{p_l(1) f_l(x)}{\sum_k f_l(x,k)} = \frac{f_l(x|k)}{\sum_k f_l(x|k)}.$$

This means $p_l(k|x) \sim 1$ if $f_l(x|k) >> f_l(x|j),\ \forall\, j \neq k$. For some other class $j_1 \neq k$ both inequalities $f_m(x|j_1) >> f_l(x|k)$ and $f_m(x|j_1) > f_m(x|k)$ may still hold for a different local model $m \neq l$ even if the posterior probability of this class in model m is smaller than those of class k in model l.

For this reason, densities may be better to be compared rather than posterior probabilities.

Classification rules may then be derived from *(maximum-density rule)*

$$\hat{k} = \arg\max_k \left(\max_l f_l(x|k)\right) \tag{4}$$

or *(average-density rule)*

$$\hat{k} = \arg\max_k \sum_l f_l(x|k). \tag{5}$$

Table 2. Frequency of chosen local models at maximum-density rule.

	chosen local models								
instrument	cel.	el.guit.	flute	french.	marim.	piano	sing.	tromb.	viol.
cello	0	0	2	0	0	1	2	2	11
elecguitar	0	3	9	0	0	2	7	7	8
flute	1	3	14	0	1	6	8	0	12
frenchhorn	0	2	4	0	0	0	0	16	14
marimba	3	2	6	0	6	1	0	0	0
piano	0	4	17	1	0	3	4	6	19
singer	1	6	40	4	0	21	29	24	28
trombone	0	1	5	0	0	2	2	6	11
violin-viola	0	0	11	0	1	0	4	5	24

Comparing the densities of the different local models is questionable if they are built on different variables. Therefore in this situation the variable selection was performed simultaneously for all local models in common (see Section 3.2).

Classifying according to the local model with maximum density (see equation 4) yields a total L1o error rate of 0.354. Table 2 shows that in many cases the wrong local model is chosen. Averaging over all local model densities results in an error rate of 0.301.

Note that comparing posteriors and densities yields in identical results in the case of known local models, hence we refer to Section 3.2 for lower error bounds (0.194).

Classifying on the whole data set of all 26 variables without any variable selection here leads to error rates of 0.368 (maximum-density rule, equation 4) or 0.366 (averaging, equation 5).

3.5 Global weighting of local models

Simply comparing posterior probabilities or densities of local models may work well if all instruments have (at least nearly) the same probability $\pi(l)$ to appear. But these measures do not take into account that on the other hand the probability of an observed object to belong to this local partition (corresponding to a certain local model) of the population can be very small. Therefore a weighting of the posterior probabilities is done in this attempt by performing a double classification. Besides predicting by the local models, a *global step* is added that classifies an observation into one of the local models, returning posterior probabilities $\pi(l)$ that the observed tone comes from instrument l.

The result can be used to choose the local model *(global model choice rule)*:

$$\hat{k} = \arg\max_k \, p_{\hat{l}}(k|x), \tag{6}$$

$$\hat{l} = \arg\max_l \, \pi(l).$$

Table 3. Summary of error rates for the various presented methods.

method	global model	idealized min error individual	idealized min error overall	local models according to equation number (1)	(2)	(3)	(4)	(5)	(6)	(7)
error	0.340	0.150	0.194	0.479	0.391	0.391	0.354	0.301	0.285	0.269

The local results (in form of local posterior probabilities) also can be weighted according to the posterior probabilities of the global modelling step *(global model weighting rule)*:

$$\hat{k} = \arg\max_k \sum_l p_l(k|x)\pi(l). \tag{7}$$

Using LDA as a classification method in both the global and the local step, and using variable selection in the global step (*width02*, *mass01*) as well as in the local step (*mass01*, *mass02*, *width01*, and *width02*) this led to an overall error rate of 0.285. But the error in choosing the right local model is 0.564. Weighting the local posteriors by the global posteriors reduces the error to be 0.269.

4 Summary

Classification of register in music is performed using timbre characteristics of the first 13 harmonics. Different aspects are shown, how local classification rules can replace global classification rules, some of them yielding ameliorations of the misclassification rate.

Table 3 shows a summary of the misclassification rates. While the global model results in an error of 0.340, lower error bounds are given for local models — assuming the correct local model is known a-priori. One lower error bound (0.150) is based on variable selection in each local model and is comparable with models 1–3. These models cannot even outperform the global model. The other lower error bound (0.194) is based on the same variable selection for all local models and is comparable with models 4–7. Using the model described in equation (7), we have considerably improved the error (0.269) compared to the global model, but there is still some space for improvements to shorten the gap to the lower bound.

The presented proceeding can be understood as some basic work allowing a more comprehensible and more effective classification of musical data by issues of local modelling. One topic of further research may take into account that variable subsets of different local models may differ. At the moment, comparison of such local models is difficult.

Acknowledgment. This work has been supported by the Collaborative Research Center 'Reduction of Complexity in Multivariate Data Structures' (SFB 475) of the German Research Foundation (DFG).

References

McADAMS, S., BEAUCHAMP, J. and MENEGUZZI, S. (1999): Discrimination of Musical Instrument Sounds Resynthesized with Simplified Spectrotemporal Parameters. *Journal of the Acoustical Society of America*, 105 (2), 882–897.

BROCKWELL, P.J. and DAVIS, R.A. (1991): *Time Series: Theory and Methods.* Springer, New York.

HORNER, A., BEAUCHAMP, J. and SO, R. (2004): A Search for Best Error Metrics to Predict Discrimination of Original and Spectrally Altered Musical Instrument Sounds. In: *Proceedings of the 2004 International Computer Music Conference*, Miami, 9–16.

KLEBER, B. (2002): *Evaluation von Stimmqualität in westlichem, klassischen Gesang.* Diploma Thesis, Fachbereich Psychologie, Universität Konstanz, Germany

LIGGES, U., WEIHS, C. and HASSE-BECKER, P. (2002): Detection of Locally Stationary Segments in Time Series. In: W. Härdle and B. Rönz (Eds.): *COMPSTAT 2002 - Proceedings in Computational Statistics - 15th Symposium held in Berlin, Germany.* Physika, Heidelberg, 285–290.

OPOLKO, F. and WAPNICK, J. (1987): McGill University Master Samples. McGill University, Quebec, Canada. URL: `http://www.music.mcgill.ca/resources/mums/html/index.htm`

REUTER, C. (1996): *Die auditive Diskrimination von Orchesterinstrumenten - Verschmelzung und Heraushörbarkeit von Instrumentalklangfarben im Ensemblespiel.* Peter Lang, Frankfurt/M.

REUTER, C. (2002): *Klangfarbe und Instrumentation - Geschichte - Ursachen - Wirkung.* Peter Lang, Frankfurt/M.

RÖVER, C., KLEFENZ, F. and WEIHS, C. (2005): Identification of Musical Instruments by Means of the Hough-Transformation. C. Weihs and W. Gaul (Eds.): *Classification: The Ubiquitous Challenge*, Springer, Berlin, 608–615.

SZEPANNEK, G., LIGGES, U., LUEBKE, K., RAABE, N. and WEIHS, C. (2005): Local Models in Register Classification by Timbre. *Technical Report 47/2005, SFB 475, Universität Dortmund.*

WEIHS, C., BERGHOFF, S., HASSE-BECKER, P. and LIGGES, U. (2001): Assessment of Purity of Intonation in Singing Presentations by Discriminant Analysis. In: J. Kunert and G. Trenkler (Eds.): *Mathematical Statistics and Biometrical Applications.* Josef Eul, Köln, 395–410.

WEIHS, C. and LIGGES, U. (2003): Voice Prints as a Tool for Automatic Classification of Vocal Performance. In: R. Kopiez, A.C. Lehmann, I. Wolther and C. Wolf (Eds.): *Proceedings of the 5th Triennial ESCOM Conference.* Hanover University of Music and Drama, Germany, 8-13 September 2003, 332–335.

WEIHS, C. and LIGGES, U. (2005): From Local to Global Analysis of Music Time Series. In: K. Morik, J.-F. Boulicaut and A. Siebes (Eds.): *Local Pattern Detection, Lecture Notes in Artificial Intelligence, 3539*, Springer, Berlin, 217–231.

WEIHS, C., LIGGES, U., LUEBKE, K. and RAABE, N. (2005a): klaR Analyzing German Business Cycles. In: D. Baier, R. Becker and L. Schmidt-Thieme (Eds.): *Data Analysis and Decision Support*, Springer, Berlin, 335–343.

WEIHS, C., REUTER, C. and LIGGES, U. (2005b): Register Classification by Timbre. In: C. Weihs and W. Gaul (Eds.): *Classification: The Ubiquitous Challenge*, Springer, Berlin, 624–631.

Part VIII

Gene and Microarray Analysis

Improving the Performance of Principal Components for Classification of Gene Expression Data Through Feature Selection

Edgar Acuña and Jaime Porras

Department of Mathematics, University of Puerto Rico at Mayaguez, Mayaguez, PR 00680, USA

Abstract. The gene expression data is characterized by its considerably great amount of features in comparison to the number of observations. The direct use of traditional statistics techniques of supervised classification can give poor results in gene expression data. Therefore, dimension reduction is recommendable prior to the application of a classifier. In this work, we propose a method that combines two types of dimension reduction techniques: feature selection and feature extraction. First, one of the following feature selection procedures: a univariate ranking based on the Kruskal-Wallis statistic test, the Relief, and recursive feature elimination (RFE) is applied on the dataset. After that, principal components are formed with the selected features. Experiments carried out on eight gene expression datasets using three classifiers: logistic regression, k-nn and rpart, gave good results for the proposed method.

1 Introduction

Some of the classical classification methods, like k-nearest-neighbors (k-nn) classifiers, do not require explicitly n(observations)$>$ p(features) but give poor classification accuracy in practice when the number of irrelevant features(variables) is too large, as in gene expression data from microarray experiments. Other methods, like classical discriminant analysis, can not be applied if $n < p$. There are three main ways to handle high-dimensional data in a supervised classification framework: i) Select a subset of relevant features and apply a classical classification method (e.g. linear discriminant analysis, decision trees, k-nn, Support Vector Machine (SVM), etc.) on this small subset of features. For gene expression data, this approach is often referred as gene selection, feature selection, or subset selection. ii) Use a dimension reduction technique, where either linear or non-linear principal components of the features are created. These components summarize the data as well as possible in certain way. The new components are then used as predictor variables for a classical classification method. Principal component analysis (PCA) and Partial least squares (PLS) are two examples of these techniques. iii) Use of a regularization method, also known as penalty-based or feature weighting method. These methods constrain the magnitudes of the parameters by assigning them a degree of relevance during the learning process.

Hence, regularization methods are an indirect way to select features. Penalized Logistic Regression is an example of regularization. The three approaches can be combined. For instance, Dettling and Buhlmann (2003) use a decision tree classifier after feature selection, Nguyen and Rocke (2002) perform PLS after feature selection, and Zhu and Hastie (2004) carried out penalized logistic regression after applying two feature selection methods: univariate ranking and recursive feature elimination. In this paper, we will combine three feature selection procedures with principal component analysis (PCA), perhaps the best known dimension reduction technique. We will show empirically that this combination gives very good results and it improves the predictive capabilities of PCA for supervised classification. This idea has also been used by Bair et al (2004) in the context of regression and survival analysis, and was called by them Supervised Principal Components. This paper is organized as follows. In section 2, a description of the feature selection methods is given. In section 3, we discuss the algorithms used for combining feature selection and principal components. In section 4, experimental results using real data are presented, and finally conclusions and future work are given in section 5.

2 Feature selection methods

Various feature selection schemes have been applied to gene expression data with a double purpose: i)As a preliminary step before classification, because the chosen classification method works only with a small subset of variables. ii)Biologists are very interested in identify genes which are associated with specific diseases. Guyon and Eliseeff (2003) classified feature selection methods into three distinct groups: filters, wrappers, and embedded. In a wrapper method, the optimality criterion of a each subset of features is based on the accuracy of decision functions built using only these features. Wrappers have two major drawbacks. First, they are often computationally intensive and difficult to set up than filter methods. Second, they generally suffer from overfitting. In this paper, we use two filter methods: The first one is a univariate ranking method based on the Kruskal-Wallis test, and the second one it is called the Relief. We have also considered a wrapper selection method called Recursive Feature Elimination (RFE) which uses SVM as a classifier. Next we will give a brief description of the methods used in this paper.

2.1 Univariate ranking methods

In these methods each feature is taken individually and a relevance score measuring the discriminating power of the feature is computed. The features are then ranked according to their score. One can choose to select only the p_1 top-ranking features (where $p_1 < p$) or the features whose score exceeds a given threshold. One of the first relevance score proposed was a variant of the F-statistic (Golub et al, 1999). The BSS/WSS ratio used by Dudoit et al (2002) equals the F-statistic up to a constant. For two classes problems, the t-statistic can be used ($t^2 = F$). Since microarray data contain a lot of outliers and few samples, some authors (e.g. Dettling and Buhlmann (2003)) prefer to use a more robust statistic such as Wilcoxon's rank sum statistic for the case $K = 2$. For multi-categorical responses ($K > 2$), one can use the Kruskal-Wallis test statistic. Considering that for each feature we have K independent samples, corresponding to the K classes, each of them of size n_k ($k = 1, \ldots, K$), then the values taken in each feature are ordered in increasing order and a rank is given to each of them. Let R_k be the sum of the ranks for the given feature in the k-th class. The Kruskal-Wallis statistic test will be given by $H = \frac{12}{n(n+1)} \sum_{k=1}^{K} \frac{R_k^2}{n_k} - 3(n+1)$. In case of ties an adjustment is applied. We must assume that the values of the features are independents among the classes. Also the measurement scale should be at least ordinal. The major drawback of univariate ranking methods is that correlations and interactions among features are omitted. In some cases, the subset of the top-ranking variables is not the best subset according to classification accuracy.

2.2 The Relief method for feature selection

The idea of this filter method is to choose among the features which distinguish most between classes. In the Relief the features are considered simultaneously. Initially all the p features of the dataset D have relevance weight $W_j = 0$ ($j = 1, \ldots, p$). Then at each step of an iterative process an instance x is chosen randomly from D and the weights are updated according to the distance of x to its *NearHit* and *Nearmiss.* The *Nearmiss* is the instance in D closest to x but that belongs to different class. The updating formula of W_j is given by $W_j = W_j - diff(x_j, Nearhit_j)^2 + diff(x_j, Nearmiss_j)^2$, where x_j is the $j - th$ component of x, and the function $diff$ computes the distance between the values of a feature for two given instances. For nominal and binary features, $diff$ is either 1 (the values are different) or 0 (the values are equal). For ordinal and continuous features, $diff$ is the difference of the values of the feature for the two instances normalized by the range of the feature. The process is repeated M times, usually M is equal to the size of D. At the end the best subset is the one that includes the features with relevance weight greater than a threshold fixed beforehand. For more details, see Acuna, (2003). Computation of the Relief was carried out using the R library *dprep* (Acuna and Rodriguez, 2005).

2.3 Recursive feature elimination (RFE)

RFE is a wrapper method described in detail in Guyon and Elisseff (2003). It is based on the recursive elimination of features. In each step of an iterative process a classifier (SVM) is used with all current features, a ranking criterion is calculated for each feature, and the feature with the smallest criterion value is eliminated. A ranking criterion commonly used is defined by $\Delta P_j = \frac{b_j}{2}\{\frac{\delta^2 P}{\delta b_j^2}\}$ where, P is a loss function computed on the training dataset and b_j is the corresponding coefficient of the j-th feature in the model. The sensitivity of P for the j-th feature is approximated by ΔP_j. For the SVM classifier and the quadratic loss function $P = ||y - b^T x||^2$, we get that $\Delta P_j = b_j^2 ||x_j||^2$. Assuming that the features have similar range then $\Delta P_j = b_j^2$ is frequently used. Due to computational reasons it might be more efficient to eliminate at the same time a large number of features, but there is a risk to degrade the performance of the classifier. In this paper, we have selected features by the RFE method using the R library RFE, developed by Ambroise and McLachlan (2002).

3 Combining feature selection and PCA

Let X be an $n \times p$ matrix, and Y the class vector, where the columns (features) of X have been centered to have mean zero. The singular value decomposition of X is given by $X = UDV^T$ where U,D,V are of order $n \times m$, $m \times m$ and $m \times p$ respectively, and $m = min(n, p)$ is the rank of X. The diagonal matrix D contains the singular values d_j of X. The matrix $Z = UD$ is called the principal components scores matrix, and $C = V^T$ is the loadings matrix. Using schemes similar to Nguyen and Rocke (2002) but with different misclassification error estimation, we are proposing two algorithms to combine feature selection and principal components.

First algorithm Repeat r times the following steps:
1. Apply a feature selection method to the whole dataset to find out the p_1 features that discriminate best the classes. This step will produce a reduced matrix X^R or order $n \times p_1$.where $p_1 < p$
2. Divide X^R and the class vector Y in a learning sample L formed by X_L^R of order $n_L \times p_1$ and $Y(n_L \times 1)$, where $n_L = 2n/3$, and a test sample T formed by the remaining $n/3$ observations. The splitting is done in such way that in each class of the learning and test sample the proportion $2/3$ and $1/3$ holds.
3. Standardize the learning sample and test sample using in both cases the mean and the standard deviation of each column from the learning sample L. Let X_L^{R*} and X_T^{R*} be the standardized learning sample and standardized test sample respectively.
4. Apply principal component analysis to X_L^{R*} to obtain the transformed data matrix Z_L (scores) of order $n_L \times m$, where m is the number of components to be used, and the loadings matrix C of order $p_1 \times m$.
5. Apply a classifier to Z_L and Y_L to find out the optimum number of principal components based on the misclassification error rate. This number of components will be used on the test sample.
6. Find the scores matrix $Z_T = X_T^{R*} C$.
7. Apply a classifier to Z_T and Y_T to obtain the misclassification error on the

Table 1. Summary of datasets used in the experiments.

Datasets	Genes	classes	Observations per class	Reference
Colon	2000	2	40 Tumor, 22 Normal	Alon et al, 1999
Leukemia	3571	2	47 ALL, 25 AML	Golub et al., 1999
Prostate	6033	2	50 Normal, 52 prostate	Singh et al, 2002
Carcinoma	7463	2	18 Normal, 18 carcinomas	Notterman et al, 2001
BRCA	3207	3	7 sporadic, 7 BRCA1, 8 BRCA2	Hedenfalk et al. 2001
Lymphoma	4026	3	42 DLBCL, 9 FL, 11 CLL	Alizadeh et al, 2001
SRBCT	2308	4	23 EWS, 20 BL, 12 NB, 8 RMS	Kahn et al, 2001
Brain	5597	5	10 medulloblastomas, 10 ATRT, 10 gliomas, 8 PNET and 4 human cerebella	Pomeroy et al., 2002

test sample.
In this work we have chosen $p_1 = 100$, and the number m of components was varied from 1 to 5. Also, the number of repetitions was taken as $r = 50$. These repetitions are done to reduce the variability of misclassification error, since in each of them different learning and test are obtained.

Second algorithm It is similar to the First algorithm, but interchanging the first two steps. Thus,
1. Divide X and the vector class Y in a learning sample L formed by X_L of order $n_L \times p$ and $Y(n_L \times 1)$, where $n_L = 2n/3$, and a test sample T formed by the remaining $n/3$ observations. The splitting is done in such way that in each class of the learning and test sample the proportion $2/3$ and $1/3$ holds.
2. Apply a feature selection method to L to find out the p_1 features that discriminate best the classes. In this work we have chosen $p_1 = 100$. This step will produce a reduced matrix X^R or order $n \times p_1$, where $p_1 < p$.
Then, the steps 3-7 from the first algorithm are applied to the matrix X^R. As in the first algorithm, the whole procedure is repeated r times. In this paper we used $r = 50$.

4 Experimental methodology

The two algorithms described in Section 3 have been applied to eight well known gene expression datasets. Table 1 shows the number of genes, the number of classes, and the number of observations per class for each dataset. The table also includes the main reference for each dataset.

Most of these datasets have been preprocessed through thresholding and filtering.

In this paper, we have used three classifiers: i) Rpart (it stands for recursive partition) which is a decision tree classifier, ii) the k-nn classifier, and iii) the multinomial logistic regression. All these classifiers are available in libraries of the R statistical software. The first one is available in the *rpart*,

Table 2. Misclassification error rates for the three classifiers using up to three principal components without feature selection.

Datasets	Rpart			3-nn			LogR		
	1PC	2PC	3PC	1PC	2PC	3C	1PC	2PC	3PC
Colon	33.2	31.0	22.4	22.8	20.6	18.6	33.8	31.6	23.9
Leukemia	18.0	10.5	12.5	16.5	4.0	4.2	21.0	4.7	0.9
Prostate	32.1	29.5	17.6	22.1	19.6	15.6	42.3	42.7	18.7
Carcinoma	52.5	47.5	49.5	29.0	4.8	9.6	38.7	0.3	0.0
BRCA	57.1	57.1	57.1	26.6	29.7	32.0	28.6	10.9	0.9
Lymphoma	15.1	15.2	14.9	6.4	2.9	1.6	5.9	0.6	0.0
SRBCT	53.5	51.1	48.9	39.9	31.4	29.6	59.6	42.6	27.2
Brain	77.5	76.7	78.7	39.0	34.4	30.1	41.0	18.6	3.3

the second in the *class* library, and the last one in the *nnet* library. We have considered $k = 3$ neighbors in our experiments.

4.1 Misclassification error estimation

The error estimation is computed by repeating N times the random splitting of the whole dataset into a training dataset and a test dataset, followed by the construction of the classifier using the training set, and computing then the misclassification error on the test sample. The sample mean of the N errors estimates will be the overall misclassification error. Increasing N decreases the variance of the sample mean. Decreasing the ratio between the size of the learning set and the size of the whole dataset generally increases the mean error rate. Increasing the ratio increases the correlation between the estimated error rates obtained with the N partitions. Common values for the ratio are 2/3, 0.7 and 9/10. Dudoit et al. (2002) recommended a ratio of 2/3. According to Braga-Neto and Dougherty (2004) the training-test splitting should be preferred to k-fold cross-validation for gene expression data. In this work, we have used $N = 50$ and ratio of 2/3.

4.2 Results

In table 2, we report misclassification error for the three classifiers based on principal components without performing feature selection. We can see that Rpart does not perform well in classifying gene expression data, specially when there are more than two classes. The 3-nn classifier seems to be the best classifier based on one principal component, but logistic regression tends to improve quickly, and when 3 components are used it becomes the best classifier. From tables 3 and 4 we can see clearly that feature selection boost the performance of principal components for supervised classification. The effect is more evident when more components are used. Also, the first algorithm seems to be better than the second algorithm since on average yields lower misclassification error. Furthermore, its computation is faster.

Table 3. Misclassification error rate for the three classifiers using the first algorithm. The optimal number of components appears between brackets.

Datasets	KW			RFE			Relief		
	Rpart	3-nn	LogR	Rpart	3-nn	LogR	Rpart	3-nn	logR
Colon	8.9 [4]	11.1 [3]	5.4 [4]	8.4 [2]	5.9 [4]	1.4 [3]	11.2 [5]	10.6 [2]	8.3 [3]
Leukemia	0.6 [1]	1.0 [1]	0.1 [2]	0.0 [1]	0.0 [1]	0.0 [1]	0.0 [1]	0.2 [1]	0.0 [1]
Prostate	9.3 [1]	6.3 [3]	3.1 [4]	1.6 [3]	0.8 [3]	0.0 [3]	3.0 [2]	3.2 [2]	3.7 [2]
Carcinoma	51.7 [2]	1.2 [3]	0.0 [1]	50.8 [1]	0.0 [1]	0.0 [1]	51.3 [3]	0.0 [1]	0.0 [1]
BRCA	57.1 [1]	12.2 [2]	0.0 [2]	57.1 [1]	3.1 [2]	0.0 [2]	57.1 [1]	16.0 [3]	0.0 [3]
Lymphoma	14.2 [2]	0.9 [5]	0.0 [3]	14.2 [2]	0.0 [1]	0.0 [1]	14.2 [2]	0.0 [2]	0.0 [1]
SRBCT	31.9 [1]	0.1 [4]	0.0 [2]	31.8 [4]	0.1 [4]	0.0 [2]	33.0 [2]	0.4 [3]	0.0 [3]
Brain	76.6 [2]	13.5 [5]	0.0 [3]	77.3 [3]	5.3 [5]	0.0 [2]	78.7 [4]	19.0 [4]	0.0 [4]

Table 4. Misclassification error rates for the three classifiers using the second algorithm. The optimal number of components appears between brackets.

Datasets	KW			RFE			Relief		
	Rpart	3-nn	LogR	Rpart	3-nn	LogR	Rpart	3-nn	logR
Colon	12.6[3]	12.4[2]	12.8[1]	14.1[4]	14.4[2]	1.7[1]	10.9[4]	12.6[2]	11.7[1]
Leukemia	1.2[4]	2.3[3]	0.0[3]	1.0[1]	1.3[1]	0.2[3]	0.8[3]	2.3[2]	0.0[4]
Prostate	11.2[4]	8.0[5]	6.7[3]	10.0[4]	6.0[2]	5.1[3]	8.1[4]	7.3[2]	6.9[3]
Carcinoma	52.1[3]	2.6[1]	0.0[1]	48.3[5]	2.8[1]	0.0[2]	48.5[1]	3.7[1]	0.0[2]
BRCA	57.1[3]	25.7[2]	3.6[3]	57.1[4]	23.5[3]	2.1[3]	57.1[2]	29.2[2]	1.4[3]
Lymphoma	14.2[2]	3.3[5]	0.7[3]	14.3[2]	0.2[2]	0.0[2]	14.3[4]	2.7[1]	0.0[2]
SRBCT	31.8[2]	0.9[3]	0.9[2]	32.0[4]	1.3[3]	4.5[2]	32.9[3]	0.7[4]	0.5[3]
Brain	76.9[5]	20.7[4]	1.9[3]	82.3[2]	18.8[4]	0.4[3]	75.3[1]	26.2[5]	4.6[3]

5 Conclusions

Our experiments strongly suggest that the classification accuracy of classifiers using principal components as predictors improves when feature selection is applied previously. Our conclusions are based on the 100 best features, but it should be interesting to see what happen if a smaller number of features is used. RFE seems to be a better procedure for feature selection than Relief and KW, but the difference diminishes when more components are used. The combination RFE with logistic regression gives the best results. Doing first feature selection followed by cross validation gives better results that doing feature selection within the cross validation procedure.

References

ACUNA, E. (2003), A Comparison of Filter and Wrapper Methods for Feature Selection in Supervised Classification, Proceedings of Interface Computing Science and Statistics, 35.

ACUNA, E., and RODRIGUEZ, C. (2005), Dprep: data preprocessing and visualization functions for classification. R package version 1.0. http://math.uprm.edu/edgar/dprep.html.

ALON, U., BARKAI, N., NOTTERMAN, D. A., GISH, K., et al. (1999). Broad patters of gene expression revealed by clustering analysis of tumor and normal colon tissues probed by oligonucleotide arrays. PNAS 96, 6745-6750.

ALIZADEH, A., EISEN, M., DAVIS, R., MA, C., LOSSOS, I., ROSENWALD, A., BOLDRICK, J., SABET, H., et al. (2000). Distinct types of diffuse large B-Cell-Lymphoma Identified by Gene Expression Profiling. Nature, 403, 503-511.

AMBROISE, C. and MCLACHLAN, G. (2002). Selection bias in gene extraction on the basis of microarray gene-expression data. PNAS vol. 99, 6562-6566.

BAIR, E., HASTIE, T., DEBASHIS, P., and TIBSHIRANI, R. (2004). Prediction by supervised principal components. Technical Report, Departament of Statistics, Stanford University

BRAGA-NETO, U. and DOUGHERTY, E. R. (2004). Is cross-validation valid for small-sample microarray classification?. Bioinformatics 20, 2465-2472.

DETTLING, M. and BUHLMANN, P. (2003). Boosting for Tumor Classification with Gene Expression Data. Bioinformatics, 19, 1061-1069.

DUDOIT, S., FRIDLYAND, J. and SPEED, T. (2002). Comparison of Discrimination Methods for the Classification of Tumors Using Gene Expression Data. JASA, 97, 77-87.

GOLUB, T., SLONIM, D., TAMAYO, P., HUARD, C., GASSENBEEK, M., et al. (1999). Molecular Classification of Cancer: Class Discovery and Class Prediction by Gene Expression Monitoring. Science, 286, 531-537.

GUYON, I. and ELISSEEFF, A. (2003). An introduction to Variable and Feature Selection. Journal of Machine Learning Research 3, 1157-1182.

HEDENFALK, I., DUGGAN, D., CHEN, Y., RADMACHER, M., BITTNER, M., SIMON, R., MELTZER, P., et al. (2001). Gene-expression profiles in hereditary breast cancer. New England Journal of medicine 344, 539-548.

KHAN, J., WEI, J., RINGNER, M., SAAL, L., LADANYI, M., et al. (2001). Classification and Diagnostic Prediction of Cancer Using Gene Expression Profiling and Artificial Neural Networks. Nature Medicine, 6, 673-679.

NGUYEN, D.V. and ROCKE, D. M. (2002). Multi-Class Cancer Classification via Partial Least Squares with Gene Expression Profiles Bioinformatics, 18, 1216-1226.

NOTTERMAN, D. A., ALON, U., SIERK, A. J., et al. (2001). Trancriptional gene expression profiles of colorectal adenoma, adenocarcinoma, and normal tissue examined by oligonucleotide arrays. Cancer Research 61, 3124-3130.

POMEROY, S., TAMAYO, P., GAASENBEEK, M., STURLA, L., ANGELO, M., MCLAUGHLIN, M., et al. (2002). Prediction of Central Nervous System EmbryonalTumor Outcome Based on Gene Expression. Nature, 415, 436-442.

SINGH, D., FEBBO, P., ROSS, K., JACKSON, D., MANOLA, J., LADD, C., TAMAYO, P., RENSHAW, A., et al. (2002). Gene Expression Correlates of Clinical Prostate Cancer Behavior. Cancer Cell, 1, 203-209.

A New Efficient Method for Assessing Missing Nucleotides in DNA Sequences in the Framework of a Generic Evolutionary Model

Abdoulaye Baniré Diallo[1], Vladimir Makarenkov[2], Mathieu Blanchette[1], and François-Joseph Lapointe[3]

[1] McGill Centre for Bioinformatics and School of Computer Science, McGill University 3775 University Street, Montreal, Quebec, H3A 2A7, Canada
[2] Département d'informatique, Université du Québec à Montréal, C.P. 8888, Succ. Centre-Ville, Montréal (Québec), H3C 3P8, Canada
[3] Département de sciences biologiques, Université de Montréal, C.P. 6128, Succ. Centre-Ville, Montréal (Québec), H3C 3J7, Canada

Abstract. The problem of phylogenetic inference from datasets including incomplete characters is among the most relevant issues in systematic biology. In this paper, we propose a new probabilistic method for estimating unknown nucleotides before computing evolutionary distances. It is developed in the framework of the Tamura-Nei evolutionary model (Tamura and Nei (1993)). The proposed strategy is compared, through simulations, to existing methods "Ignoring Missing Sites" (IMS) and "Proportional Distribution of Missing and Ambiguous Bases" (PDMAB) included in the PAUP package (Swofford (2001)).

1 Introduction

Incomplete datasets can arise in a variety of practical situations. For example, this is often the case in molecular biology, and more precisely in phylogenetics, where an additive tree (i.e. phylogenetic tree) represents an intuitive model of species evolution. The fear of missing data often deter systematists from including in the analysis the sites with missing characters (Sanderson et al. (1998), Wiens (1998)). Huelsenbeck (1991) and Makarenkov and Lapointe (2004) pointed out that the presence of taxa comprising big percentage of unknown nucleotides might considerably deteriorate the accuracy of the phylogenetic analysis. To avoid this, some authors proposed to exclude characters containing missing data (e.g. Hufford (1992) and Smith (1996)). In contrast, Wiens (1998) argued against excluding characters and showed a benefit of "filling the holes" in a data matrix as much as possible. The popular PAUP software (Swofford (2001)) includes two methods for computing evolutionary distances between species from incomplete sequence data. The first method, called IMS ("Ignoring missing sites"), is the most commonly used strategy. It proceeds by the elimination of incomplete sites while computing evolutionary distances. According to Wiens (2003), such an approach represents a viable solution only for long sequences because of the presence of a sufficient number of known nucleotides. The second method included in PAUP, called PDMAB ("Proportional distribution of missing and ambiguous bases"), computes evolutionary distances taking into account missing bases. In this paper we propose a new method, called PEMV ("Probabilistic estimation of missing values"), which estimates the identities of all missing bases prior to computing pairwise distances between taxa. To estimate a missing base, the new method proceeds by computing a similarity score between the sequence comprising the missing base and all other sequences. A probabilistic approach is used to determine the likelihood of an unknown base to be either A, C, G or T for DNA sequences. We show how this method can be applied in the framework of Tamura-Nei evolutionary model (Tamura and Nei (1993)). This model is considered as a further extension of the Jukes-Cantor (Jukes and Cantor (1969)), Kimura 2-parameter (Kimura, (1980)), HKY (Hasegawa et al. (1985)), and F84 (Felsenstein and Churchill (1996)) models. In the next section we introduce the new method for estimating missing entries in sequence data. Then, we discuss the results provided by the methods IMS, PDMAB and PEMV in a Monte Carlo simulation study carried out with DNA sequences of various lengths, containing different percentages of missing bases.

2 Probabilistic estimation of missing values

The new method for estimating unknown bases in nucleotide sequences, PEMV, is described here in the framework of the Tamura-Nei (Tamura and Nei (1993)) model of sequence evolution. This model assumes that the equilibrium frequencies of nucleotides (π_A, π_C, π_G and π_T) are unequal and substitutions are not equally likely. Furthermore, it allows for three types of nucleotide substitutions: from purine (A or G) to purine, from pyrimidine (C or T) to pyrimidine and from purine to pyrimidine (respectively, from pyrimidine to purine). To compute the evolutionary distance between a pair of sequences within this model, the following formula is used:

$$\begin{aligned} D = & -\frac{2\pi_A\pi_G}{\pi_R} ln\left(1 - \frac{\pi_R}{2\pi_A\pi_G}P_R - \frac{1}{2\pi_R}Q\right) \\ & -\frac{2\pi_C\pi_T}{\pi_Y} ln\left(1 - \frac{\pi_Y}{2\pi_C\pi_T}P_Y - \frac{1}{2\pi_Y}Q\right) \\ & -\left(\pi_R\pi_Y - \frac{\pi_A\pi_G\pi_Y}{\pi_R} - \frac{\pi_C\pi_T\pi_R}{\pi_Y}\right) ln\left(1 - \frac{1}{2\pi_R\pi_Y}Q\right), \end{aligned} \tag{1}$$

where P_R,P_Y and Q are respectively the transitional difference between purines, the transitional difference between pyrimidines and the transversional difference involving pyrimidine and purine; π_R and π_Y are respectively the frequencies of purines ($\pi_A + \pi_G$) and pyrimidines ($\pi_C + \pi_T$).

Assume that **C** is a matrix of aligned sequences, the base k,denoted as X, in the sequence i is missing and X is either A, C, G or T. To compute the distance between the sequence i and all other considered sequences, PEMV estimates, using Equation 2 below, the probabilities $P_{ik}(X)$, to have the nucleotide X at site k of the sequence i. The probability that an unknown base at site k of the sequence i is a specific nucleotide depends on the number of sequences having this nucleotide at this site as well as on the distance (computed ignoring the missing sites) between i and all other considered sequences having known nucleotides at site k. First, we calculate the similarity score δ between all observed sequences while ignoring missing data. For any pair of sequences, this score is equal to the number of matches between homologous nucleotides divided by the number of comparable sites.

$$P_{ik}(X) = \frac{1}{N_k}\left(\sum_{\{j|C_{jk}=X\}} \delta_{ij} + \frac{1}{3}\sum_{\{j|C_{jk}\neq X\}} (1-\delta_{ij})\right), \tag{2}$$

where N_k is the number of known bases at site k (i.e. column k) of the considered aligned sequences, and δ_{ij} is the similarity score between the sequences i and j computed ignoring missing sites. The following theorem characterizing the probabilities $P_{ik}(A)$, $P_{ik}(C)$, $P_{ik}(G)$ and $P_{ik}(T)$, can be stated:

Theorem 1. *For any sequence i, and any site k of the matrix C, such that C_{ik} is a missing nucleotide, the following equality holds: $P_{ik}(A) + P_{ik}(C) + P_{ik}(G) + P_{ik}(T) = 1$.*

Due to space limitation the proof of this theorem is not presented here.

Once the different probabilities P_{ik} are obtained, we can compute for any pair of sequences i and j, the evolutionary distance using Equation 1. First, we have to calculate the nucleotide frequencies (Equation 3), the transitional differences P_R and P_Y(Equation 4), and the transversional difference Q (Equation 5). Let π_X be the new frequency of the nucleotide X:

$$\pi_X = \frac{\Lambda_X^i + \sum_{\{k|C_{ik}=?\}} P_{ik}(X) + \Lambda_X^j + \sum_{\{k|C_{jk}=?\}} P_{jk}(X)}{2L}, \tag{3}$$

where X denotes the nucleotide A, C, G or T;Λ_X^i is the number of nucleotides X in the sequence i; symbol ? represents a missing nucleotide; L is the total number of sites compared.

$$P(i,j) = \frac{P'(i,j) + \sum_{\{k|(C_{ik}=?orC_{jk}=?)\}} P'(i,j,k)}{L}, \tag{4}$$

$$Q(i,j) = \frac{Q'(i,j) + \sum_{\{k|(C_{ik}=?orC_{jk}=?)\}} Q'(i,j,k)}{L}, \tag{5}$$

where $P'(i,j)$ is the number of transitions of the given type (either purine to purine P'_R, or pyrimidine to pyrimidine P'_Y) between the sequences i and j computed ignoring missing sites; $P'(i,j,k)$ is the probability of transition of the given type between the sequences i and j at site k when the nucleotide at site

k is missing either in i or in j (e.g. if the nucleotide at site k of the sequence i is A and the corresponding nucleotide in j is missing, the probability of transition between purines is the probability that the missing base of the sequence j is G, whereas the probability of transition between pyrimidines is 0); $Q'(i,j)$ is the number of transversions between i and j computed ignoring missing sites; $Q'(i,j,k)$ is the probability of transversion between i and j at site k when the nucleotide at site k is missing either in i or in j.

When both nucleotides at site k of i and j are missing, we use similar formulas as those described in Diallo et al. (2005). It is worth noting that PEMV method can be used to compute the evolutionary distance independently of the evolutionary model (Equation 6):

$$d^*_{ik} = \frac{N^c_{ij} - N^m_{ij} + \sum_{\{k|(C_{ik}=?orC_{jk}=?)\}}(1 - P^k_{ij})}{L}, \tag{6}$$

where N^m_{ij} is the number of matches between homologous nucleotides in the sequences i and j; N^c_{ij} is the number of comparable pairs of nucleotides in i and j (i.e. when both nucleotides are known in the homologous sites of i and j); P^k_{ij} is the probability to have a pair of identical nucleotides at site k of i and j.

3 Simulation study

A Monte Carlo study has been conducted to test the ability of the new method to compute accurate distances matrices that can be used as input of distance-based methods of phylogenetic analysis. We examined how the new PEMV method performed, compared to the PAUP strategies, testing them on random phylogenetic data with different percentages of missing nucleotides. The results were obtained from simulations carried out with 1000 random binary phylogenetic trees with 16 and 24 leaves. In each case, a true tree topology, denoted T, was obtained using the random tree generation procedure proposed by Kuhner and Felsenstein (1994). The branch lengths of the true tree were computed using an exponential distribution. Following the approach of Guindon and Gascuel (2002), we added some noise to the branches of the true phylogeny to create a deviation from the molecular clock hypothesis. The source code of our tree generation program, written in C, is available at the following website: http://www.labunix.uqam.ca/~makarenv/tree_generation.cpp.

The random trees were then submitted to the SeqGen program (Rambault and Grassly (1997)) to simulate sequence evolution along their branches. We used SeqGen to obtain the aligned sequences of the length l (with 250, 500, 750, and 1000 bases) generated according to the HKY evolutionary model (Hasegwa et al. (1985)) which is a submodel of Tamura-Nei. According to Takashi and Nei (2000), the following equilibrium nucleotide frequencies were chosen: $\pi_A = 0.15$, $\pi_C = 0.35$, $\pi_G = 0.35$, and $\pi_T = 0.15$. The transition/transversion rate was set to 4. To simulate missing data in the sequences, we used one of the two strategies described by Wiens (2003). This strategy consists of the random elimination of blocks of nucleotides of different sizes. This elimination is certainly more realistic from a biological point of view. Here, we generated data with 0 to 50% of missing bases. The obtained sequences were submitted to the three methods for computing evolutionary distances. For each distance matrix provided by IMS, PDMAB and PEMV, we inferred a phylogeny T' using the BioNJ algorithm (Gascuel (1997)). The

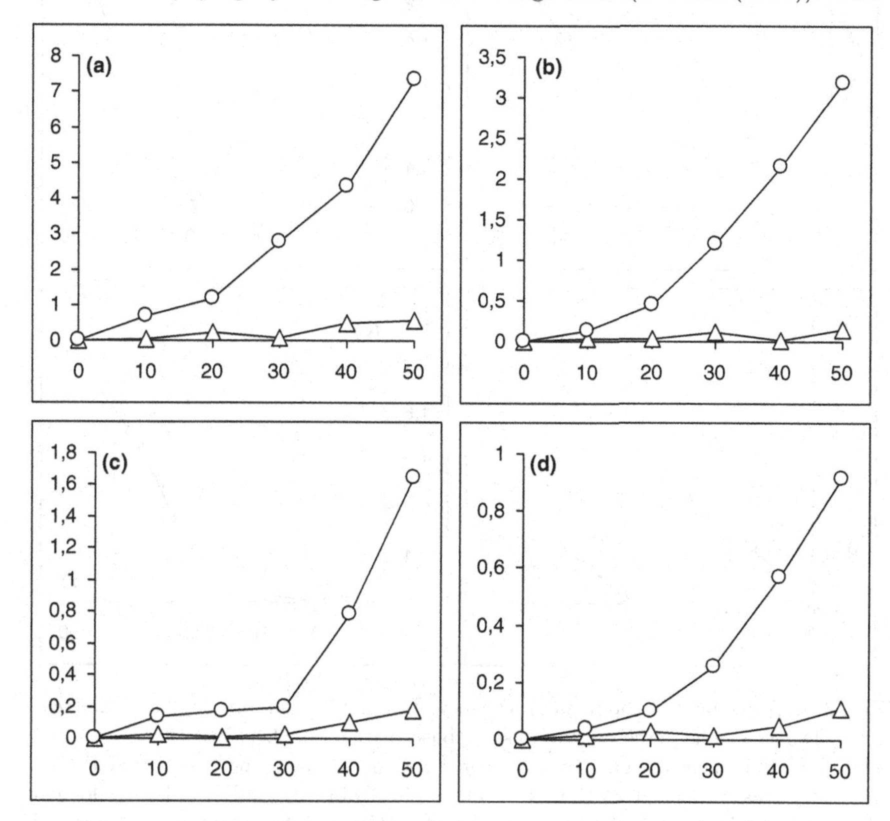

Fig. 1. Improvement in topological recovery obtained for random phylogenetic trees with 16 species. The percentage of missing bases varies from 0 to 50% (abscissa axis). The curves represent the gain (in %) against the less accurate method of PAUP. The difference was measured as the variation of the Robinson and Foulds topological distance between the less accurate method of PAUP and the most accurate method of PAUP (△) and PEMV (○).The sequences with (a) 250 bases, (b) 500 bases, (c) 750 bases, and (d) 1000 bases are represented.

phylogeny T' was then compared to the true phylogeny T using the Robinson and Foulds (1981) topological distance. The Robinson and Foulds distance between two phylogenies is the minimum number of operations, consisting of merging and splitting internal nodes, which are necessary to transform one tree into another. This distance is reported as percentage of its maximum value ($2n$-6 for a phylogeny with n leaves). The lower this value is, the closer the obtained tree T' to the true tree T.

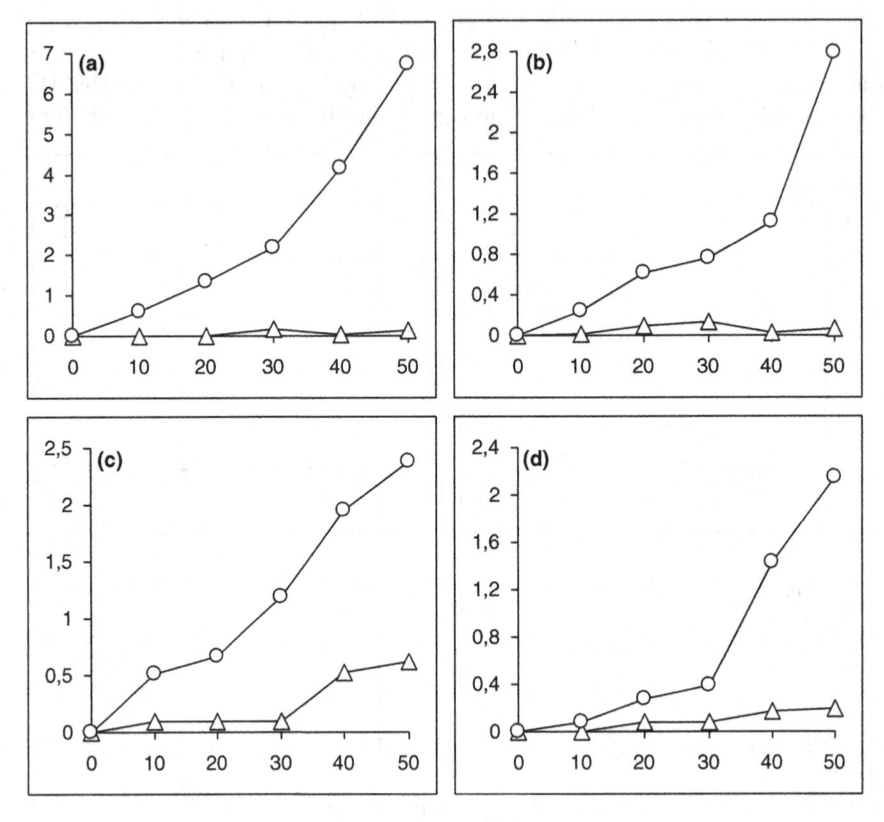

Fig. 2. Improvement in topological recovery obtained for random phylogenetic trees with 24 species. The percentage of missing bases varies from 0 to 50% (abscissa axis). The curves represent the gain (in %) against the less accurate method of PAUP. The difference was measured as the variation of the Robinson and Foulds topological distance between the less accurate method of PAUP and the most accurate method of PAUP (△) and PEMV (○). The sequences with (a) 250 bases, (b) 500 bases, (c) 750 bases, and (d) 1000 bases are represented.

For each dataset, we tested the performance of the three methods depending on the sequence length. Figures 1 and 2 present the results given by the three competing methods for the phylogenies with 16 and 24 leaves. First, for the phylogenies of both sizes PEMV clearly outperformed the PAUP methods (IMS and PDMAB) when the percentage of missing data was large (30% to 50%). Second, the results obtained with IMS were very similar to those given by PDMAB. Third, the gain obtained by our method was decreasing while the sequences length was increasing. At the same time, the following trend can be observed: the impact of missing data decreases when sequence length increases. Note that the same tendency has been pointed out by Wiens (2003).

4 Conclusion

The PEMV technique introduced in this article is a new efficient method that can be applied to infer phylogenies from nucleotide sequences comprising missing data. The simulations conducted in this study demonstrated the usefulness of PEMV in estimating missing bases prior to phylogenetic reconstruction. Tested in the framework of the Tamura-Nei model (Tamura and Nei (1993)), the PEMV method provided very promising results. The deletion of missing sites, as it is done in the IMS method, or their estimation using PDMAB (two methods available in PAUP) can remove important features of the data at hand. In this paper, we presented PEMV in the framework of the Tamura-Nei (Tamura and Nei (1993)) model which can be viewed as a generalization of the popular F84 (Felsenstein and Churchill (1996) and HKY85 (Hasegawa et al. (1985)) models. It would be interesting to extend and test this probabilistic approach within Maximum Likelihood and Maximum Parsimony models. It is also important to compare the results provided by BioNJ to those obtained using other distance-based methods of phylogenetic reconstruction, as for example, NJ (Saitou and Nei (1987)), FITCH (Felsenstein (1997)) or MW (Makarenkov and Leclerc (1999)).

References

DIALLO, Ab. B., DIALLO, Al. B. and MAKARENKOV, V. (2005): Une nouvelle mthode efficace pour l'estimation des données manquantes en vue de l'inférence phylogénétique. In: *Proceeding of the 12th meeting of Société Francophone de Classification.*Montréal, Canada, *121–125.*

FELSENSTEIN, J. and CHURCHILL, G.A. (1996): A hidden Markov model approach to variation among sites in rate of evolution. *Molecular Biology Evolution, 13, 93–104.*

FELSENSTEIN, J. (1997): An alternating least squares approach to inferring phylogenies from pairwise distances. *Systematic Biology, 46, 101–111.*

GASCUEL, O. (1997): An improved version of NJ algorithm based on a simple model of sequence Data. *Molecular Biology Evolution, 14, 685–695.*

GUINDON, S. and GASCUEL, O. (2002): Efficient biased estimation of evolutionary distances when substitution rates vary across sites. *Molecular Biology Evolution, 19, 534–543.*

HUELSENBECK, J. P. (1991): When are fossils better than existent taxa in phylogenetic analysis? *Systematic Zoology, 40, 458–469.*

HASEGAWA, M., KISHINO, H. and YANO, T.(1985): Dating the humanape split by a molecular clock of mitochondrial DNA. *Journal of Molecular Evolution, 22, 160–174.*

HUFFORD, L. (1992): Rosidaea and their relationships to other nonmagnoliid dicotyledons: A phylogenetic analysis using morphological and chemical data. *Annals of the Missouri Botanical Garden, 79, 218–248.*

JUKES, T. H. and CANTOR, C. (1969): Mammalian Protein Metabolism, chapter Evolution of protein molecules. *Academic Press, New York, 21–132.*

KIMURA, M. (1980): A simple method for estimating evolutionary rate of base substitutions through comparative studies of nucleotide sequence. *Journal of Molecular Evolution, 16, 111–120.*

KUHNER, M. and FELSENSTEIN. J.: A simulation comparison of phylogeny algorithms under equal and unequal evolutionary rates. *Molecular Biology Evolution, 11, 459–468.*

MAKARENKOV, V. and LECLERC, B. (1999): An algorithm for the fitting of a phylogenetic tree according to a weighted least-squares criterion. *Journal of Classification, 16, 3–26.*

MAKARENKOV, V. and LAPOINTE, F-J. (2004): A weighted least-squares approach for inferring phylogenies from incomplete distance matrices. *Bioinformatics, 20, 2113–2121.*

RAMBAULT, A. and GRASSLY, N. (1997): SeqGen: An application for the Monte Carlo simulation of DNA sequences evolution along phylogenetic trees. *Bioinformatics, 13, 235–238.*

ROBINSON, D. and FOULDS, L. (1981): Comparison of phylogenetic trees. *Mathematical Biosciences, 53, 131–147.*

SAITOU, N. and NEI, M.(1987): The neighbor-joining method: A new method for reconstructing phylogenetic trees. *Molecular Biology Evolution, 4, 406–425.*

SANDERSON, M.J., PURVIS, A. and HENZE, C. (1998): Phylogenetic supertrees: Assembing the tree of life. *Trends in Ecology and Evolution, 13, 105–109.*

SMITH, J.F.(1997): Tribal relationships within Gesneriaceae: A cladistic analysis of morphological data. *Systematic Botanic, 21, 497–513.*

SWOFFORD, D. L. (2001): PAUP*. Phylogenetic Analysis Using Parsimony (*and Other Methods). Version 4. *Sinauer Associates*, Sunderland, Massachusetts.

TAKAHASHI, K. and NEI, M. (2000): Efficiencies of fast algorithms of phylogenetic inference under the criteria of maximum parsimony, minimum evolution, and maximum likelihood when a large number of sequences are used. *Molecular Biology and Evolution, 17, 1251–1258.*

TAMURA, N. and NEI, M. (1993): Estimation of the number of nucleotide substitutions in the control region of mitochondrial DNA in humans and chimpanzees. *Molecular Biology and Evolution, 10/3, 512–526.*

WIENS, J. J. (1998): Missing data, incomplete taxa, and phylogenetic accuracy. *Systematic Biology, 52, 528–538.*

WIENS, J. J. (2003): Does adding characters with missing data increase or decrease phylogenetic accuracy. *Systematic Biology, 47, 625–640.*

New Efficient Algorithm for Modeling Partial and Complete Gene Transfer Scenarios

Vladimir Makarenkov[1], Alix Boc[1], Charles F. Delwiche[2], Alpha Boubacar Diallo[1], and Hervé Philippe[3]

[1] Département d'informatique, Université du Québec à Montréal, C.P. 8888, Succ. Centre-Ville, Montréal (Québec), H3C 3P8, Canada,
[2] Cell Biology and Molecular Genetics, HJ Patterson Hall, Bldg. 073, University of Maryland at College Park, MD 20742-5815, USA.
[3] Département de biochimie, Faculté de Médecine, Université de Montréal, C.P. 6128, Succ. Centre-ville, Montréal, QC, H3C 3J7, Canada.

Abstract. In this article we describe a new method allowing one to predict and visualize possible horizontal gene transfer events. It relies either on a metric or topological optimization to estimate the probability of a horizontal gene transfer between any pair of edges in a species phylogeny. Species classification will be examined in the framework of the complete and partial gene transfer models.

1 Introduction

Species evolution has long been modeled using only phylogenetic trees, where each species has a unique most recent ancestor and other interspecies relationships, such as those caused by horizontal gene transfers (HGT) or hybridization, cannot be represented (Legendre and Makarenkov (2002)). HGT is a direct transfer of genetic material from one lineage to another. Bacteria and Archaea have sophisticated mechanisms for the acquisition of new genes through HGT, which may have been favored by natural selection as a more rapid mechanism of adaptation than the alteration of gene functions through numerous mutations (Doolittle (1999)). Several attempts to use network-based models to depict horizontal gene transfers can be found (see for example: Page (1994) or Charleston (1998)). Mirkin et al (1995) put forward a tree reconciliation method that combines different gene trees into a unique species phylogeny. Page and Charleston (1998) described a set of evolutionary rules that should be taken into account in HGT models. Tsirigos and Rigoutsos (2005) introduced a novel method for identifying horizontal transfers that relies on a gene's nucleotide composition and obviates the need for knowledge of codon boundaries. Lake and Rivera (2004) showed that the dynamic deletions and insertions of genes that occur during genome evolution, including those introduced by HGT, may be modeled using techniques similar to those used to model nucleotide substitutions (e.g. general Markov models). Moret et al (2004) presented an overview of the network modeling in phylogenetics. In this paper we continue the work started in Makarenkov et al (2004), where we described an HGT detection algorithm based on the least-squares optimization. To design a detection algorithm which is mathematically and biologically sound we will consider two possible approaches allowing for complete and partial gene transfer scenarios.

2 Two different ways of transferring genes

Two HGT models are considered in this study. The first model, assumes partial gene transfer. In such a model, the original species phylogeny is transformed into a connected and directed network where a pair of species can be linked by several paths (Figure 1a). The second model assumes complete transfer; the species phylogenetic tree is gradually transformed into the gene tree by adding to it an HGT in each step. During this transformation, only tree structures are considered and modified (Figure 1b).

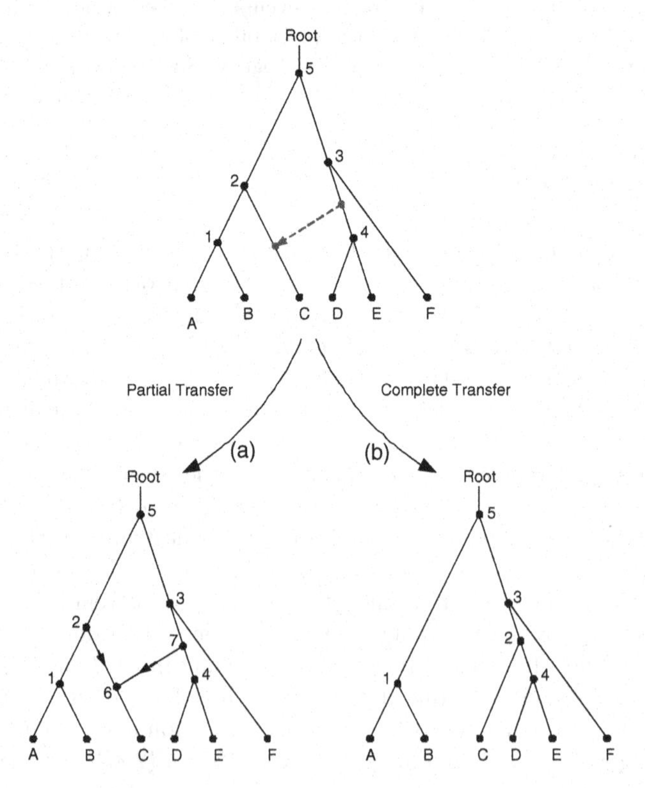

Fig. 1. Two evolutionary models, assuming that either a partial (a) or complete (b) HGT has taken place. In the first case, only a part of the gene is incorporated into the recipient genome and the tree is transformed into a directed network, whereas in the second, the entire donor gene is acquired by the host genome and the species tree is transformed into a different tree.

3 Complete gene transfer model

In this section we discuss the main features of the HGT detection algorithm in the framework of the complete gene transfer model. This model assumes that the entire transferred gene is acquired by the host (Figures 1b). If the homologous gene was present in the host genome, the transferred gene can supplant it. Two optimization criteria will be considered. The first of them is the least-squares (LS) function Q:

$$Q = \sum_{i}\sum_{j}(d(i,j) - \delta(i,j))^2, \tag{1}$$

where $d(i,j)$ is the pairwise distance between the leaves i and j in the species phylogenetic tree T and $\delta(i,j)$ the pairwise distance between i and j in the gene tree T_1. The second criterion that can be useful to assess the incongruence between the species and gene phylogenies is the Robinson and Foulds (RF) topological distance (1981). When the RF distance is considered, we can use it as an optimization criterion as follows: All possible transformations (Figure 1b) of the species tree, consisting of transferring one of its subtrees from one edge to another, are evaluated in a way that the RF distance between the transformed species tree T' and the gene tree T_1 is computed. The subtree transfer providing the minimum of the RF distance between T' and T_1 is retained as a solution. Note that the problem asking to find the minimum number of subtree transfer operations necessary to transform one tree into another has been shown to be NP-hard but approximable to within a factor of 3 (Hein et al (1996)).

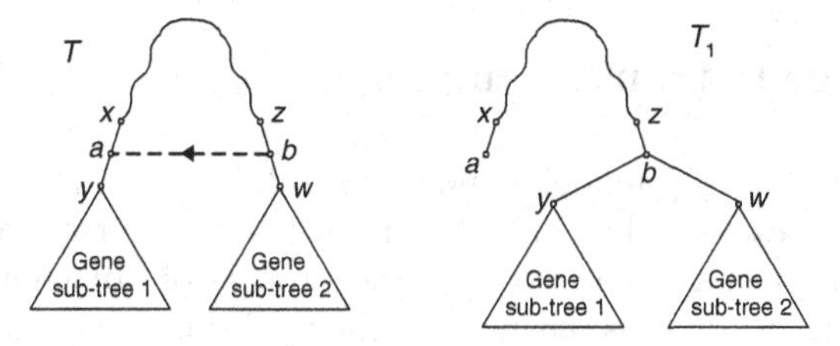

Fig. 2. Timing constraint: the transfer between the edges (z,w) and (x,y) of the species tree T can be allowed if and only if the cluster regrouping both affected subtrees is present in the gene tree T_1.

Several biological rules have to be considered in order to synchronize the way of evolution within a species phylogeny (Page and Charleston (1998)). For instance, transfers between the species of the same lineage must be prohibited. In addition, our algorithm relies on the following timing constraint: The cluster combining the subtrees rooted by the vertices y and w must be present in the gene tree T_1 in order to allow an HGT between the edges (z,w) and (x,y) of the species tree T (Figure 2). Such a constraint enables us,

first, to arrange the topological conflicts between T and T_1 that are due to the transfers between single species or their close ancestors and, second, to identify the transfers that have occurred deeper in the phylogeny. The main steps of the HGT detection algorithm are the following:
Step 0. This step consists of inferring the species and gene phylogenies denoted respectively T and T_1 and labeled according to the same set X of n taxa (e.g. species). Both species and gene trees should be explicitly rooted. If the topologies of T and T_1 are identical, we conclude that HGTs are not required to explain the data. If not, either the RF difference between them can be used as a phylogeny transformation index, or the gene tree T_1 can be mapped into the species tree T fitting by least-squares the edge lengths of T to the pairwise distances in T_1 (see Makarenkov and Leclerc (1999)).
Step 1. The goal of this step is to obtain an ordered list L of all possible gene transfer connections between pairs of edges in T. This list will comprise all different directed connections (i.e. HGTs) between pairs of edges in T except the connections between adjacent edges and those violating the evolutionary constraints. Each entry of L is associated with the value of the gain in fit, computed using either LS function or RF distance, found after the addition of the corresponding HGT connection. The computation of the ordered list L requires $O(n^4)$ operations for a phylogenetic tree with n leaves. The first entry of L is then added to the species tree T.
Steps 2 ... k. In the step k, a new tree topology is examined to determine the next transfer by computing the ordered list L of all possible HGTs. The procedure stops when the RF distance equals 0 or the LS coefficient stops decreasing (ideally dropping to 0). Such a procedure requires $O(kn^4)$ operations to add k HGT edges to a phylogenetic tree with n leaves.

4 Partial gene transfer model

The partial gene transfer model is more general, but also more complex and challenging. It presumes that only a part of the transferred gene has been acquired by the host genome through the process of homologous recombination. Mathematically, this means that the traditional species phylogenetic tree is transformed into a directed evolutionary network (Figure 1a). Figure 3 illustrates the case where the evolutionary distance between the taxa i and j may change after the addition of the edge (b,a) representing a partial gene transfer from b to a.

From a biological point of view, it is relevant to consider that the HGT from b to a can affect the distance between the taxa i and j if and only if a is located on the path between i and the root of the tree; the position of j is assumed to be fixed. Thus, in the network T (Figure 3) the evolutionary distance $dist(i,j)$ between the taxa i and j can be computed as follows:

$$dist(i,j) = (1-\mu)d(i,j) + \mu(d(i,a) + d(j,b)), \tag{2}$$

where μ indicates the fraction (unknown in advance) of the gene being transferred and d is the distance between the vertices in T before the addition of the HGT edge (b,a). A number of biological rules, not discussed here due to the space limitation, have to be incorporated into this model (see Makarenkov et al (2004) for more details). Here we describe the main features of the network-building algorithm:

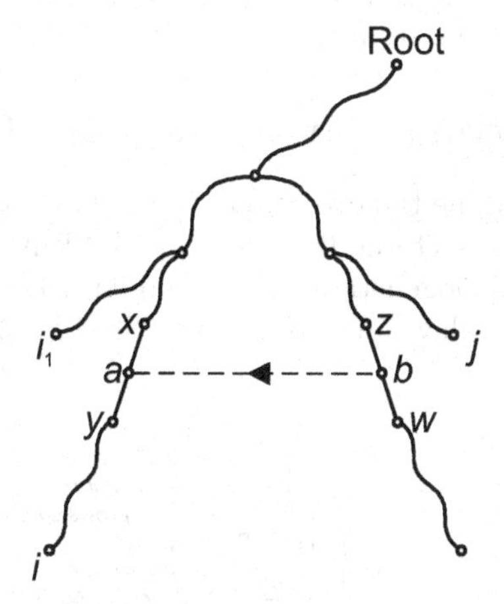

Fig. 3. Evolutionary distance between the taxa i and j can be affected by the addition of the edge (b,a) representing a partial HGT between the edges (z,w) and (x,y). Evolutionary distance between the taxa i_1 and j cannot be affected by the addition of (b,a).

Step 0. This step corresponds to Step 0 defined for the complete gene transfer model. It consists of inferring the species and gene phylogenies denoted respectively T and T_1. Because the classical RF distance is defined only for tree topologies, we use the LS optimization when modeling partial HGT.
Step 1. Assume that a partial HGT between the edges (z,w) and (x,y) (Figure 3) of the species tree T has taken place. The lengths of all edges in T should be reassessed after the addition of (b,a), whereas the length of (b,a) is assumed to be 0. To reassess the edge lengths of T, we have first to make an assumption about the value of the parameter μ (Equation 2) indicating the gene fraction being transferred. This parameter can be estimated either by comparing sequence data corresponding to the subtrees rooted by the vertices y and w or by testing different values of μ in the optimization problem. Fixing this parameter, we reduce to a linear system the system of equations establishing the correspondence between the experimental gene distances and the path-length distances in the HGT network. This system having generally more variables (i.e. edge lengths of T) than equations (i.e. pairwise distances in T; number of equations is always $n(n\text{-}1)/2$ for n taxa)

can be solved by approximation in the least-squares sense. All pairs of edges in T can be processed in this way. The HGT connection providing the smallest value of the LS coefficient and satisfying the evolutionary constraints will be selected for the addition to the tree T transforming it into a phylogenetic network.

Steps 2 ... k. In the same way, the best second, third and other HGT edges can be added to T, improving in each step the LS fit of the gene distance. The whole procedure requires $O(kn^5)$ operations to build a reticulated network with k HGT edges starting from a species phylogenetic tree with n leaves.

5 Detecting horizontal transfers of PheRS synthetase

In this section, we examine the evolution of the PheRS protein sequences for 32 species including 24 Bacteria, 6 Archaea, and 2 Eukarya (see Woese et al (2000)). The PheRS phylogenetic tree inferred with PHYML (Guindon and Gascuel (2003)) using G-law correction is shown in Figure 4.

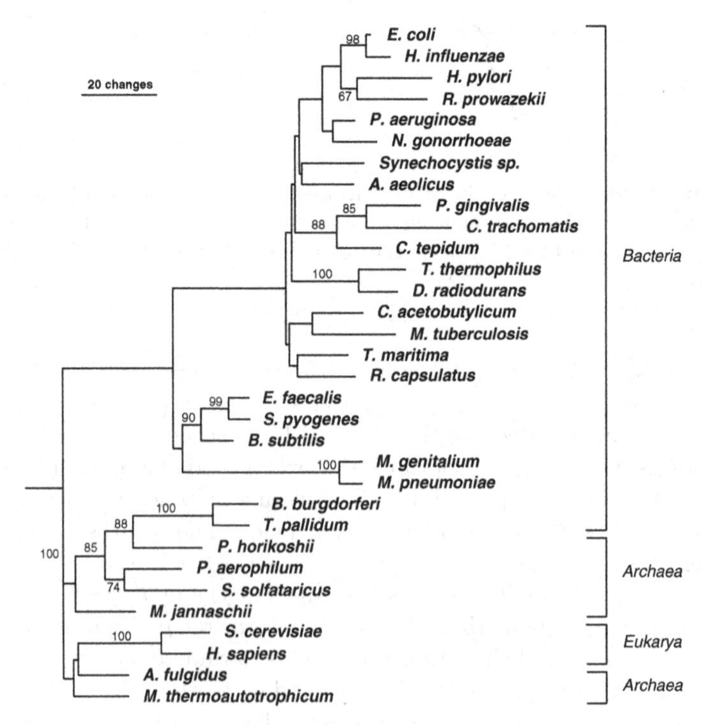

Fig. 4. Phylogenetic tree of PheRS sequences (i.e. gene tree). Protein sequences with 171 bases were considered. Bootstrap scores above 60% are indicated.

This tree is slightly different from the phylogeny obtained by Woese et al (2000, Fig. 2); the biggest difference involves the presence of a new cluster formed by two Eukarya (*H. sapiens* and *S. cerevisiae*) and two Archaea (*A. fulgidus* and *M. thermoautotrophicum*). This 4-species cluster with a low bootstrap support is probably due to the reconstruction artifacts. Otherwise, this tree shows the canonical pattern, the only exception being the spirochete PheRSs (i.e. *B. bugdorferi* and *T. pallidum*). They are of the archaeal, not the bacterial genre, but seem to be specifically related to *P. horokoshii* within that grouping (Figure 4). The species tree corresponding to the NCBI taxonomic classification was also inferred (Figure 5, undirected lines). The computation of HGTs was done in the framework of the complete gene transfer model. The five transfers with the biggest bootstrap scores are represented.

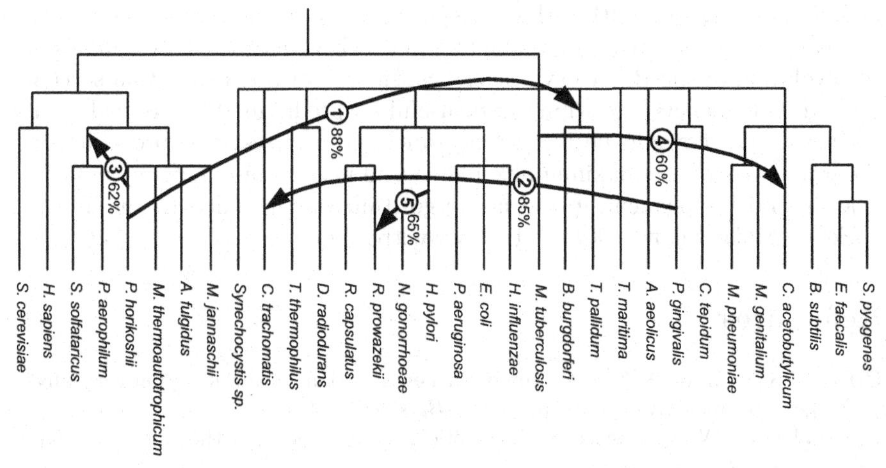

Fig. 5. Species phylogeny corresponding to the NCBI taxonomy for the 32 species in Figure 4. HGTs with bootstrap scores above 60% are depicted by arrows. Numbers on the HGT edges indicate their order of appearance in the transfer scenario.

The bootstrap scores for HGT edges were found fixing the topology of the species tree and resampling the PheRS sequences used to obtain the gene tree. The transfer number 1, having the biggest bootstrap support, 88%, links *P. horokoshii* to the clade of spirochetes. This bootstrap score is the biggest one that could be obtained for this HGT, taking into account the identical 88% score of the corresponding 3-species cluster in the PheRS phylogeny (Figure 4). In total, 14 HGTs, including 5 trivial connections, were found; trivial transfers occur between the adjacent edges. Trivial HGTs are necessary to transform a non-binary tree into a binary one. The non-trivial HGTs with low bootstrap score are most probably due to the tree reconstruction artifacts. For instance, two HGT connections (not shown in Figure 5) linking the cluster of Eukarya to the Archaea (*A. fulgidus* and *M. thermoautotrophicum*) have a low bootstrap support (16% and 32%, respectively). In this example, the solution found with the RF distance was represented. The usage of the LS function leads to the identical scenario differing from that shown in Figure 5 only by the bootstrap scores found for the HGT edges 3 to 5.

6 Conclusion

We described a new distance-based algorithm for the detection and visualization of HGT events. It exploits the discrepancies between the species and gene phylogenies either to map the gene tree into the species tree by least-squares or to compute a topological distance between them and then estimate the probability of HGT between each pair of edges of the species phylogeny. In this study we considered the complete and partial gene transfer models, implying at each step either the transformation of a species phylogeny into another tree or its transformation into a network structure. The examples of the evolution of the PheRS synthetase considered in the application section showed that the new algorithm can be useful for predicting HGT in real data. In the future, it would be interesting to extend and test this procedure in the framework of the maximum likelihood and maximum parsimony models. The program implementing the new algorithm was included to the T-Rex package (Makarenkov (2001), http://www.trex.uqam.ca).

References

CHARLESTON, M. A. (1998): Jungle: a new solution to the host/parasite phylogeny reconciliation problem. *Math. Bioscience, 149, 191-223.*

DOOLITTLE, W. F. (1999): Phylogenetic classification and the universal tree. *Science, 284, 2124-2129.*

GUINDON, S. and GASCUEL, O. (2003): A simple, fast and accurate algorithm to estimate large phylogenies by maximum likelihood. *Syst. Biol., 52, 696-704.*

LAKE, J. A. and RIVERA, M. C. (2004): Deriving the genomic tree of life in the presence of horizontal gene transfer: conditioned reconstruction. *Mol. Biol. Evol., 21, 681-690.*

LEGENDRE, P. and V. MAKARENKOV. (2002): Reconstruction of biogeographic and evolutionary networks using reticulograms. *Syst. Biol., 51, 199-216.*

MAKARENKOV, V. and LECLERC, B. (1999): An algorithm for the fitting of a tree metric according to a weighted LS criterion. *J. of Classif., 16, 3-26.*

MAKARENKOV, V. (2001): reconstructing and visualizing phylogenetic trees and reticulation networks. *Bioinformatics, 17, 664-668.*

MAKARENKOV, V., BOC, A. and DIALLO, A. B. (2004): Representing lateral gene transfer in species classification. Unique scenario. In: D. Banks, L. House, F. R. McMorris, P. Arabie, and W. Gaul (eds.): *Classification, Clustering and Data Mining Applications.* Springer Verlag, proc. IFCS 2004, Chicago 439-446

MIRKIN, B. G., MUCHNIK, I. and SMITH, T.F. (1995): A Biologically Consistent Model for Comparing Molecular Phylogenies. *J. of Comp. Biol., 2, 493-507.*

MORET, B., NAKHLEH, L., WARNOW, T., LINDER, C., THOLSE, A., PADOLINA, A., SUN, J. and TIMME, R. (2004): Phylogenetic Networks: Modeling, Reconstructibility, Accuracy. *Trans. Comp. Biol. Bioinf., 1, 13-23.*

PAGE, R. D. M. (1994): Maps between trees and cladistic analysis of historical associations among genes, organism and areas. *Syst. Biol., 43, 58-77.*

PAGE, R. D. M. and CHARLESTON, M. A. (1998): Trees within trees: phylogeny and historical associations. *Trends Ecol. Evol., 13, 356-359.*

ROBINSON, D. R. and FOULDS, L. R. (1981): Comparison of phylogenetic trees. *Math. Biosciences, 53, 131-147.*

TSIRIGOS, A. and RIGOUTSOS, I. (2005): A Sensitive, Support-Vector-Machine Method for the Detection of Horizontal Gene Transfers in Viral, Archaeal and Bacterial Genomes. *Nucl. Acids Res., 33, 3699-3707.*

WOESE, C., OLSEN, G., IBBA, M. and SÖLL, D. (2000): Aminoacyl-tRNA synthetases, genetic code, evolut. process. *Micr. Mol. Biol. Rev., 64, 202-236.*

List of Reviewers

Vladimir Batagelj, University of Ljubljana, Slovenia
Lynne Billard, University of Georgia, USA
Hans–Hermann Bock, RWTH Aachen University, Germany
Slavka Bodjanova, Texas A&M University, Kingsville, USA
Marko Bohanec, Jožef Stefan Institute, Slovenia
Hamparsum Bozdogan, The University of Tennessee, USA
Paula Brito, University of Porto, Portugal
Jonathan G. Campbell, Letterkenny Institute of Technology, Ireland
William H.E. Day, Canada
Edwin Diday, Université Paris Dauphine, France
Patrick Doreian, University of Pittsburgh, USA
Katherine Faust, University of California, Irvine, USA
Bernard Fichet, University of Aix Marseille, France
Herwig Friedl, Technical University Graz, Austria
Klaus Frieler, University of Hamburg, Germany
Gerard Govaert, Université de Technologie de Compiègne, France
John C. Gower, The Open University, UK
Michael Greenacre, Universitat Pompeu Fabra, Spain
Patrick J.F. Groenen, Erasmus University Rotterdam, The Netherlands
Alain Guénoche, Les Universités à Marseille, France
Jacques A.P. Hagenaars, Tilburg University, The Netherlands
David Hand, Imperial College London, UK
Richard J. Hathaway, Georgia Southern University, USA
Christian Hennig, University College London, UK
Thomas Hirschberger, Ludwig–Maximilians–Universität, Germany
Henk A.L. Kiers, University of Groningen, The Netherlands
Katarina Košmelj, University of Ljubljana, Slovenia
Wojtek Krzanowski, University of Exeter, UK
Berthold Lausen, University of Erlangen–Nuremberg, Germany
Nada Lavrač, Jožef Stefan Institute, Slovenia
Yves Lechevallier, INRIA, Le Chesnay, France
Bruno Leclerc, CAMS–EHESS, France
Jung Jin Lee, Soong Sil University, Korea
Pierre Legendre, Université de Montréal, Canada
Daniel Müllensiefen, University of Hamburg, Germany
Fionn Murtagh, University of London, UK
Siegfried Numberger, Ludwig–Maximilians–Universität, Germany
Jean-Paul Rasson, University of Namur, Belgium
Gunter Ritter, University of Passau, Germany
Tamas Rudas, Eotvos Lorand University, Hungary

Alexander Schliep, Max Planck Inst. for Molecular Genetics, Germany
Lars Schmidt-Thieme, University of Freiburg, Germany
Janez Stare, University of Ljubljana, Slovenia
Gerhard Tutz, Ludwig–Maximilians–Universität, Germany
Rosanna Verde, Seconda Universitá degli Studi di Napoli, Italy
Maurizio Vichi, University of Rome 'La Sapienza', Italy
Claus Weihs, Universität Dortmund, Germany
Djamel Zighed, Université Lumière Lyon 2, France

Key words

Authors